T0191941

Quantum Science and Technology

Aims and Scope

The book series Quantum Science and Technology is dedicated to one of today's most active and rapidly expanding fields of research and development. In particular, the series will be a showcase for the growing number of experimental implementations and practical applications of quantum systems. These will include, but are not restricted to: quantum information processing, quantum computing, and quantum simulation; quantum communication and quantum cryptography; entanglement and other quantum resources; quantum interfaces and hybrid quantum systems; quantum memories and quantum repeaters; measurement-based quantum control and quantum feedback; quantum nanomechanics, quantum optomechanics and quantum transducers; quantum sensing and quantum metrology; as well as quantum effects in biology. Last but not least, the series will include books on the theoretical and mathematical questions relevant to designing and understanding these systems and devices, as well as foundational issues concerning the quantum phenomena themselves. Written and edited by leading experts, the treatments will be designed for graduate students and other researchers already working in, or intending to enter the field of quantum science and technology.

More information about this series at http://www.springer.com/series/10039

Dimitris G. Angelakis
Editor

Quantum Simulations with Photons and Polaritons

Merging Quantum Optics with Condensed Matter Physics

Editor
Dimitris G. Angelakis
Centre for Quantum Technologies, National
University of Singapore
Singapore
Singapore

ISSN 2364-9054 ISSN 2364-9062 (electronic)
Quantum Science and Technology
ISBN 978-3-319-84799-3 ISBN 978-3-319-52025-4 (eBook)
DOI 10.1007/978-3-319-52025-4

Printed on acid-free paper

This Springer imprint is published by Springer Nature
The registered company is Springer International Publishing AG
The registered company address is: Gewerbestrasse 11, 6330 Cham, Switzerland

To my family and all the families of the contributing scientists for being patient with them.

Preface

One of the most important problems in the interdisciplinary area spanned by Physics, Computing, Engineering, and Nanotechnology is the efficient simulation of the dynamics of quantum many-body systems. Quantum simulators offer a new perspective allowing also for the implementation of quantum computation and the design of new materials with specific properties for technological applications. Initial approaches focused in the use of cold atom in optical lattices and ions in ion traps for mimicking a variety of models from condensed matter. More recent technological developments in the fields of interfacing light and matter, and especially in quantum nonlinear optics, have motivated complementary proposals for quantum simulators based on strongly correlated photons and polaritons. The latter are generated in hybrid light-matter systems ranging from coupled QED resonator arrays in superconducting circuits to slow light setups and semiconductor-based photonic chips. In this edited volume, several world experts in the area review some of the most important works on emulations of phenomena ranging from Mott transitions and Luttinger liquid physics to interacting relativistic theories and gauge fields with photons and polaritons. The aim was to review both the major theory proposals and the ongoing experimental efforts to realize photonic simulators in the laboratory in circuit QED, semiconductor, and integrated photonic lattice structures.

The book should be useful as an introduction to this area for graduate students working in quantum physics, quantum optics, implementations of quantum simulation, and strongly correlated physics. It can also serve as a reference material of the major developments at the time of writing to experts in the field.

We acknowledge the support of Technical University of Crete and the Centre for Quantum Technologies (CQT) Singapore while preparing this volume. We also like to thank CQT Ph.D. student Jirawat Tangpatinanon for his help in processing the latex files and the staff in Springer for their patience while this volume was being prepared.

Crete
October 2016

Dimitris G. Angelakis

Contents

Introduction

Computer simulation has led to huge advances in technology, from the engineering of smart materials to the design of drugs. Supercomputers are now a mainstream tool of science. However, studying many-body effects in condensed matter physics such as high-temperature superconductivity, topological states, and quantum magnetism, or even phenomena predicted by quantum field theories, which are otherwise only accessible in high-energy accelerators still poses tremendous challenges. The main obstacle is usually the computational complexity of solving even the most simplified models.

Quantum simulators present a powerful alternative, as originally suggested by Feynman. The recent advances in manipulating quantum particles and their environment have allowed demonstrations of quantum simulators of many-body and exotic physics under tabletop laboratory conditions. The latter includes realizing Mott-insulator–superfluid transitions, BEC-BCS crossovers, frustrated spin models, and artificial gauge fields in disordered systems. The main experimental platforms used so far consist of cold atoms in optical lattices and cold ions in electromagnetic traps.

More recently, a new approach for quantum simulation based on hybrid light-matter systems has been slowly emerging. Here, the main role is played by photons and hybrid light-matter excitations, also known as polaritons. In these ideas, the accessibility to local observables, and the ability to probe out-of-equilibrium many-body phenomena in driven systems with efficient optical measurements, offers several advantages over other approaches based on purely atomic or photonic setups.

The purpose of this book was to review the progress in this area in terms of both physical platforms and phenomena. In the following chapter, experts in this emerging area review some of the most important ideas so far. Among others, we see how photons and polaritons generated in circuit QED, integrated linear, and nonlinear quantum optics setups can be used to prepare and simulate novel out-of-equilibrium phases of matter, to probe topological physics and to implement phenomena from the realm of exotic physics.

In more detail, in Chap. 1, the equilibrium phase diagram of the Jaynes–Cummings–Hubbard (JCH) model is reviewed. The latter is done both within mean field theory and using sophisticated techniques such as DMRG. Aspects of the non-equilibrium behavior of coupled resonator arrays (CRAs), especially in terms of time-dependent quenches, are also presented.

Chapter 2 reviews some of the analytical as well as the numerical approaches to treat the JCH model, including the Green function and slave boson methods to calculate both elementary excitations and critical exponents of the equilibrium case. The fate of these results in open dissipative systems is briefly discussed, and an outline of schemes for their experimental verifiability is presented.

Chapter 3 reviews some of the exotic phases predicted in driven dissipative CRAs, including photon fermionization, crystallization, and photonic quantum Hall states in out-of-equilibrium setting. Possible experimental candidates to realize coupled resonator arrays along with the two theoretical models that capture their physics, the Jaynes–Cummings–Hubbard and Bose–Hubbard Hamiltonians are also discussed.[1]

Chapter 4 discusses the possibility to simulate synthetic gauge fields for photons and reviews the early observation of topological features such as photonic edge states in linear coupled resonator systems. After discussing the basics of the transfer matrix and coupled mode approaches, the cases of interacting regimes are also explored including proposal to simulate quantum hall parent Hamiltonians.

Chapter 5 reviews the basics of solid-state quantum simulator platforms based on exciton-polaritons. After introducing the microcavity exciton-polaritons basics, a review of the works on exciton-polariton condensation is discussed along with the efforts for simulating lattice models.

In Chap. 6, many-body quantum photonics in semiconductor photonic crystal structures is discussed. After an introduction to nanophotonics platforms and the basics of light propagation and confinement there, developments in achieving strong light-matter coupling are reviewed. The chapter closes with a discussion of the efforts to realize strongly correlated states of photons in nanophotonics chips with a focus on driven dissipative approaches.

Chapter 7 deals with quantum simulation using strongly interacting microwave excitations in superconducting circuits. The basic notions of circuit quantization are introduced followed by a review of some of the theoretical and experimental works for analog and digital quantum simulations with superconducting quantum circuits.

The last chapter deals with classical analogue simulations in integrated photonic chips. The chapter starts with a review of the femtosecond laser writing of

[1]At the time of writing this introduction, after the completion of the main chapters, a preprint of an experiment involving 72 coupled CCQED resonators interacting with artificial atoms in a superconducting circuit has been reported in arXiv:1607.06895. In this work, a realization of the proposal for a one-dimensional Jaynes–Cummings lattice or Jaynes–Cummings–Hubbard Hamiltonian as in 2007 by Angelakis et al. in Phys. Rev. A(R), 76, 031805 2007 is reported. The experiment took place in the group of A Houck in Yale. Among other effects, signatures of a dissipative phase transition taking place is reported.

waveguides in silica and proceeds with discussing the simulation of Dirac physics in one-dimensional photonic lattices. Works in two-dimensional lattices are then presented where topological physics including edge states, effective gauge fields, and Floquet topological insulators states were implemented.

Dimitris G. Angelakis
Centre for Quantum Technologies,
National University of Singapore, Singapore, Singapore

and

School of Electrical and Computer Engineering,
Technical University of Crete, Chania, Crete, Greece

Chapter 1
Strongly Correlated Polaritons in Nonlinear Cavity Arrays

Andrea Tomadin, Davide Rossini and Rosario Fazio

Abstract Arrays of coupled QED cavities have been proposed as promising candidates to study hybrid many-body states of light and matter in a controlled way. The rich scenario emerging in these systems stems from the interplay between intra-cavity light-matter interaction and inter-cavity photon hopping. Coherent light-matter interaction generates polaritonic excitations with physical properties resembling those of bosonic particles in a lattice. We review the most salient features of the zero-temperature equilibrium phase diagram of a polaritonic lattice model, focusing on a quantitative analysis for the one-dimensional case. A judicious analysis of the system, however, cannot neglect the effect of losses and decoherence of both light and matter components of the polaritonic excitation. External driving is typically needed to counteract such losses. In this case, the knowledge of the equilibrium phase diagram is not sufficient to describe the state of the system during its time-evolution, and its possible approach to the steady state. For this reason, we also discuss the nonequilibrium dynamics resulting from the interplay between losses and external driving in some relevant cases.

1.1 Introduction

Collective behavior emerges in many-body systems as a result of the interaction between their constituents. Strong interactions are believed to be at the root of still not completely understood phenomena as high temperature superconductivity or in the fractional quantum Hall effect. To the aim of understanding these features,

A. Tomadin · D. Rossini · R. Fazio
NEST Istituto Nanoscienze-CNR and Scuola Normale Superiore, 56126 Pisa, Italy
e-mail: andrea.tomadin@sns.it

D. Rossini (✉)
Dipartimento di Fisica, Università di Pisa and INFN, Largo Pontecorvo 3, 56127 Pisa, Italy
e-mail: davide.rossini@unipi.it

R. Fazio
Abdus Salam ICTP, Strada Costiera 11, 34151 Trieste, Italy
e-mail: fazio@sns.it

D.G. Angelakis (ed.), *Quantum Simulations with Photons and Polaritons*,
Quantum Science and Technology, DOI 10.1007/978-3-319-52025-4_1

artificial many-body systems started to be explored as quantum simulators, i.e. to simulate other quantum systems. Together with more mature implementations as cold atoms in optical lattices or trapped ions, interest is growing in using interacting photonic systems. Remarkable progresses in tailoring light-matter interaction have fuelled the interest towards the realization of many-body states of photons [1–5]. A paradigm to study these effects is an array of coupled QED-cavities, a periodic arrangement of cavities mutually interacting through photon hopping and with strong local photon nonlinearities.

In these few years significant progress has been achieved on the theoretical side. Recently some experiments started to appear (see for example in [4]). The aim of this chapter is to briefly review some of our activity on this field. We will concentrate in particular on the transition from the Mott to superfluid state. We will discuss the properties of the phase diagram in the absence of damping, and we will analyze how to detect this "equilibrium" quantum phase transition in the more realistic case of a cavity array in the presence of leakage. We will finally describe an interesting possibility in which cavity arrays can be exploited to implement models with engineered dissipation.

The chapter is organized as follows. In the next section we will introduce cavity arrays, the model Hamiltonian which is commonly used to describe their properties, and the relevant mechanism of dissipation. We will then proceed by describing the equilibrium phase diagram, i.e. the one obtained ignoring any coupling to the external world. The latter is obviously an idealized situation, which is impossible to be realized. We will therefore discuss how it would be possible to detect some features of the nonequilibrium diagram through a quench experiment in which the array is brought out of equilibrium by a laser pulse. We will finally conclude this chapter by discussing how to engineer dissipation in cavity arrays. As already mentioned, this is intended as a review of our work, where we provide some more details as compared to what has been published already. The list of references is therefore not complete. We refer to the existing reviews for an extensive account of the current literature.

1.2 Modelling of Nonlinear Cavity-Arrays

1.2.1 Effective Photon-Photon Interaction in Nonlinear Cavities

Photons hardly interact in open space. Nonetheless, if a resonator is used to trap photons and atoms, it is possible to substantially increase the interaction between them. The simplest conceivable model of resonator takes into account only one mode of the electromagnetic field, with frequency ω, which can interact with one or more identical atoms located in the cavity. We assume that only one atomic transition, between two atomic levels $\{|g\rangle, |e\rangle\}$, with energy difference ω_{q} (hereafter we are using units such that $\hbar = 1$), is almost resonant with the cavity mode. The frequency difference $\Delta = \omega - \omega_q$ between the cavity mode and the atomic transition is called the cavity detuning. In the jargon of quantum information, the two atomic levels coupled to the electromagnetic field are called a *qubit*. In the dipole and rotating-wave approximations, the Hamiltonian of the system is given by the Jaynes-Cummings (JC) model [6]:

$$\hat{\mathcal{H}}_{\text{JC}} = \omega_q \hat{S}^z + \omega \hat{a}^\dagger \hat{a} + g(\hat{S}^+ \hat{a} + \hat{S}^- \hat{a}^\dagger), \tag{1.1}$$

where g is the atom-field coupling strength, and $\hat{a}^\dagger$ ($\hat{a}$) is the creation (annihilation) operator of the photonic mode. Here, $\hat{S}^\alpha$ is a collective pseudospin operator for atoms (with $\alpha = x, y, z$; $\hat{S}^\pm = \hat{S}^x \pm i\hat{S}^y$), defined by $\hat{S}^\alpha = \frac{1}{2}\sum_{j=1}^{N} \hat{\sigma}_j^\alpha$, with N being the number of atoms inside the cavity and $\hat{\sigma}_j^\alpha$ the Pauli matrices for the j-th qubit.

The JC Hamiltonian (1.1) conserves the total number m of atomic and photonic excitations, which are the eigenvalues of the polariton number operator

$$\hat{m} = \hat{a}^\dagger \hat{a} + \sum_{j=1}^{N} \hat{\sigma}_j^+ \hat{\sigma}_j^- . \tag{1.2}$$

When restricted to the subspace of a fixed number of excitations, $\hat{m}$ can be written in a tridiagonal form and easily diagonalized (see Sect. 1.3.2 for a detailed discussion).

The spectrum of the Hamiltonian in Eq. (1.1) is anharmonic and thus the JC model describes a *nonlinear* cavity [6]. Such anharmonicity is at the origin of an effective repulsion between photons. To understand this fact, let us suppose that one photon is injected into the cavity and, as a consequence, the system is in its first excited state. Adding a second excitation requires an energy larger than that of two photons in the same system with vanishing light-matter coupling. The strength of this effective interaction depends on the detuning Δ and on the number N of atoms in the cavity. If the interaction is strong enough, the excitation of the atom-cavity system by a first photon may block the transmission of a second photon, thereby converting an incident thermal stream of photons into an antibunched stream. This nonlinear effect has been named *photon blockade* [7], after the Coulomb blockade effect, in which charge transport through a device occurs on an electron-by-electron basis.

The Hamiltonian in Eq. (1.1) generates the unitary evolution of a simplified atom-cavity system. However, even within a simple atom-cavity model, two major loss and decoherence processes exist and are responsible for the energy transfer with the surrounding environment: i) spontaneous emission with rate γ_a from the higher- to the lower-energy atomic state; ii) photon propagation at a rate κ from the cavity mode to the continuum of modes outside of the cavity. The rate κ is inversely proportional to the quality factor of the cavity. The ratio between the interaction strength squared g^2 and the algebraic average $\sqrt{\kappa\gamma_a}$ of the two loss rates is called the *cooperativity factor* η. For $\eta \gg 1$, the typical time-scale of the coherent evolution generated by the Hamiltonian is much shorter than the time-scale of the non-unitary processes that cause loss and decoherence. In this parameter regime, the Hamiltonian model (1.1) is sufficient to describe the dynamics of the system. For longer times, or for smaller cooperativity factors, it is necessary to take into account the loss and decoherence processes as well (see Sect. 1.2.3).

1.2.2 Photon-Mediated Coupling Between Nonlinear Cavities

The coupling between different cavities may be due to photon propagation between the cavity modes, due to the finite quality factor of the cavities. This process can be understood in terms of photon *hopping* between neighboring cavities, as massive particles do between neighboring wells due to quantum tunneling. Eventually, photons propagate through the whole cavity array and delocalized modes can be defined in the array, adopting the same formalism used to describe electrons traveling in a solid-state crystal [8]. The Bloch theorem can be applied to the electromagnetic vector potential $\mathbf{A}(\mathbf{r})$, in the Coulomb gauge, so that a Wannier function [8] $\xi_\ell(\mathbf{r})$ can be defined in the ℓ-th site of the lattice (i.e. the ℓ-th cavity of the array) and interpreted as the wave function of the corresponding cavity mode. As Wannier functions are not entirely localized within a lattice site (i.e. the cavity volume), the finite overlap between the Wannier functions of different cavities is responsible for finite photon hopping amplitude. In terms of Wannier functions, and annihilation operators for the cavity modes, the non-relativistic second-quantized vector potential operator [9] reads:

$$\hat{\mathbf{A}}(\mathbf{r}) = \sum_{\ell=1}^{L} \xi_\ell(\mathbf{r}) \hat{a}_\ell \,. \tag{1.3}$$

Choosing a basis of maximally localized Wannier functions [10] allows to neglect overlap between cavities which are not nearest neighbors in the array. In this *nearest-neighbor approximation*, photon propagation through the cavity array contributes through the following “kinetic” term to the cavity array Hamiltonian:

$$\hat{\mathcal{H}}_{\text{kin}} = -J \sum_{\langle \ell, \ell' \rangle} \hat{a}_\ell^\dagger \hat{a}_\ell \,, \tag{1.4}$$

where the symbol $\langle \ldots \rangle$ restricts the summation over couples of indices ℓ, ℓ' labelling neighboring cavities. The hopping amplitude J is given by $J = \int d\mathbf{r}\, \epsilon(\mathbf{r}) \xi_\ell(\mathbf{r}) \xi_{\ell'}(\mathbf{r})$, where $\epsilon(\mathbf{r})$ is the inhomogeneous dielectric constant in the cavity array.

Adding the kinetic term (1.4) to the Jaynes-Cummings Hamiltonian (1.1), one obtains the so-called Jaynes-Cummings-Hubbard (JCH) Hamiltonian:

$$\hat{\mathcal{H}}_{\text{JCH}} = \sum_{\ell=1}^{L} \hat{\mathcal{H}}_{\text{JC}}^{(\ell)} - J \sum_{\langle \ell, \ell' \rangle} \hat{a}_\ell^\dagger \hat{a}_{\ell'} - \mu \sum_{\ell=1}^{L} \hat{m}_\ell. \tag{1.5}$$

Here, the Hamiltonian is written in its "gran canonical" form [11], which includes a term proportional to the chemical potential μ of polaritons (neglecting losses, the total number of polaritons is conserved). The JCH Hamiltonian describes a system of L coupled identical cavities, each of them containing N atoms. Its structure closely matches that of the Bose-Hubbard (BH) Hamiltonian

$$\hat{\mathcal{H}}_{\mathrm{BH}} = \frac{U}{2}\sum_{\ell=1}^{L}\hat{n}_\ell(\hat{n}_\ell - 1) - J\sum_{\langle \ell,\ell'\rangle}\hat{b}_\ell^\dagger \hat{b}_{\ell'} - \mu\sum_{\ell=1}^{L}\hat{n}_\ell\,, \tag{1.6}$$

which is the paradigmatic model to describe ultracold bosonic neutral atoms trapped in periodic electromagnetic potentials created by counterpropagating laser beams, called optical lattices [12]. In Eq. (1.6), $\hat{b}_\ell^\dagger$ ($\hat{b}_\ell$) is the atomic creation (annihilation) operator in the ℓ-th optical lattice site and $\hat{n}_\ell = \hat{b}_\ell^\dagger \hat{b}_\ell$. The onsite JC Hamiltonian is substituted by a nonlinear term, proportional to the interaction strength U, which originates from a zero-range repulsion potential between the neutral atoms. It has been shown that the phase diagram and the low-energy excitations of the two Hamiltonians (1.5) and (1.6) are qualitatively similar. However, at higher energies, substantial differences arise between the spectra of the two models, due to the composite nature of the polaritons. Although the BH Hamiltonian naturally arises in the context of ultracold atoms in optical lattices, a nonlinear cavity array setup has been devised [13–15] where a family of polaritons obeys the BH Hamiltonian in a specific parameter regime. In Sects. 1.3 and 1.4 we discuss the equilibrium and nonequilibrium dynamics of a cavity array resorting both to the JCH and to the BH Hamiltonian.

1.2.3 Photon-Induced Losses and Decoherence

Despite being described by similar Hamiltonians (1.5) and (1.6), polaritons in a nonlinear cavity array and ultracold atoms in an optical lattice are substantially different systems for the following reason. Optical lattices are able to trap a number of ultracold atoms on a time-scale which is very long compared to the typical time-scales of the BH Hamiltonian [12]. This important feature guarantees that the number of atoms is practically constant during the whole time-evolution and the BH Hamiltonian can be solved both in the gran-canonical and in the canonical ensemble. On the contrary, polaritons in the JCH Hamiltonian are composite excitations, which eventually disappear from the cavity array due to coupling of the cavity modes to the continuum of electromagnetic modes outside the cavity. State-of-the-art implementations of a cavity array are such that the time-scale $\propto \kappa^{-1}$ for photon propagation outside of the array (often referred to as "photon leakage") cannot be assumed to be much longer than the time-scale g^{-1} for light-matter coherent energy exchange. In other words, a cavity array is an inherently *open quantum system* [16]. Moreover, the detection of leaked photons is the only way to probe the properties of the polaritons inside the cavity array. An external driving is thus necessary to measure a nonvanishing output signal. While this measurement scheme is active, the number of polaritons is not well-defined and the effects of the driving and of the environment on the cavity array must be taken into consideration.

The driving of a cavity array can be achieved by shining a coherent light source (i.e. a laser beam) onto the cavities. On the ℓ-th cavity, it is represented by the Hamiltonian term [9]

$$\hat{\mathcal{H}}^{(\ell)}_{\text{pump}}(t) = \Omega(t)^* \, \hat{a}_\ell + \Omega(t) \, \hat{a}^\dagger_\ell \,, \tag{1.7}$$

where $\Omega(t)$ is proportional to the amplitude of the electromagnetic field of the coherent light source. If the frequency of the driving field is ω_d and the rotating-wave approximation [9] is used, we have $\Omega(t) = \Omega e^{-i\omega_\text{d} t}$.

The inclusion of photon leakage into the model cannot be done within the Hamiltonian formalism. The profound reason for this difficulty is that losses *necessarily* induce decoherence (as a broad consequence of the fluctuation-dissipation theorem [17]) and thus require a more general dynamical framework than a coherent unitary evolution. In general, the time-evolution of a mixed state described by the density matrix $\rho(t)$ obeys the master equation [6]

$$\partial_t \rho(t) = -i[\hat{\mathcal{H}}(t), \rho(t)] + \hat{\mathcal{L}}[\rho](t) \,. \tag{1.8}$$

For the JCH and BH models, in the presence of external driving by a coherent light source, the time-dependent Hamiltonian reads:

$$\hat{\mathcal{H}}_{\text{JCH}-\text{pump}}(t) = \hat{\mathcal{H}}_{\text{JCH}} + \sum_{\ell=1}^{L} \hat{\mathcal{H}}^{(\ell)}_{\text{pump}}(t), \quad \hat{\mathcal{H}}_{\text{BH}-\text{pump}}(t) = \hat{\mathcal{H}}_{\text{BH}} + \sum_{\ell=1}^{L} \hat{\mathcal{H}}^{(\ell)}_{\text{pump}}(t) \,. \tag{1.9}$$

The Liouvillian superoperator $\hat{\mathcal{L}}[\rho](t)$ includes the contributions of losses and decoherence and, in general, depends on the state $\rho(t')$ at $t' < t$. Under the assumption that the external environment (i) is weakly coupled to the cavity array and (ii) has a smooth continuous density of states at the frequency ω, the system dynamics can be shown to be Markovian [18]. In this case, the Liouvillian superoperator is a linear function $\hat{\mathcal{L}}[\rho(t)]$ of the state $\rho(t)$ at time t only, which can be written in the Lindblad form [18]:

$$\hat{\mathcal{L}}[\rho(t)] = \Gamma \sum_{\ell=1}^{L} (2\, \hat{K}_\ell \, \rho \, \hat{K}^\dagger_\ell - \hat{K}^\dagger_\ell \hat{K}_\ell \, \rho - \rho \, \hat{K}^\dagger_\ell \, \hat{K}_\ell) \,, \tag{1.10}$$

where $\hat{K}_\ell$ is a so-called Kraus operator. Photon leakage corresponds to local Kraus operators $\hat{K}_\ell = \hat{a}_\ell$ with coefficient $\Gamma = \kappa$. For a cavity array operating in the microwave spectral range at non-cryogenic temperatures, propagation of photons from the environment to the cavity mode has also to be taken into account, and corresponds to $\hat{K}_\ell = \hat{a}^\dagger_\ell$.

The system is in a *steady state* ρ_{ss} when the time-derivative of the density matrix vanishes. The concept of steady state is the closest analogue of the ground state that can be devised in a nonequilibrium system. The steady state of a cavity array described by the JCH Hamiltonian with photon leakage, in the absence of external driving, is the vacuum state. It is also useful to define a *dark state* ρ_{ds} such that $\hat{K}_\ell \rho_{ds} = 0$ for each ℓ. It is easy to see that if a dark state is an eigenstate of the Hamiltonian, it is a steady state of the system. Otherwise a much more complicated dynamics arises and the steady state results from the interplay between the Hamiltonian and the Liouvillian. Nonlocal and nonlinear Kraus operators can be engineered to drive the system towards a desired steady state (see Sect. 1.4.2).

1.3 Equilibrium Phase Diagram of Cavity-Array Models

1.3.1 The Mott-Insulator—Superfluid Quantum Phase Transition

The crucial property of both the JCH and the BH Hamiltonian is the competition between the kinetic energy (i.e. the photon or atom hopping, respectively) and the onsite interactions (i.e. the light-matter coherent interaction or the atomic repulsive interaction, respectively). Consequently, the phase portrait of both models presents substantial similarities. In the following discussion we focus on the JCH Hamiltonian. The competition between the kinetic and the interaction terms opens up the possibility to stabilize two qualitatively different phases at zero temperature, where the polaritonic excitations have opposite behaviors. If the kinetic term dominates, the photonic component of the polaritons is delocalized across the cavity array and the system demonstrates long-range spatial coherence. The property of long-range spatial coherence is typical of superfluid (SF) liquids and gases, so this phase of the polaritonic system is called superfluid phase. In the opposite limit where the interaction term is much larger then the kinetic term, polaritons are localized in the cavity volume. The system is incompressible and photon propagation through the array is inhibited by onsite interactions, so that the polaritonic system realizes a Mott insulating (MI) phase. The two phases are separated by a quantum phase transition, which takes place at a well-defined *critical* value of the ratio between the hopping amplitude J and the atom-field coupling strength g, for a given number of atoms N in each cavity.

The phase diagram of the JCH Hamiltonian has been first obtained by means of a mean-field approach, applied for one [14] as well as many [19] atoms in each cavity, and later confirmed with a strong-coupling theory [20]. Subsequently, accurate numerical techniques, such as variational cluster approximations [21], quantum Monte Carlo [22–24], and density matrix renormalization group (DMRG) algorithms [25, 26], have yielded a more precise picture, which includes the role of dimensionality and the information on the critical exponents at the quantum phase

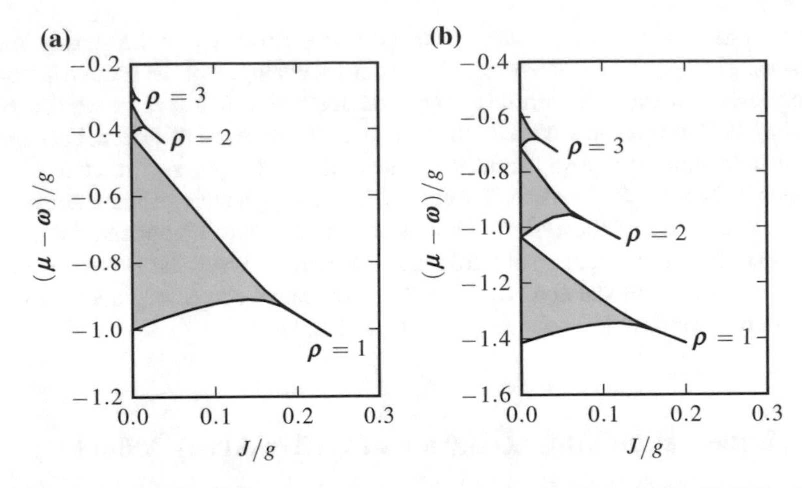

Fig. 1.1 Phase diagram of the 1D JCH Hamiltonian at $\Delta = 0$, obtained with DMRG calculations [25]. *Dark shaded* areas denote the MI lobes at fixed integer density ρ, while in the surrounding white region the system is in a SF phase. The *left and right panel* correspond to $N = 1$ and $N = 2$, respectively

transition. In Fig. 1.1 we show the phase diagram for a one-dimensional (1D) cavity array. However, the following discussion applies to higher dimensionalities as well.

The phase diagram in the plane spanned by the hopping amplitude J and the chemical potential μ is characterized by several lobes extending from the $J = 0$ axis. (The energy ω of the cavity mode is subtracted from the chemical potential in the plot of Fig. 1.1.) The system is in the MI phase when the parameters μ, J correspond to a point inside the lobes, and in the SF phase otherwise. Inside the ρ-th lobe, the ground state of the system has ρ polaritons localized in each cavity and the fluctuations of the polariton number operator vanish. The excitation spectrum is gapped, thus the MI phase is incompressible. In the SF phase, the global $U(1)$ symmetry of the Hamiltonian, realized by the transformation $\hat{a}_\ell \to \hat{a}_\ell e^{i\phi}$, is broken. In other words, the expectation value of the electromagnetic field in the cavity array acquires a uniform, constant phase. Such state of light is called "coherent" and its properties are markedly differ from light emitted by a standard thermal source [9]. The number of polaritons in each cavity is not well-defined, since quantum fluctuations $\langle (\hat{m}_\ell - \langle \hat{m}_\ell \rangle)^2 \rangle$ of the photonic number operator are finite; moreover they considerably increase if one considers a block of cavities with an enlarging dimension [26]. Local fluctuations of the particle number correspond to the property, mentioned above, that the photonic component of the polaritons is delocalized over the whole lattice. The SF phase is characterized by a finite stiffness and compressibility, and the energy gap closes.

In the BH model (1.6), the SF-MI transition has been studied theoretically in great detail [27]. Most importantly, an experimental implementation of the BH model using ultracold bosonic neutral atoms in optical lattices [28] has been achieved by several groups around the world and has been the target of constant efforts for more than a decade. On the contrary, complete experimental implementations of the JCH model have not been reported yet. Several implementation schemes have been put forward [2, 5] and first experiments started to appear recently [29].

In the rest of this section we discuss in detail the zero-temperature phase diagram of the 1D JCH Hamiltonian. We consider first the case of vanishing photon hopping amplitude, $J = 0$, in Sect. 1.3.2 and then the general case, $J \neq 0$, in Sect. 1.3.3. A comparison with the BH Hamiltonian is discussed at a quantitative level.

1.3.2 Vanishing Photon Hopping Amplitude

When the photon hopping is completely suppressed, the cavities of the array are decoupled and the problem is exactly solvable. Let us also suppose that the ground state contains ρ polaritons in each cavity. The system is homogeneous, therefore it contains a total number ρL of polaritons. To create (annihilate) a polariton in a cavity requires a positive (negative) energy $\Delta E_+ = E_+ - E_0$ ($\Delta E_- = E_- - E_0$). The minimum between $+\Delta E_+$ and $-\Delta E_-$ is the energy gap of the spectrum of the cavity array ("Mott gap"). The existence of an energy gap in each cavity implies that the ground state is incompressible and insulating.

In the parameter space, the phase boundary between ground states with a different number ρ of polaritons per cavity is determined by the condition that the Mott gap vanishes: $\Delta E_\pm = 0$. To calculate the Mott gap it is sufficient to diagonalize the JC Hamiltonian matrix, reduced to the subspace with ρ polaritons:

$$\hat{\mathcal{H}}_{\mathrm{JC}}\Big|_{\rho} = \begin{pmatrix} \rho\omega & g\sqrt{N\rho} & & & \\ g\sqrt{N\rho} & \omega_q + (\rho-1)\omega & g\sqrt{2(N-1)(\rho-1)} & & \\ & g\sqrt{2(N-1)(\rho-1)} & 2\omega_q + (\rho-2)\omega & g\sqrt{3(N-2)(\rho-2)} & \\ & & g\sqrt{3(N-2)(\rho-2)} & 3\omega_q + (\rho-3)\omega & \cdots \\ & & & \cdots & \cdots \end{pmatrix} . \tag{1.11}$$

The Hamiltonian matrix is written in the basis $\{|0, \rho\rangle, |1, \rho-1\rangle, \cdots, |N, \rho-N\rangle\}$, where the first index gives the number of excited atoms, and the second index gives the number of photons in the cavity mode. Note that, for $\rho < N$, the above matrix has to be truncated to the first $(\rho + 1) \times (\rho + 1)$ rows and columns.

In Fig. 1.2 we plot the boundaries $\mu_c^\pm(\rho, J = 0)$ between ground states with different number of polaritons in each cavity, as a function of the detuning Δ, for different number of atoms N in each cavity. In the grand canonical ensemble, the boundaries correspond to the extra energies to create or annihilate a polariton,

$$\mu_c^\pm(\rho, 0) = \pm\Delta E_\pm(\rho, 0) . \tag{1.12}$$

One can identify a critical value $\Delta^\star(N)$ of the detuning such that, for $\Delta > \Delta^\star(N)$, the Mott lobes (corresponding to segments at $J = 0$) tend to group into two sets, with $\rho > N$ and $\rho < N$, respectively. The width of the Mott lobe with $\rho = N$, separating the two sets, increases as the detuning increases past $\Delta^\star(N)$. We estimated

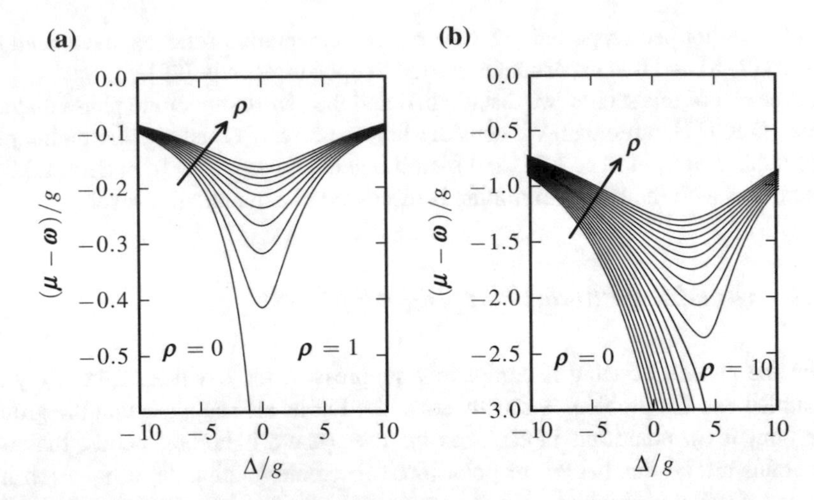

Fig. 1.2 Boundaries between MI phases with different polariton density ρ in each cavity, as a function of the detuning Δ, for $J = 0$. The value of ρ increases along the arrow and is indicated explicitly in some regions. The *left and right panel* correspond to $N = 1$ and $N = 10$, respectively. Note that the absolute value of ω simply shifts the values of the critical chemical potentials

$\Delta^\star$ numerically, taking the value of the detuning at which the curve $\mu_c^+(\rho = N, 0)$ in Fig. 1.2 has a minimum, and found a scaling $\Delta^\star(N) \sim \sqrt{N}$.

At a fixed polariton density ρ, one can give an estimate of the effective interaction strength U_{eff} between polaritons by taking the second-order difference quotient of the ground-state energy $E_0(\rho)$ with respect to the number of polaritons:

$$U_{\text{eff}}(\rho) = \left.\frac{\Delta^2 E_0(m)}{\Delta m^2}\right|_{m=\rho} = \mu_c^+(\rho, 0) - \mu_c^-(\rho, 0)\,. \tag{1.13}$$

An important difference with respect to the BH model is the fact that the effective repulsion $U_{\text{eff}}(\rho)$ for the JC model explicitly depends on the number of polaritons in a cavity. At a fixed ρ, this dependence weakens when the number N of atoms inside each cavity increases (see Fig. 1.3). We also notice that, for large ρ, the effective repulsion drops as a power law $U_{\text{eff}} \sim \rho^{-3/2}$, for each value of N and Δ. The ρ-independent effective interaction, mimicking the BH situation, can be thus recovered in the limit of many atoms $N \gg \rho$.

1.3.3 Finite Photon Hopping Amplitude

When a finite hopping amplitude is taken into account, $J \neq 0$, all the cavities in the array are coupled and it is not practical to straightforwardly diagonalize the JCH Hamiltonian. Moreover, it can be shown that the JCH Hamiltonian is non-

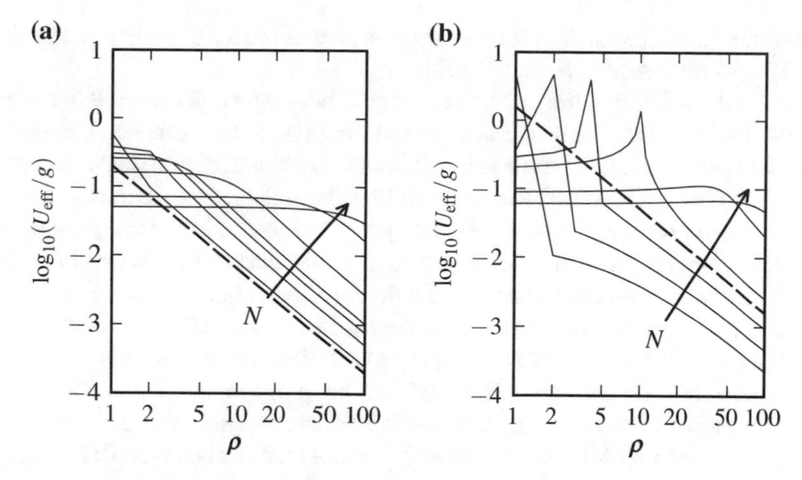

Fig. 1.3 Onsite effective interaction strength U_{eff} in the JCH model, as a function of the number ρ of polaritons in each cavity, for $J = 0$. Different data sets correspond to $N = 1, 2, 4, 10, 50, 100$, increasing as indicated by the *arrow*. On the horizontal axis, ρ can attain only integer values, however, the data is shown using *solid lines* to make the graphics clearer. The *dashed line* corresponds to $U_{\text{eff}} \sim \rho^{-3/2}$. The *left and the right panel* correspond to $\Delta = 0$ and $\Delta = 5$, respectively. For our convenience, here and in the next Figs. 1.4 and 1.5 we set $\omega = 0$

integrable, so that an analytical solution of the model in its whole complexity is not possible. To overcome these difficulties in a 1D geometry, the DMRG algorithm has been employed and accurate results on the phase diagram have been obtained. The strategy of DMRG [30] is to build up a portion of the system (the system's block) and then recursively enlarge it, until the desired system size is reached. At every step, the basis of the corresponding Hamiltonian is truncated, so that the size of the Hilbert space is kept manageable as the physical system grows. The truncation of the Hilbert space is performed by retaining only some eigenstates, corresponding to the highest eigenvalues of the block's reduced density matrix.

The phase boundaries between MI and SF phases have been obtained by calculating the Mott gap as function of the hopping amplitude and finding the values of the chemical potential $\mu_c^{\pm}$ at which the Mott gap vanishes [25, 26]. In the same way as for the zero-hopping case, Eq. (1.12), along the μ axis and at fixed J, the upper and lower extreme of the ρ-th Mott lobe are given by

$$\mu_c^{\pm}(\rho, J) = \pm \Delta E_{\pm}(\rho, J)\,. \tag{1.14}$$

More precisely, on a chain of finite length L, the Mott gap for a generic value of J can be evaluated by performing three iterations of the DMRG procedure in the canonical ensemble at integer density ρ, with projections on different polariton number sectors ρL, $\rho L \pm 1$. The corresponding ground states give the desired energies E_0, $E_{\pm} = E_0 + \Delta E_{\pm}$, from which the chemical potential $\mu_c^{\pm}$ is extrapolated through Eq. (1.14).

The thermodynamic limit is recovered by means of a finite-size scaling with a linear fit in $1/L$, as discussed in Refs. [31, 32].

The most significant effect of a finite photon hopping amplitude is that the Mott lobes shrink with increasing J and superfluid regions appear between neighboring Mott lobes (see Fig. 1.1). At some value $J^\star(\rho)$ the two boundaries of the MI lobe meet, $\mu_c^+[\rho, J^\star(\rho)] = \mu_c^-[\rho, J^\star(\rho)]$ and the ρ-th Mott lobe features a sharp tip, as shown in Fig. 1.1. (In two- and three-dimensional geometry the tip is generally smoother.)

Photon hopping between neighboring cavities can also be interpreted in terms of effective *polariton hopping*, with amplitude $J_{\rm eff}$. To find the relation between $J_{\rm eff}$ and J, we consider the energy of a *defect state*, i.e. a state that is obtained from the MI ground state by creating or annihilating a polariton. We also calculate the energy of a defect state in an auxiliary BH model with hopping amplitude $J_{\rm eff}$. By requiring the two energies to be the same, we find the desired relation. For definiteness, we focus here on the first MI lobe, with $\rho = 1$, for both the JCH and the BH model. An analogous calculation can be performed for any other lobe.

Let us start out with the calculation of the defect state energy in the BH model. We follow the route of a perturbative expansion in $J_{\rm eff}$, which can also be understood as a strong-coupling expansion, valid in the limit $J_{\rm eff} \ll U$ [33]. The energy of the defect state is then given by the average of the kinetic operator over the unperturbed ground state. To zeroth order in $J_{\rm eff}/U$, the MI ground state of the BH model, with $\rho = 1$ boson per lattice site, is given by:

$$|\psi_0\rangle \propto \prod_{\ell=1}^{L} \hat{b}_\ell^\dagger |{\rm vac}\rangle \ , \tag{1.15}$$

where $|{\rm vac}\rangle$ is the vacuum state. Let us consider the two defect states

$$|\psi_+\rangle \propto \sum_{\ell=1}^{L} \hat{b}_\ell^\dagger |\psi_0\rangle \, , \quad |\psi_-\rangle \propto \sum_{\ell=1}^{L} \hat{b}_\ell |\psi_0\rangle \ , \tag{1.16}$$

where one boson is created (annihilated) with the same probability and phase on each cavity of the array. The energy of the defect states (with respect to the ground state energy), at first order in $J_{\rm eff}$, is given by $E_+ - E_0 \sim -4J_{\rm eff}$ and $E_- - E_0 \sim -2J_{\rm eff}$, respectively.

Following the same procedure for the JCH model, we can also explicitly calculate the expectation value of the photon hopping term $-J\sum_\ell \hat{a}_\ell^\dagger \hat{a}_\ell$ on the same one-defect states (1.16), thus finding

$$E_+ - E_0 \sim -4J \times \left|[\hat{a}_\ell, \hat{b}_{\ell,1}^\dagger]\right|^2, \quad E_- - E_0 \simeq -2J \times \left|[\hat{a}_\ell, \hat{b}_{\ell,1}^\dagger]\right|^2 . \tag{1.17}$$

Here, $\hat{b}_{\ell,1}^\dagger$ is the operator that, applied to the ground state, gives the state with one polariton in the ℓ-th site of the lattice, and reads

$$\hat{b}^{\dagger}_{\ell,1} = \frac{1}{\sqrt{2}} \cdot \frac{\left(\sqrt{4Ng^2 + \Delta^2} - \Delta\right)\hat{a}^{\dagger}_{\ell} - \hat{S}^{+}_{\ell}}{\sqrt{4Ng^2 + \Delta^2 - \Delta\sqrt{4Ng^2 + \Delta^2}}} . \tag{1.18}$$

We remark that polaritons do not have a well-defined statistics, hence they cannot be associated to standard bosonic creation and annihilation operators. (It is easy to check that $[\hat{b}_{\ell,1}, \hat{b}^{\dagger}_{\ell,1}] \neq 1$, in general.)

Equating the energies of the defect states in the BH and JCH model, we find

$$J_{\text{eff}} \sim J \times \left|[\hat{a}_{\ell}, \hat{b}^{\dagger}_{\ell,1}]\right|^2 . \tag{1.19}$$

Using the expression (1.18), and the commutation relations $[\hat{a}_{\ell}, \hat{a}^{\dagger}_{\ell}] = 1$ and $[\hat{a}_{\ell}, \hat{S}^{+}_{\ell}] = 0$, we finally obtain:

$$J_{\text{eff}} \sim J \times \frac{1}{2}\left(1 - \frac{\Delta}{\sqrt{4Ng^2 + \Delta^2}}\right) . \tag{1.20}$$

The ratio J_{eff}/J is shown in Fig. 1.4 (solid line) as a function of the detuning, and compared to numerical results obtained with the DMRG algorithm (symbols). In the following we elucidate the origin of the discrepancies, by working out more explicitly the comparison with the BH model.

As a matter of fact, the effective interaction strength U_{eff} and the hopping amplitude J_{eff} for the polaritons allow us to define a mapping from the JCH to the BH

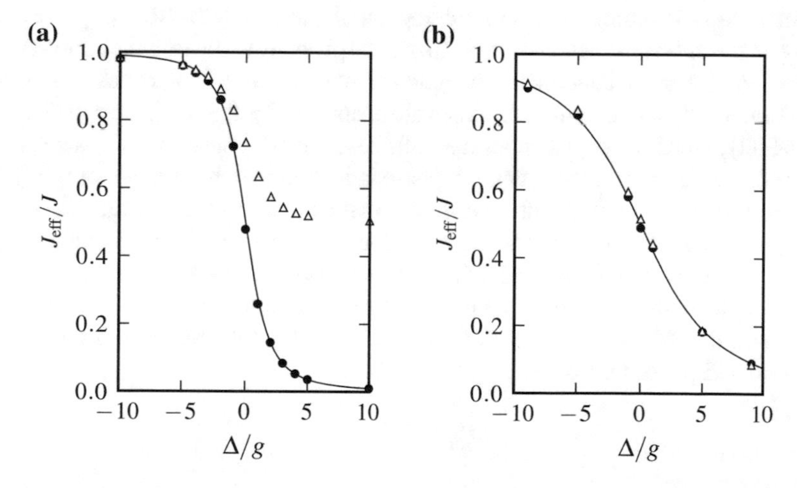

Fig. 1.4 Ratio between the effective polaritonic hopping amplitude J_{eff} and the photonic hopping amplitude J as a function of the detuning Δ. The *solid line* corresponds to the analytical estimate in Eq. (1.20). The filled (*empty*) symbols correspond to numerical data obtained from the lower (*upper*) slope of the MI lobes. The *left and right panel* correspond to $N = 1$ and $N = 10$, respectively

model, where polaritons are approximately represented as bosonic particles and the effective parameters are used in the Hamiltonian (1.6).

The robustness of this mapping can be checked by comparing the analytical expressions for $J_{\rm eff}$ given above with results of DMRG calculations. The first step in this direction is to show that a relation exists between the slope of the boundaries of the MI lobes for $J \gtrsim 0$ and the ratio $J_{\rm eff}/J$. Specializing Eq. (1.14) to the first lobe and using Eqs. (1.17) and (1.19), we find

$$\frac{d\mu_{\rm c}^-}{dJ} = 2\frac{J_{\rm eff}}{J}, \quad \frac{d\mu_{\rm c}^+}{dJ} = -4\frac{J_{\rm eff}}{J} . \tag{1.21}$$

Numerical results for both the lower and the upper boundary of the $\rho = 1$ MI lobe are reported in Fig. 1.4 and compared to the analytical expression (1.20). We point out that the ratio $J_{\rm eff}/J$ is very different for the lower and the upper boundary in the case $N = 1$ [panel (a)]. In particular, Eq. (1.20) agrees only with the numerical results corresponding to the lower boundary. The discrepancy indicates that the polariton hopping depends on the occupation of the cavity where this happens [the form of polariton operators depend on the excitation subspace they live in; in particular, $\hat{b}^\dagger_{\ell,2}$ is generally different from $\hat{b}^\dagger_{\ell,1}$ in Eq. (1.18)]. We conclude that, in this limit, the mapping to an effective BH model is less accurate, and cannot be used to extract information about the phase transition. On the other hand, when N is increased, the mapping becomes more accurate [panel (b)]. This trend is supported by the observation that, only in this situation, one can unequivocally define a *constant* effective interaction strength $U_{\rm eff}$ for the mapping to the BH model, while for $N \leq \rho$ the interaction strength depends strongly on the density (see Fig. 1.3).

An accurate estimate of the critical hopping $J^\star$, at which the Mott gap closes and the system exhibits the MI-SF transition, can be given by analyzing the compressibility $\kappa = \partial\rho/\partial\mu$ in the canonical ensemble at fixed number of polaritons ρ per cavity. In the thermodynamic limit, MI states are identified by the vanishing of the compressibility, which is, on the other hand, finite in the SF phase. A finite-size scaling of κ is thus able to detect the quantum phase transition with high accuracy[1] [25]. The results for $\rho = 1$, obtained with DMRG calculations, are plotted in Fig. 1.5. When N is large, it is possible to extrapolate such value with large precision by exploiting the mapping with the BH model. Indeed, it is well known that the MI-SF transition in the BH model with one boson per lattice site in the ground state occurs at $J/U \simeq 0.3$ [31, 32]. Given the effective onsite polariton repulsion $U_{\rm eff}$, the transition is thus located at the critical polariton hopping

$$J^\star_{\rm eff} \simeq 0.3\, U_{\rm eff} . \tag{1.22}$$

[1]This procedure is formally equivalent to finding the critical hopping $J^\star$ at which, in the grand canonical ensemble, the critical chemical potentials $\mu_{\rm c}^\pm(\rho, J)$ of Eq. (1.14) coincide.

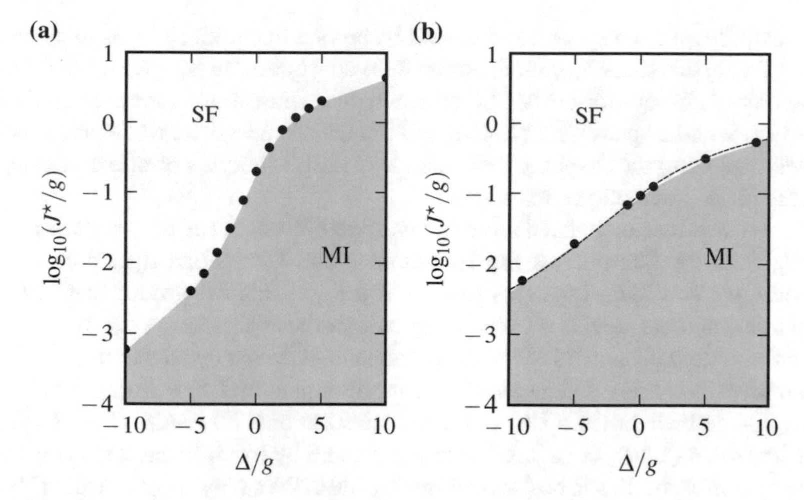

Fig. 1.5 Critical hopping in the first MI lobe as a function of the detuning Δ. Symbols are obtained by means of DMRG calculations. The *gray shading* is a guide to the eye connecting the symbols. The *left and right panel* correspond to $N = 1$ and $N = 10$, respectively. In panel (**b**), the *dashed line* is an analytical approximation discussed in the main text, following from $J^\star_{\text{eff}} \simeq 0.3 U_{\text{eff}}$

From this analytic estimate it is possible to draw the dashed line in the panel (b) of Fig. 1.5 (in the case of $N = 10$ atoms per cavity), which compares very well with the numerical results.

Finally we observe that, for large positive detuning $\Delta = \omega - \omega_q$, the superfluid is stabilized for increasing values of the photon hopping, thus meaning that it becomes suppressed with respect to the MI state. (On the contrary, for large negative detuning the opposite situation occurs.) This can be intuitively understood by the fact that, when $\Delta \gg 1$, the single-photon cavity frequency ω is much larger than the excitation frequency of the atomic transition, therefore the photonic (delocalized) part of the polaritons is energetically suppressed. Note also that, in such circumstance, the onsite repulsion energy U_{eff} exhibits a marked peak around $\rho = N$ [see panel (b) in Fig. 1.3], thus enhancing the stabilization of a polaritonic MI state at large detunings if the number of atoms in each cavity matches the density ρ. In Fig. 1.5 we observe indeed that, at fixed large Δ values, the first MI lobe is more robust for $N = 1$.

1.4 Nonequilibrium Dynamics of Cavity-Array Models

1.4.1 Nonequilibrium Signatures of the Equilibrium Phase Diagram

In Sect. 1.3 we have discussed the equilibrium MI-SF quantum phase transition in the JCH and BH models. A cavity array implementation of either model, however, is necessarily an open system and cannot be probed in equilibrium, as explained in Sect. 1.2.3. In this section, we address the following question: is it possible to identify signatures of the equilibrium MI-SF quantum phase transition in the nonequilibrium steady-state of the open JCH or BH model? For definiteness, here we focus on the BH model, exploiting the mapping discussed in Sect. 1.3 in terms of effective hopping amplitude and interaction strength.

We remark that our goal is not to investigate a novel set of quantum phases, arising from the interplay of the Hamiltonian and Liouvillian dynamics. On the contrary, we consider signatures proper of the well-known ground-state MI and SF phases, and we devise a protocol by which these signatures can be accessed when the system is coupled to the environment and driven by an external field. The dynamics of the system is generated by the master equation (1.8) with the Hamiltonian $\hat{\mathcal{H}}_{\mathrm{BH-pump}}$ defined in Eq. (1.6) (with the substitutions $J \mapsto J_{\mathrm{eff}}$, $U \mapsto U_{\mathrm{eff}}$) and the Liouvillian (1.10). Here, the losses are defined by local, linear Kraus operators $\hat{K}_\ell = \hat{b}_\ell$, corresponding to bosons leaking outside of the cavity array, at a rate $\Gamma \mapsto \kappa$.

The crucial observable to distinguish the MI from the SF phase is the "order parameter" $\langle \hat{b}_\ell(t) \rangle$ [14, 34], where $\langle \hat{O}(t) \rangle = \mathrm{Tr}[\hat{O}\rho(t)]$ is the quantum average of the operator $\hat{O}$ at time t. The order parameter vanishes in the MI and is finite in the SF phase, indicating the breaking of the $U(1)$ symmetry associated to the latter phase (see Sect. 1.3.1) and, more tangibly, emission of coherent light [6, 9] from the cavity array. To search for a signature of the MI-SF transition in the amplitude of the emitted coherent light requires to carefully devise a driving scheme. Indeed, it is easy to understand that a continuous and coherent driving of the cavity array washes out signatures of the coherent light generated in the cavity array *spontaneously* due to light-matter interactions. To overcome this difficulty, it has been proposed [35] to drive the cavity array periodically, using a pulsed light source. The dynamics of the light coherence *between* pulses is unaffected by the driving and is distinctly different in the parameter regime corresponding to the MI and SF phases [35]. A convenient observable is thus the time-average of the absolute value of the coherent amplitude, $\overline{|\langle \hat{b}_\ell(t) \rangle|}$.

The solution of the equations of motion for the cavity array in these nonequilibrium conditions, with many-body interactions between the bosonic particles, poses a formidable task. In general, explicitly solvable models of master equations (1.8) are very limited, and support from extensive numerical simulations is desirable. In the Lindblad formalism, the most common approach relies on the quantum trajectories technique, a method that is widely used in quantum optics [18]. This is based on a stochastically evolving state vector which can occasionally jump to other states according to the Lindblad operators. A more recent method is based on time-dependent DMRG in the formalism of the matrix product operators (MPO): the density matrix of the system is represented as a MPO, while the time evolution is performed by employing a time-evolving block-decimation (TEBD) algorithm that integrates the master equation (1.8) in time until a steady state is reached [36, 37].

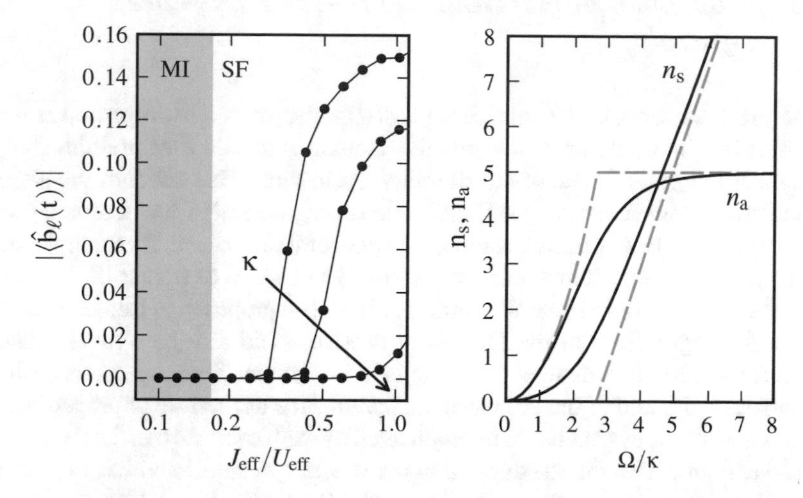

Fig. 1.6 The *leftmost panel* shows the time-averaged absolute value of the order parameter of the MI-SF quantum phase transition in the BH model as a function of the ratio between the effective hopping amplitude and interaction strength. Data sets correspond to different values $\kappa/U_{\text{eff}} = 5 \times 10^{-3}$, 10^{-1}, and 2×10^{-2}, increasing as indicated by the *arrow*. The *shaded area* marks the parameter range where the equilibrium Mott phase is expected for $\rho = 1$. The *right panel* shows the steady-state value of the average photonic occupations n_s, n_a in the two-cavities symmetric and antisymmetric mode, respectively, for the model discussed in Sect. 1.4.2. The *dashed lines* correspond to the semiclassical solution of Eq. (1.26). The *solid line* has been obtained with the exact diagonalization of the two-cavity system

In Ref. [35] a mean-field scheme was used [14, 34], where the kinetic part of the Hamiltonian is approximated as follows:

$$\hat{\mathcal{H}}_{\text{kin}}(t) = -J \sum_{\langle \ell,\ell' \rangle} \langle \hat{b}_{\ell'}(t) \rangle \hat{b}_\ell^\dagger + \text{H.c.}\,. \tag{1.23}$$

This approximation corresponds to neglecting nonlocal correlation functions beyond the connected term. In other words, quantum correlations between different sites are lost, while onsite quantum correlations are considered. We point out that the "semi-classical" approximation [9], often used in quantum optics, neglects local correlation and can be recovered from the mean-field scheme by substituting the creation and annihilation operators with their time-dependent quantum averages.

Panel (a) of Fig. 1.6 shows that the time-averaged coherent amplitude increases as the ratio $J_{\text{eff}}/U_{\text{eff}}$ increases, corresponding to a transition from a localized (MI) to a delocalized (SF) dynamics. The increase is particularly sharp around a critical value, which shifts towards the boundary of the equilibrium phase transition as the decay rate becomes smaller. We conclude that the MI-SF quantum phase transition, which characterizes the BH model in the absence of losses and decoherence, could be demonstrated experimentally by detecting light from an open, driven cavity array.

1.4.2 Liouvillian Engineering and Driven-Dissipative Dynamics

In the previous section, we considered a dissipative mechanism (photon leakage) which drives the cavity array towards the vacuum state and thus prevents the persistence of the ground state of the Hamiltonian in time. This effect of the losses is particularly detrimental when the Hamiltonian is *engineered* to have a certain ground state, which may be of practical interest as a platform to perform, for example, quantum computation algorithms, or to study correlated states of matter [2, 3]. A very general approach to overcome this difficulty has been proposed in Ref [38] and consists in the engineering of the Liouvillan itself to yield a desired steady state. A practical route to Liouvillian engineering is to implement Kraus operators such that the dark state, instead of the ground state, features the desired set of properties. The Hamiltonian of the system has to be engineered as well to ensure that the dark state of the Liouvillian is also the steady state of the whole system. Liouvillian engineering generally requires some external driving applied to the system, which connects the low-energy states of the lattice to a set of detuned higher-energy states, which can be adiabatically eliminated. In this case, the driving is integral part of the dissipation process, and it is instrumental to achieving a desired Liouvillian. For this reason, these systems are called *driven-dissipative*.

The first proposal of a many-body driven-dissipative system [38] considered a cold-atom setup in an optical lattice, described by the BH model. In this case, the effective dynamics is not unitary but there is no loss of particles from the system. The same principle, however, has been applied to cavity arrays [39] where photon losses and driving, as discussed in Sect. 1.2.3, must be taken into account in addition to the engineered Liouvillian.

We consider now a cavity array which is driven towards a state consisting of a Bose-Einstein condensate (BEC) of photons. This state is interesting because photons cannot condense in a black body, due to their vanishing chemical potential. Only recently alternative routes to photon condensation have been realized [40]. To achieve a BEC steady state, the Kraus operators are nonlocal and nonlinear, and read [39]:

$$\hat{K}_\ell = \frac{1}{2}(\hat{b}_\ell^\dagger + \hat{b}_{\ell+1}^\dagger)(\hat{b}_\ell - \hat{b}_{\ell+1}) \, . \tag{1.24}$$

It is easy to see that a BEC state with M photons and zero wavevector

$$|\mathrm{BEC}\rangle \propto \left(\sum_{\ell=1}^{L} \hat{b}_\ell^\dagger \right)^M |\mathrm{vac}\rangle \, , \tag{1.25}$$

is a dark state of the Liouvillian.

The physical implementation of the system [39] consists of an array of stripline resonators, coupled to each other by pairs of interacting superconducting qubits. Photons in the stripline resonators are capacitively coupled to the qubits. One transition in the four-levels spectrum of the interacting qubits is driven, two transitions are quasi-resonant with the cavity modes, and a fourth transition includes a dominant dissipative process. The cavity array is driven by a coherent source which creates photons and photon leakage has to be considered as well.

For simplicity, let us first consider an array with only two cavities, $L = 2$. It is convenient to introduce annihilation operators for the symmetric and antisymmetric superposition of the cavity modes, $\hat{b}_{\rm s} = (\hat{b}_1 + \hat{b}_2)/\sqrt{2}$, $\hat{b}_{\rm a} = (\hat{b}_1 - \hat{b}_2)/\sqrt{2}$, respectively. The equations of motion for the expectation values of the annihilation operators reads

$$\partial_t \langle \hat{b}_{\rm s} \rangle = [-i(\delta_{\rm d} - 2J) - \kappa]\langle \hat{b}_{\rm s} \rangle + \Gamma \langle \hat{b}_{\rm s} \hat{b}^\dagger_{\rm a} \hat{b}_{\rm a} \rangle,$$
$$\partial_t \langle \hat{b}_{\rm a} \rangle = (-i\delta_{\rm d} - \kappa)\langle \hat{b}_{\rm a} \rangle - i\Omega - \Gamma \langle (\hat{b}^\dagger_{\rm s} \hat{b}_{\rm s} + 1)\hat{b}_{\rm a} \rangle \, , \qquad (1.26)$$

where the time-dependence of the expectation values has been dropped from the notation, and $\delta_{\rm d}$ is the detuning of the antisymmetric cavity mode with respect to the frequency of the driving field. Here, Γ is the strength of the Lindblad terms in the engineered part of the Liouvillian and Ω is the amplitude of the driving field. The Liouvillian is engineered to generate a BEC in the symmetric mode. For this reason, external driving pumps photons in the antisymmetric mode only: all the photonic population in the symmetric mode is a direct consequence of the Liouvillian dynamics.

The solution of Eqs. (1.26) in the steady state can be obtained numerically by setting a limit to the maximum number of photons which can be present in the cavity modes. The photon occupation $n_{\rm s,a} = \langle \hat{b}^\dagger_{\rm s,a} \hat{b}_{\rm s,a} \rangle$ in the steady state is shown in the panel (b) of Fig. 1.6. An analytical solution is also easy to obtain within a semiclassical approximation, where each operator is substituted with its time-dependent quantum average: $\hat{b}_{\rm s,a} \mapsto b_{\rm s,a}(t)$. In this approximation, we find that a critical value of the driving strength Ω exists

$$\Omega_{\rm c} = \kappa \sqrt{\frac{\delta_{\rm d}^2 + (\kappa + \Gamma)^2}{\kappa \Gamma}} \, , \qquad (1.27)$$

such that: (i) for $\Omega < \Omega_{\rm c}$ the symmetric mode is empty and the population in the antisymmetric mode increases quadratically with Ω; (ii) for $\Omega > \Omega_{\rm c}$ the population in the antisymmetric mode is independent of Ω, and the population in the symmetric mode increases linearly with Ω. In other words, above thresholds the action of the engineered Liouvillian is to scatter photons from the antisymmetric to the symmetric mode. The numerical solution allows to compute the second-order, equal time correlation function $g^{(2)}$ of photons in the symmetric mode and shows that increasing the pumping strength produces a change in the light statistics from super-Poissonian to Poissonian. Poissonian statistics is a feature expected for a coherent state of photons in the same mode [6, 9], i.e. a BEC of photons. The numerical solution (which takes into account quantum fluctuations) and the semiclassical solution give comparable results, as shown in Fig. 1.6.

The time-evolution of an extended array with L cavities can be understood in terms of the two-cavities model. In this case, one introduces a family of modes with annihilation operators $\hat{b}_q \propto \sum_{\ell=1}^{L} e^{iq\ell}\hat{b}_\ell$, corresponding to Bloch waves [8] with quasimomentum q in the Brillouin zone of the lattice [8]. The analogous of the symmetric (antisymmetric) mode is the $q = 0$ ($q = \pi$) Bloch wave. Condensation takes place in the $q = 0$ mode, i.e. the mode obtained by the symmetric superposition of all the cavity modes. The external driving couples to all Bloch modes with amplitude Ω_q. By tuning the spatial dependence of the phase of the external driving along the lattice, one can change the profile Ω_q. Let us take $\Omega_q = \Omega\delta_{q,\pi}$ for analogy with the two-cavity case. (Here, $\delta_{i,j}$ is the Kronecker symbol.) It can be shown [39] that all the Bloch modes with $q \neq 0$ and $q \neq \pi$ are effectively decoupled from the dynamics and can be neglected. The cavity array then maps precisely into the two-cavity setup considered above. Exact diagonalization of a small array with $L = 4$ cavities supports these conclusions [39].

Acknowledgements We acknowledge the EU (through IP-SIQS and STREP-TermiQ) and the Italian MIUR through FIRB (Project RBFR12NLNA) and PRIN (Project 2010LLKJBX), for financial support. We would also acknowledge a fruitful collaboration on the topics discussed in this chapter with I. Carusotto, S. Diehl, D. Gerace, V. Giovannetti, A. Imamoğlu, D. Marcos, P. Rabl, G. Santoro, and H. Türeci.

References

1. C. Ciuti, I. Carusotto, Rev. Mod. Phys. **85**, 299 (2013)
2. M.J. Hartmann, F.G.S.L. Brandão, M.B. Plenio, Laser Photon. Rev. **2**, 527 (2008)
3. A. Tomadin, R. Fazio, J. Opt. Soc. Am. B **27**, A130 (2010)
4. A.A. Houck, H.E. Türeci, J. Koch, Nature Phys. **8**, 292 (2012)
5. S. Schmidt, J. Koch, Ann. der Physik **525**, 395 (2013)
6. S.M. Barnett, P.M. Radmore, *Methods in Theoretical Quantum Optics* (Clarendon Press, Oxford, 1997)
7. A. Imamoğlu, H. Schmidt, G. Woods, M. Deutsch, Phys. Rev. Lett. **79**, 1467 (1997)
8. N.W. Ashcroft, N.D. Mermin, *Solid State Physics* (Harcourt, Orlando, 1976)
9. D.F. Walls, G.J. Milburn, *Quantum Optics* (Springer-Verlag, Berlin, 1994)
10. N. Marzari, A.A. Mostofi, J.R. Yates, I. Souza, D. Vanderbilt, Rev. Mod. Phys. **84**, 1419 (2012)
11. A.L. Fetter, J.D. Walecka, *Quantum Theory of Many-Particle Systems* (Dover, New York, 2003)
12. M. Lewenstein, A. Sanpera, V. Ahufinger, *Ultracold Atoms in Optical Lattices* (Oxford University Press, USA, 2012)
13. M.J. Hartmann, F.G.S.L. Brandão, M.B. Plenio, Nature Phys. **2**, 849 (2006)
14. A.D. Greentree, C. Tahan, J.H. Cole, L. Hollenberg, Nature Phys. **2**, 856 (2006)
15. D. Angelakis, M. Santos, S. Bose, Phys. Rev. A **76**, 031805 (2007)
16. H.-P. Breuer, F. Petruccione, *The Theory of Open Quantum Systems* (Oxford University Press, USA, 2002)
17. L.D. Landau, E.M. Lifshitz, *Statistical Physics* (Pergamon Press, Oxford, 1980)
18. H. Carmichael, *An Open Systems Approach to Quantum Optics* (Springer-Verlag, Berlin, 1993)
19. N. Na, S. Utsunomiya, L. Tian, Y. Yamamoto, Phys. Rev. A **77**, 031803(R) (2007)
20. S. Schmidt, G. Blatter, Phys. Rev. Lett. **103**, 086403 (2009)
21. M. Aichhorn, M. Hohenadler, C. Tahan, P. Littlewood, Phys. Rev. Lett. **100**, 216401 (2008)
22. P. Pippan, H. Evertz, M. Hohenadler, Phys. Rev. A **80**, 033612 (2009)

23. M. Hohenadler, M. Aichhorn, S. Schmidt, L. Pollet, Phys. Rev. A **84**, 041608(R) (2011)
24. M. Hohenadler, M. Aichhorn, L. Pollet, S. Schmidt, Phys. Rev. A **85**, 013810 (2012)
25. D. Rossini, R. Fazio, Phys. Rev. Lett. **99**, 186401 (2007)
26. D. Rossini, R. Fazio, G. Santoro, Europhys. Lett. **83**, 47011 (2008)
27. M. Lewenstein, A. Sanpera, V. Ahufinger, B. Damski, A. Sen(De), and U. Sen. Adv. Phys. **56**, 243 (2007)
28. I. Bloch, J. Dalibard, W. Zwerger, Rev. Mod. Phys. **80**, 885 (2008)
29. K. Toyoda, Y. Matsuno, A. Noguchi, S. Haze, S. Urabe, Phys. Rev. Lett. **111**, 160501 (2013)
30. U. Schollwöck, Rev. Mod. Phys. **77**, 259 (2005)
31. T.D. Kühner, H. Monien, Phys. Rev. B **58**, 14741(R) (1998)
32. T.D. Kühner, S.R. White, H. Monien. Phys. Rev. B **61**, 12474 (2000)
33. J.K. Freericks, H. Monien, Phys. Rev. B **53**, 2691 (1996)
34. M.P.A. Fisher, P.B. Weichman, G. Grinstein, D.S. Fisher, Phys. Rev. B **40**, 546 (1989)
35. A. Tomadin, V. Giovannetti, R. Fazio, D. Gerace, I. Carusotto, H.E. Türeci, A. Imamoğlu, Phys. Rev. A **81**, 061801(R) (2010)
36. F. Verstraete, J.J. García-Ripoll, J.I. Cirac, Phys. Rev. Lett. **93**, 207204 (2004)
37. M. Zwolak, G. Vidal, Phys. Rev. Lett. **93**, 207205 (2004)
38. S. Diehl, A. Micheli, A. Kantian, B. Kraus, H.P. Büchler, P. Zoller, Nature Phys. **4**, 878 (2008)
39. D. Marcos, A. Tomadin, S. Diehl, P. Rabl, New J. Phys. **14**, 055005 (2012)
40. J. Klaers, J. Schmitt, F. Vewinger, M. Weitz, Nature **468**, 545 (2010)

Chapter 2
Phase Diagram and Excitations of the Jaynes-Cummings-Hubbard Model

Sebastian Schmidt and Gianni Blatter

Abstract The Jaynes-Cummings-Hubbard model (JCHM) has emerged as a fundamental model at the interface of quantum optics and condensed matter physics. It describes strongly correlated photons in a coupled qubit-cavity array and predicts a superfluid-Mott insulator transition of polaritons under quasi-equilibrium conditions. Here, we review recent analytical as well as numerical results for the phase diagram, elementary excitations and critical exponents of the JCHM and compare them to closely related models such as the Bose-Hubbard and the Dicke model. We comment on the fate of these results in open dissipative systems and outline schemes for their experimental verifiability.

2.1 Introduction

Since its proposal by Greentree [1] and Angelakis [2], the study of the Jaynes-Cummings-Hubbard model (JCHM) has become an active and versatile research field bridging two areas of modern physics: quantum optics and condensed matter. The JCHM describes photons hopping in an array of coupled cavities and locally interacting with qubits. Its Hamiltonian is given by

$$H = \sum_i h_i^{\mathrm{JC}} - J \sum_{\langle ij \rangle} a_i^\dagger a_j \,, \tag{2.1}$$

where h_i^{JC} denotes the local Jaynes-Cummings Hamiltonian

$$h_i^{\mathrm{JC}} = \omega_c \, a_i^\dagger a_i + \omega_x \sigma_i^+ \sigma_i^- + g(\sigma_i^+ a_i + \sigma_i^- a_i^\dagger) \tag{2.2}$$

S. Schmidt (✉) · G. Blatter
Institute for Theoretical Physics, Eth Zurich, Zürich, Switzerland
e-mail: schmidts@phys.ethz.ch

G. Blatter
e-mail: blatterj@phys.ethz.ch

D.G. Angelakis (ed.), *Quantum Simulations with Photons and Polaritons*, Quantum Science and Technology, DOI 10.1007/978-3-319-52025-4_2

with site index i, photon creation (annihilation) operators $a_i^{(\dagger)}$ and qubit raising (lowering) operators $\sigma_i^{+(-)}$. The cavity mode frequency is ω_c, the two qubit levels are separated by the energy ω_x and g denotes the light-matter coupling strength (we set $\hbar = 1$). The second term in (2.1) describes photon hopping between nearest neighbour sites with amplitude J. The light-matter interaction mediated by g leads to the formation of polariton quasiparticles and introduces a nonlinearity into the spectrum of the JCHM. Depending on the perspective, the JCHM can thus be viewed as a lattice model for dressed photons, which gain an effective mass through cavity confinement and are interacting due to the nonlinearity mediated by the qubits or as an effective spin lattice model, where effective spin-spin interactions are mediated by photon exchange between distant cavities [2–5].

The physics of coupled cavity arrays is very rich reaching from strong correlation phenomena and driven dissipative phase transitions, e.g., superfluid-Mott insulator type transitions [1, 2, 6–28], self-trapping and Josephson-like phenomena [29–33], crystalline-like phases of light [34–37] or fermionization of photons [38–40] to topological phenomena associated with the engineering of artificial gauge fields [41–48], edge or Majorana-like modes [48–52] or the Fractional quantum Hall effect [53–56]. For recent reviews on the topic see Refs. [57–61].

One of the major appeals of the JCHM is the possibility to engineer this model bottom-up in solid-state cavity or circuit QED architectures. Photonic lattices (without the coupling to qubits) have already been engineered on a large scale with only little disorder in cavity frequencies (as compared to the hopping rate) [43, 62]. Effective photon-photon interactions in a coupled qubit-cavity array were realised very recently as well: In Ref. [33] a non-equilibrium delocalization-localization transition (self-trapping) of photons has been observed as originally predicted in Ref. [31]. The experiment was based on a two site version of the JCHM, i.e., a so-called Jaynes Cummings dimer. It demonstrates a new scalable architecture for quantum simulation of non-equilibrium many-body systems with a high-level of control over coherent and dissipative dynamics.

A chemical potential for polaritons?
The global $U(1)$ symmetry of the JCHM preserves the total number of polaritons $N = \sum_i (a_i^\dagger a_i + \sigma_i^+ \sigma_i^-)$. It is thus convenient to introduce a chemical potential μ, which fixes the number of polaritons [1, 2], i.e.,

$$H \to H - \mu N \,, \tag{2.3}$$

Introducing such a chemical potential assumes the existence of a grand-canonical ensemble and thus thermalisation. However, photonic systems are driven dissipative systems and typically settle in a non-equilibrium steady state, which can be far from equilibrium. It is thus important to identify situations under which calculations in the grand-canonical ensemble are meaningful and experimentally verifiable. Below we briefly discuss two possible experimental schemes.

The first situation considers incoherent pumping of a polaritonic system with weak dissipation. For example, photons confined in a dye-filled optical micro-cavity condense into an equilibrium Bose-Einstein condensate with grand-canonical number statistics [63–65]. In such a system, effective thermalisation of photons occurs through emission and absorption by dye-molecules, which themselves are coupled to a large thermal phonon reservoir. Possible extensions of photonic condensates to periodic photonic lattices are discussed in Ref. [66]. Note, that such incoherent pumping mechanisms can also be realised in strongly interacting light-matter systems based on exciton-polaritons [67] and superconducting qubits [68].

The second situation considers quenched dynamics. For example, a coupled qubit-cavity array can be initialized in a Mott-like state using a large coherent pulse [19]. Observing the subsequent decay of such a special initial state under weak dissipation reveals the quantum phase diagram of the Hamiltonian under equilibrium conditions, i.e. the coherence of the system at long times approaches zero for $J < J_c$, i.e., in the Mott phase, while it is nonzero for $J > J_c$, i.e., in the superfluid phase (J_c denotes the critical hopping strength of the quantum phase transition in equilibrium). Note, that the distinction between Mott and superfluid phases becomes sharp only when dissipation vanishes, but otherwise is smoothened and washed out (similar to finite temperature effects for ultra cold atoms in optical lattices as described by the Bose-Hubbard model (BMH) [69]).

In this review we focus on the superfluid-Mott insulator transition of polaritons under (quasi-) equilibrium conditions. In Chap. 2, we discuss the atomic limit ($J = 0$) and simple perturbative estimates for the phase diagram of the JCHM. In Chap. 3, we show that a linked-cluster expansion of the photonic Matsubara Green's function can be utilised to calculate the quantum phase diagram and elementary excitations of the JCHM in the Mott phase. In Sect. 2.3 we generalise such a strong coupling approach to the superfluid phase using a slave-boson technique. In Chap. 4 we briefly summarise the results of exact Quantum-Monte-Carlo simulations determining the universality class and critical exponents of the JCHM. Finally, in Chap. 5 we discuss an intricate relation between the SF-MI transition in the JCHM and the super-radiance phase transition in a multi-mode Dicke model. In particular, we show that a weak-coupling mean-field theory predicts the existence of a single Mott lobe even in the absence of an underlying cavity array.

2.2 Degenerate Perturbation Theory

In the atomic limit ($J = 0$) the eigenstates of the Hamiltonian (2.1) are the dressed polariton states $|n\sigma\rangle$ labelled by the polariton number n and upper/lower branch index $\sigma = \pm$. For $n > 0$ they can be written as a superposition of a Fock state with n photons plus atomic ground state $|n, g\rangle$ and $(n-1)$ photons with the atom in its excited state $|(n-1), e\rangle$,

$$\begin{aligned} |n+\rangle &= \sin\theta_n |n\,,\,g\rangle + \cos\theta_n |(n-1)\,,\,e\rangle\,, \\ |n-\rangle &= \cos\theta_n |n\,,\,g\rangle - \sin\theta_n |(n-1)\,,\,e\rangle\,, \end{aligned} \tag{2.4}$$

with the angle $\tan\theta_n = 2g\sqrt{n}/(\delta + 2\chi_n)$, $\chi_n = \sqrt{g^2 n + \delta^2/4}$ and the detuning parameter $\delta = \omega_c - \omega_x$. The corresponding eigenvalues are

$$\epsilon_n^\sigma = -(\mu - \omega_c)n - \delta/2 + \sigma\,\chi_n\,, \quad \sigma = \pm\,. \tag{2.5}$$

The zero polariton state $|0-\rangle = |0\,, g\rangle$ is a special case with $\epsilon_0^- = 0$. Note, that upper and lower polariton energies are separated by the so-called Rabi splitting $\Omega_n = 2\sqrt{g^2 n + \delta^2/4}$.

Starting from the atomic limit result in Eqs. (2.4) and (2.5) we can obtain a rough estimate for the quantum phase diagram of the JCHM at small tunnelling $J \ll g$ by calculating the excitation energies in straightforward degenerate perturbation theory. To first order, the chemical potentials at which the addition/removal of a lower polariton costs no energy is given by (we assume a hypercubic lattice in D dimensions)

$$\begin{aligned}\mu_p - \omega_c &= \chi_n - \chi_{n+1} - 2DJ(f_{n+1}^{--})^2 + \mathcal{O}(J^2/g)\,,\\ \mu_h - \omega_c &= \chi_{n-1} - \chi_n + 2DJ(f_n^{--})^2 + \mathcal{O}(J^2/g)\,.\end{aligned} \tag{2.6}$$

The two equations in (2.6) define the upper and lower phase boundary in the quantum phase diagram in Fig. 2.1 for small values of the hopping parameter J/g. The point where the two lines meet (i.e., $\mu_p = \mu_h$) is $J_c \sim 0.1g$ for $n = 1$ and represents an upper limit for the size of the first Mott lobe. By going to higher order in the perturbative expansion for the ground-state energy and subsequent resummation of

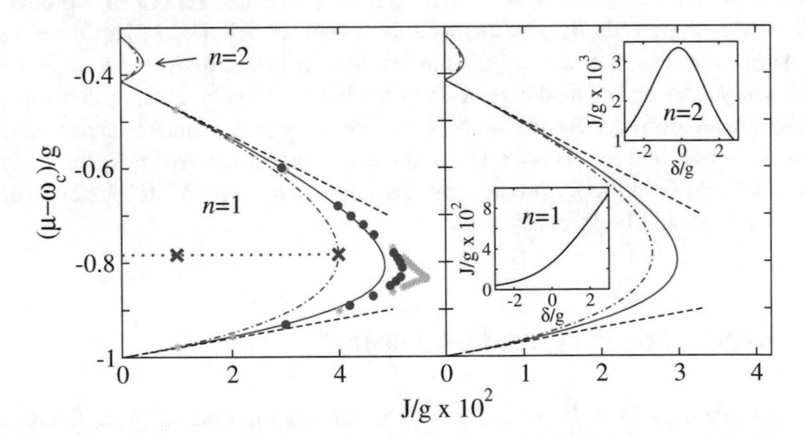

Fig. 2.1 Quantum phase diagram for a hypercubic lattice in $D = 2$ (*left figure*) and $D = 3$ (*right figure*). We show two Mott lobes for $n = 1, 2$ at zero detuning $\delta = 0$. We compare first-order perturbation theory (*dashed*), RPA (*dot-dashed*) and quantum fluctuations (*solid*) with recent results from a quantum Monte-Carlo (*filled dots*) [11] and variational cluster (stars) [9] approach. The two crosses (*left figure*) connected by the *dotted line* mark the parameter values used in Fig. 2.2. The insets show the critical hopping strength J_c/g at the tip of the lobe as a function of the detuning δ for $n = 1$ and $n = 2$ calculated within strong-coupling RPA in $D = 3$ dimensions. Figure taken with permission from Ref. [12]

the diverging strong-coupling series, e.g., via variational perturbation theory (VPT) [70, 71], one could in principle determine the exact location of the phase boundary. This has recently been achieved for the BHM [72, 73].

2.3 Greens Function Approach

In this section, we study the photonic Matsubara Green's function

$$G_{ij}(\tau;\tau') = -\langle \mathcal{T} a_i(\tau)\bar{a}_j(\tau')\rangle \tag{2.7}$$

with the time-ordering operator $\mathcal{T}$ and the Heisenberg operator $\bar{a}_j(\tau') = e^{H\tau'}a_j^{\dagger}e^{-H\tau'}$. A suitable method for the evaluation of the Matsubara Green's function is a linked-cluster expansion in terms of local cumulants originally developed by Metzner et al. [74] for the Fermi-Hubbard model (FHM) and more recently applied to the BHM [75].

First, we study the atomic limit ($J = 0$). For $J = 0$ the Matsubara Green's function is local and given by

$$G_{0ij}(\tau;\tau') = G_{0i}(\tau;\tau')\,\delta_{ij} = -\langle \mathcal{T} a_i(\tau)\bar{a}_i(\tau')\rangle_0\,\delta_{ij}\,, \tag{2.8}$$

where the average $\langle\ldots\rangle_0$ is taken with respect to the eigenstates of the local Hamiltonian in Eq. (2.4). We consider a spatially homogeneous system, drop the site index i and obtain after a Fourier transformation

$$G_0(\omega_m)=\sum_{n,\sigma,\mu}\frac{e^{-\beta\epsilon_n^{\sigma}}}{Z}\left(\frac{z_{n+1}^{\mu\sigma}}{\Delta_{n+1}^{\mu\sigma}-i\omega_m}-\frac{z_n^{\sigma\mu}}{\Delta_n^{\sigma\mu}-i\omega_m}\right) \tag{2.9}$$

with the partition function $Z = \sum_{n\sigma} e^{-\beta\epsilon_n^{\sigma}}$ and bosonic Matsubara frequencies $\omega_m = 2\pi m/\beta$ (here, $\beta = 1/(k_B T)$ with temperature T and Boltzmann constant k_B). The spectral weights in (2.9) are defined as $z_n^{\mu\sigma} = (f_n^{\mu\sigma})^2$ with the matrix elements $f_n^{\sigma\nu} = \langle n\,\sigma|a^{\dagger}|(n-1)\,\nu\rangle$. At zero detuning ($\delta = 0$) they are given by $f_n^{\sigma\nu} = \left(\sqrt{n}+\sigma\,\nu\,\sqrt{n-1}\right)/2$ for $n > 1$ ($f_1^{\sigma -} = 1/\sqrt{2}$).

Following the recipe in Metzner et al. [74], each term of the linked-cluster expansion can be written diagrammatically in terms of n-particle cumulants represented by 2n-leg vertices and tunneling matrix elements symbolized by propagating lines connecting two vertices. The strong-coupling expansion provided by the linked-cluster method is applicable to the JCHM because (i) the atomic limit Hamiltonian is local, and (ii) anomalous averages of the photon operator with respect to the eigenstates of the local Hamiltonian vanish, i.e., $\langle n\sigma|(a^{\dagger})^k|n\sigma\rangle = 0$ for $k \in N$ (since a single photon excitation always changes the polariton number). In this case an infinite set of diagrams can be summed by calculating the irreducible part of the Green's function $K(\mathbf{k}, \omega_m)$ which is connected to the full Green's function via the equation

$$G(\mathbf{k}, \omega_m) = K(\mathbf{k}, \omega_m)/[1 - J(\mathbf{k})K(\mathbf{k}, \omega_m)] \tag{2.10}$$

with the lattice dispersion $J(\mathbf{k})$, e.g., for a hypercubic lattice $J(\mathbf{k}) = 2J \sum_{i=1}^{D} \cos \mathbf{k} \cdot \mathbf{a_i}$ ($\mathbf{a}_i$ denotes a lattice vector). To second order in J we obtain

$$\begin{aligned} K(\omega_m) &= \rightarrow\!\bullet\!\rightarrow \; + \; \overset{\circlearrowright}{\rightarrow\!\bullet\!\rightarrow} \\ &= G_0(\omega_m) + 2DJ^2\, Q(\omega_m) \end{aligned} \tag{2.11}$$

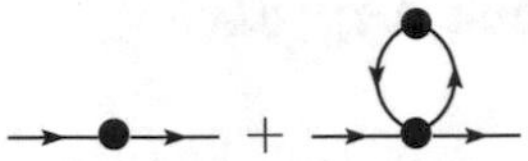

with

$$Q(\omega_m) = \int_0^\beta d\tau d\tau_1 d\tau_2\, C^{(2)}(\tau_1 0; \tau_2 \tau) G_0(\tau_2; \tau_1) e^{i\omega_m \tau}. \tag{2.12}$$

The quantum fluctuation correction $Q(\omega_m)$ involves the two-particle cumulant $C^{(2)}(\tau_1 0; \tau_2 \tau)$, which is related to the local (atomic limit) two-particle Green's function $G_0^{(2)}(\tau_1 0; \tau_2 \tau) = \langle T a_i(\tau_1) a_i(0) \bar{a}_i(\tau_2) \bar{a}_i(\tau) \rangle_0$ via

$$C^{(2)}(\tau_1 0; \tau_2 \tau) = G_0^{(2)}(\tau_1 0; \tau_2 \tau) - G_0(\tau_1; \tau_2) G_0(0; \tau) - G_0(\tau_1; \tau) G_0(0; \tau_2). \tag{2.13}$$

The algebraic expressions for the two-particle cumulant and the quantum correction $Q(\omega_n)$ are lengthy, but can be calculated in a straightforward manner [12].

The object of interest is the inverse Greens function which tells us immediately about the phase boundary $G^{-1}(\mathbf{0}, 0)|_{J_c(\mu)} = 0$ and the dispersion relation $G^{-1}(\mathbf{k}, i\omega_m \to \omega + i0^+) = 0$. It is thus convenient to introduce the strong-coupling self-energy $\Sigma(\mathbf{k}, \omega_m)$ via $G(\mathbf{k}, \omega_m)^{-1} = G_0(\omega_m)^{-1} - \Sigma(\mathbf{k}, \omega_m)$. From (2.11) we obtain to second order

$$\Sigma(\mathbf{k}, \omega_m) = J(\mathbf{k}) + 2DJ^2 Q(\omega_m)/G_0(\omega_m). \tag{2.14}$$

The first term on the r.h.s. is usually called the strong-coupling random-phase approximation (RPA), whereas the second term denotes the leading correction due to quantum fluctuations. The RPA corresponds to a summation of all self-avoiding walks (chain diagrams) through the lattice. The leading quantum correction includes in addition all one-time forward/backward hopping processes (bubble diagrams) between two neighbored sites.

Although we have carried out our calculations at finite temperature, we will only consider the $T = 0$ case from now on. At the quantum phase transition the energy of long wavelength fluctuations vanishes and we obtain an explicit expression for the critical hopping strength

$$J_c^{-1} = D\, G_0(0) \big(1 + \sqrt{1 + 2Q(0)/(D\, G_0^3(0))}\big). \tag{2.15}$$

If we ignore the second term under the square root in (2.15) we obtain the RPA phase boundary $1/J_c = 2D\, G_0(0)$ shown as a dashed-dotted line in Fig. 2.1. This result agrees exactly with the phase boundary as obtained from a numerical decoupling mean-field approach [1, 2]. The inclusion of quantum fluctuations in (2.15) leads to an improved phase boundary (solid line in Fig. 2.1). In two dimensions the result in Eq. (2.15) agrees well with Monte-Carlo calculations [11, 20, 22]. There is a small deviation near the tip of the lobe, where quantum corrections are most important. In the right panel of Fig. 2.1 we present quantitatively accurate results for the phase diagram in $D = 3$ (note, that the strong-coupling expansion becomes more accurate in higher dimensions).

We now discuss the elementary excitations of the JCHM in the Mott phase based on the RPA, i.e., by neglecting the second term in (2.14). An analytic continuation of the Matsubara Green's function via $i\omega_n \to \omega + i0^+$ yields the retarded real-time Green's function. Its poles and residues provide us with the dispersion relations and mode strengths of the fundamental excitations. In general, we obtain four poles, the conventional (Bose-Hubbard like) lower polariton particle (hole) modes $\omega^-_{(p,h)}$ and two modes $\omega^+_{(p,h)}$ which correspond to an upper polariton particle (hole) excitation (conversion modes). The presence of the latter signals a clear deviation from the usual Bose-Hubbard like physics and is due to the composite quasi-particle nature of polaritons.

The conversion modes exist already in the atomic limit, but have been overlooked in early numerical approaches. One reason might be that their bandwidth and strengths are very small as compared to the conventional modes. However, extended numerical studies have confirmed the prediction in [14]. The weakness of the conversion modes throughout the Mott lobe allows us to set their dispersion relations $\omega^+_{(p,h)}$ and mode strength $s^+_{(p,h)}$ approximately equal to $\omega^+_{(p,h)} \approx \Delta^+_{(p,h)}$ and $s^+_{(p,h)} \approx z^+_{(p,h)}$ with the atomic-limit particle (hole) gaps $\Delta^\sigma_p \equiv \Delta^{\sigma-}_{n+1}$ $(\Delta^\sigma_h \equiv \Delta^{-\sigma}_n)$ and mode strengths $z^\sigma_p \equiv z^{\sigma-}_{n+1}$ $(z^\sigma_h \equiv -z^{-\sigma}_n)$, respectively. If we neglect their contribution to the one-particle cumulant (2.9), we can derive simple analytic formulas for the dispersion relations of the conventional modes, i.e.,

$$\omega^-_{(p,h)} = (\Delta_+ \;-\; J(\mathbf{k})\, z_+ \,\pm\, \Omega)\,/2 \tag{2.16}$$

with $\Omega = \sqrt{\Delta_-^2 + J(\mathbf{k})^2\, z_+^2 - 2J(\mathbf{k})\, z_-\, \Delta_-}$ and the abbreviations $\Delta_\pm = \Delta^-_p \pm \Delta^-_h$ and $z_\pm = z^-_p \pm z^-_h$. The strength of the modes are given by

$$s^-_{(p,h)} = \frac{z_+\, \omega^-_{(p,h)} - z^-_{(p,h)} \Delta^-_{(h,p)} - z^-_{(h,p)} \Delta^-_{(p,h)}}{\omega^-_{(p,h)} - \omega^-_{(h,p)}}\,. \tag{2.17}$$

The dispersions in (2.16) are plotted in Fig. 2.2 deep inside the Mott regime and at the tip of the lobe with $n = 1$. As shown in the inset of Fig. 2.2 the energy needed for a conventional excitation is an order of magnitude smaller than for a conversion excitation. The strengths of the conventional modes $s^-_{(p,h)}$ grow with increasing tunneling

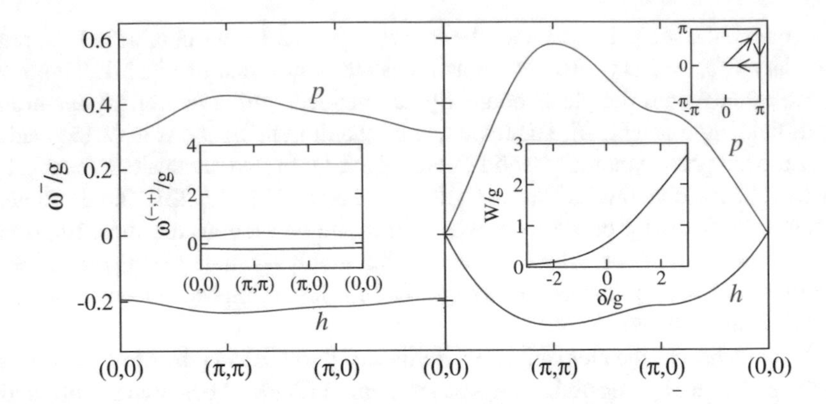

Fig. 2.2 Particle/hole dispersion of the conventional modes $\omega^{-}_{(p,h)}$ for $D = 2$ and $n = 1$ at zero detuning $\delta = 0$. *Left figure* Deep inside the Mott insulator (see cross in Fig. 2.1) for $(\mu - \omega_c)/g = -0.78$ and $J/g = 0.01$. The inset shows the conventional modes $\omega^{-}_{(p,h)}$ together with the conversion mode ω^{+}_{p}. *Right figure* At the phase boundary (see cross in Fig. 2.1) for $(\mu - \omega_c)/g = -0.78$ and $J/g = 0.04$. The inset shows the bandwidth W of the conventional particle mode ω^{-}_{p} as a function of detuning δ at the tip of the lobe. Figure taken with permission from Ref. [12]

strength while the strengths of the conversion modes $s^{+}_{(p,h)}$ stay approximately constant. If we are only interested in low energy excitations we can thus indeed neglect the contributions from upper polaritons throughout the Mott phase. They will always remain gapped even at the phase boundary. We will make use of this result in the next section, where we will also calculate the excitation spectra in the superfluid phase using a slave-boson approach.

2.4 Slave-Boson Approach

In the previous section we derived an analytic strong-coupling theory for the phase diagram and the elementary excitations in the Mott phase of the JCHM based on a linked-cluster expansion (LCA) of the Matsubara Greens function. Here, we generalise such a strong-coupling approach to the superfluid phase using a slave-boson technique [18], which was previously applied to the BHM [76].

2.4.1 Slave-Boson Formulation

A convenient starting point is the polariton representation [13] of the boson operator

$$a_i = \sum_{n\sigma\nu} f^{\sigma\nu}_{n} P^{\nu\dagger}_{in-1} P^{\sigma}_{in} \tag{2.18}$$

in terms of standard algebra operators $P_{in}^{\sigma\dagger} = |n\sigma\rangle_{ii}\langle 0|$ and matrix elements $f_n^{\sigma\nu}$ defined below Eq. (2.9). In this new basis the JCHM becomes

$$H = \sum_i \sum_{n=0}^{\infty} \sum_{\sigma} \epsilon_n^{\sigma} P_{in}^{\sigma\dagger} P_{in}^{\sigma} - J \sum_{\langle ij\rangle} \sum_{n,n'=1} \sum_{\substack{\sigma,\sigma' \\ \nu,\nu'}} f_n^{\sigma\sigma'} f_{n'}^{\nu\nu'} P_{in}^{\sigma\dagger} P_{in-1}^{\sigma'} P_{jn'-1}^{\nu'\dagger} P_{jn'}^{\nu} . \tag{2.19}$$

The upper polariton branch with $\sigma, \sigma' = +$ leads to additional high energy conversion modes in the Mott phase with small spectral weight and bandwidth as discussed in the previous section. We thus neglect the upper branch as well as particle conversion hopping from now on and drop the branch index σ. This leads to the simplified Hamiltonian

$$H = \sum_i \sum_{n=0}^{\infty} \epsilon_n P_{in}^{\dagger} P_{in} - J \sum_{\langle ij\rangle} \sum_{n,n'=1} f_n f_{n'} P_{in}^{\dagger} P_{in-1} P_{jn'-1}^{\dagger} P_{jn'} . \tag{2.20}$$

with $\epsilon_n \equiv \epsilon_n^-$ and $f_n \equiv f_n^{--}$.

2.4.2 Gutzwiller Ansatz

In order to calculate the phase boundary and static observables in the superfluid phase near a Mott lobe with filling $n \geq 1$, we restrict the Hilbert space to states with n and $n \pm 1$ bosons and make a Gutzwiller Ansatz for the ground-state wave function

$$|\psi\rangle = \prod_i [\cos(\theta) P_{i0}^{\dagger} + \sin(\theta)(\sin(\chi) P_{i-1}^{\dagger} + \cos(\chi) P_{i1}^{\dagger})]|0\rangle ,$$

where we also dropped the index n and changed the notation to $P_{i\alpha}^{\dagger} \equiv P_{in+\alpha}^{\dagger}$, $\epsilon_{n+\alpha} \equiv \epsilon_{\alpha}$, and $f_{n+\alpha} \equiv f_{\alpha}$. Note, that this variational wave function is normalized to unity and satisfies the completeness relation, i.e.

$$\sum_{n\sigma} P_{in}^{\sigma\dagger} P_{in}^{\sigma} = 1 . \tag{2.21}$$

It is straightforward to show that if the constraint (2.21) is fulfilled, the polariton operators also obey bosonic commutation relations.

The expectation value $\epsilon_{\text{var}} = \langle\psi|H|\psi\rangle$ yields the variational energy

$$\begin{aligned} \epsilon_{\text{var}} = {} & \epsilon_0 \cos(\theta)^2 + \sin(\theta)^2 \left[\epsilon_{-1} \sin(\chi)^2 + \epsilon_1 \cos(\chi)^2\right] \\ & - JD/2 \sin(2\theta)^2 \left[f_0 \cos(\chi) + f_{-1} \sin(\chi)\right]^2 , \end{aligned} \tag{2.22}$$

which has to be minimized with respect to the variational parameters θ and χ (here, D denotes the dimension of a hypercubic lattice), yielding

$$\tan(2\chi) = \frac{4JDf_0 f_1 \cos(\theta)^2}{\epsilon_{-1} - \epsilon_1 + 2JD(f_1^2 - f_0^2)\cos(\theta)^2} \tag{2.23}$$

and

$$\cos(2\theta) = \frac{1}{2JD} \frac{\epsilon_1 \cos(\chi)^2 + \epsilon_{-1}\sin(\chi)^2 - \epsilon_0}{[f_0 \sin(\chi) + f_1 \cos(\chi)]^2} \,. \tag{2.24}$$

The lobe boundaries are then determined by the vanishing of the order parameter

$$\phi_c = \langle\psi|a|\psi\rangle = \sin(\theta)\,[f_0 \sin(\chi) + f_1 \cos(\chi)]^2\,/2. \tag{2.25}$$

Setting $\phi_c = 0$ (i.e., $\theta = 0$) in (2.23) and (2.24) and eliminating χ yields the relation

$$\epsilon_{-1} - \epsilon_1 = -Jz(f_1^2 - f_0^2) \pm \sqrt{Q} \tag{2.26}$$

with

$$Q = U^2 - 2Jz(f_0^2 + f_1^2)U + J^2 z^2 (f_1^2 - f_0^2)^2 \tag{2.27}$$

and

$$U = \epsilon_{-1} - 2\epsilon_0 + \epsilon_1\,. \tag{2.28}$$

Equation (2.26) constitutes an expression for the mean-field boundaries of the Mott lobes in the JCHM shown in Fig. 2.3 (the energies ϵ_α contain the chemical potential). Note, that they agree exactly with those obtained from the RPA in the previous section.

2.4.3 Quadratic Fluctuations

In order to find the elementary excitations we define a new set of operators $\mathbf{R}^\dagger = (G_i^\dagger, E_{1i}^\dagger, E_{2i}^\dagger)^T$, which is obtained from the original polariton basis $\mathbf{P}^\dagger = (P_{i0}^\dagger, P_{i-1}^\dagger, P_{i1}^\dagger)^T$ via a unitary transformation $\mathbf{R}^\dagger = T\mathbf{P}^\dagger$ with

$$T = \begin{pmatrix} \cos(\theta) & \sin(\theta)\cos(\chi) & \sin(\theta)\sin(\chi) \\ -\sin(\theta) & \cos(\theta)\cos(\chi) & \cos(\theta)\sin(\chi) \\ 0 & -\sin(\chi) & \cos(\chi) \end{pmatrix} . \tag{2.29}$$

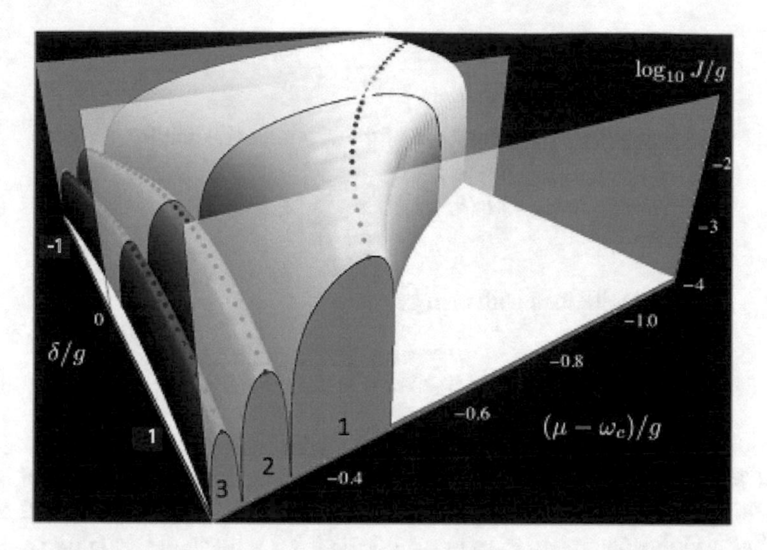

Fig. 2.3 Quantum phase diagram for the JCHM as obtained from slave-boson theory, i.e., Eq. (2.26). Shown are the lowest three Mott lobes with polariton numbers $N = 1, 2, 3$. *Dotted lines* represent the critical hopping strength's J_c/g for which the chemical potential μ and detuning δ are chosen such as to fullfill particle-hole symmetry. Finite detuning $|\delta| > 0$ decreases the critical hopping strength J_c/g for $N > 1$, but the lowest Mott lobe ($N = 1$) steadily increases when tuning through the resonance ($\delta = 0$). Figure taken with permission from [18] (with minor modifications)

The operator $G^\dagger$ creates a new vacuum state, i.e., the mean-field ground state $|\psi\rangle = \prod_i G_i^\dagger |0\rangle$, and $E_{1i}^\dagger$, $E_{2i}^\dagger$ are orthogonal operators creating excitations above the ground-state. We express the Hamiltonian in terms of these new operators and eliminate G_i by using the constraint (2.21) in the restricted Hilbert space

$$G_i \approx \sqrt{1 - E_{1i}^\dagger E_{1i} - E_{2i}^\dagger E_{2i}} \tag{2.30}$$

Expanding the square root everywhere in the Hamiltonian to quadratic order in $E_{(1,2)i}^{(\dagger)}$ yields, after a Fourier transformation, an effective quadratic Hamiltonian

$$H_{\text{eff}} = \epsilon_{\text{var}} + \sum_{\mathbf{k}} \mathbf{E}_{\mathbf{k}}^\dagger \, h_{\text{eff},\mathbf{k}} \, \mathbf{E}_{\mathbf{k}} \tag{2.31}$$

where $\mathbf{E} = (E_{1\mathbf{k}}, E_{2\mathbf{k}}, E_{1-\mathbf{k}}^\dagger, E_{2-\mathbf{k}}^\dagger)^T$ and $h_{\text{eff},\mathbf{k}}$ is a 4×4 matrix

$$h_{\text{eff},\mathbf{k}} = \begin{pmatrix} g & f \\ f & g \end{pmatrix}, \tag{2.32}$$

with f, g denoting 2×2 matrices defined in the appendix. The sum over $\mathbf{k}$ runs over the first Brioullin zone. The effective Hamiltonian can be diagonalized by a bosonic Bogoliubov transformation yielding

$$H_{\text{eff}} = \epsilon_{\text{var}} + \epsilon_{\text{fluct}} + \sum_{\alpha=\pm} \sum_{\mathbf{k}} \epsilon_\alpha(\mathbf{k}) d^\dagger_{\alpha\mathbf{k}} d_{\alpha\mathbf{k}} \tag{2.33}$$

with a fluctuation-generated correction of the ground-state energy

$$\epsilon_{\text{fluct}} = \mathcal{E}(\theta, \chi) + \sum_{\alpha=\pm} \sum_{\mathbf{k}} \epsilon_\alpha(\mathbf{k})/2 \tag{2.34}$$

and $d^\dagger_{\alpha\mathbf{k}}$ creating excitations with energy

$$\epsilon_\pm(\mathbf{k}) = \sqrt{A(\mathbf{k}) \pm \sqrt{A(\mathbf{k})^2 - B(\mathbf{k})}} \tag{2.35}$$

with rather lengthy expressions for $\mathcal{E}(\theta, \chi)$, $A(\mathbf{k})$, and $B(\mathbf{k})$ given in the appendix of Ref. [26]. In the Mott phase, these spectra agree with the expressions for particle/hole like modes in Eq. (2.16). In the superfluid phase, we obtain a gapless, linear Goldstone mode $\epsilon_-(\mathbf{k}) = c_s|\mathbf{k}| + \mathcal{O}(\mathbf{k}^2)$ with sound velocity c_s. A second mode, the so-called amplitude or Higgs mode, generally remains gapped with $\epsilon_+(\mathbf{k}) = \Delta_a + \mathcal{O}(\mathbf{k}^2)$ (except for the tip of the lobe). For a more detailed discussion of the excitation spectra we refer to the caption in Fig. 2.4.

2.5 Critical Exponents

In the previous two sections we have calculated the dispersion relations for the elementary excitations of the JCHM in the Mott and superfluid phases. The long wavelength behavior of the dispersion at $\mathbf{k} \to 0$ right at the phase boundary determines the dynamical critical exponent z defined by $\omega \sim \xi^{-z} \sim k^z$ with the diverging correlation length $\xi \sim |J - J_c|^{-\nu}$ and its associated critical exponent ν. If the phase boundary is approached away from the tip of the lobe, either the particle (upper phase boundary) or the hole (lower phase boundary) gap vanishes linearily $\Delta \sim |J - J_c|$ and the dispersion remains quadratic $\omega \sim k^2$. The situation changes at the tip of the lobe, where particle and hole gaps vanish simultaneously with a square-root behavior $\Delta \sim |J - J_c|^{1/2}$, while their dispersions become linear $\omega \sim k$ (see Fig. 2.4). This indicates a special transition at the tip of the lobe, reminiscent of an emergent particle-hole symmetry.

Consequently, according to the analytical results of the previous sections the dynamical critical exponent has the generic value $z = 2$ everywhere in the phase diagram except for the special critical point at the tip of the lobe where it changes to $z = 1$. At $\mathbf{k} = \mathbf{0}$ the gap vanishes as $\Delta \sim |J - J_c|^{z\nu}$ when the tunneling strength approaches its critical value J_c. This leads to a mean-field exponent $\nu = 1/2$ everywhere in the phase diagram. Thus on a mean-field level, the JCHM has the same critical exponents as the BHM [69] with a change of its universality class along the

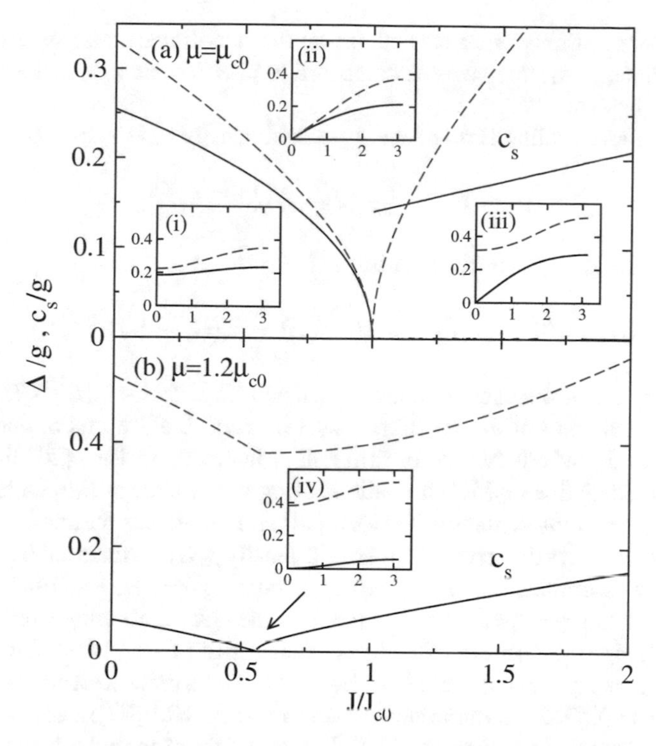

Fig. 2.4 Elementary excitations of the JCHM as a function of the effective hopping strength J/J_{c0} at zero detuning $\delta/g = 0$ and for **a** $\mu = \mu_{c0}$ where μ_{c0} denotes the critical chemical potential at the tip of the lobe with critical hopping strength $J_c = J_{c0}$ (*top figure*) and **b** away from the tip at $\mu = 1.2\mu_{c0}$ with $J_c = 0.566J_{c0}$ (*bottom figure*). Shown are the gaps of particle (*dashed*) and hole (*solid*) modes in the Mott phase ($J < J_c$) and as well as the gaps of the Amplitude mode (*dashed*) and the sound velocity of the Goldstone mode (*solid*). The insets show the corresponding excitation spectra at (i) $J = 0.5J_{c0}$ (in the Mott phase) (ii) $J = J_{c0}$ (at the tip of the lobe) (iii) $J = 1.5J_{c0}$ (in the superfluid phase) (iv) $J = 0.566J_{c0}$ (at the phase boundary away from the tip of the lobe). At the phase boundary, the particle and hole mode of the Mott phase are identical with the Goldstone and Amplitude modes of the superfluid phase. At the tip of the lobe (ii), where the polariton density can remain constant during the superfluid-insulator transition, the Amplitude mode becomes gapless and linear (its mass vanishes). The sound velocity of the Goldstone mode remains non-zero, confirming a special point in the phase diagram with dynamical critical exponent $z = 1$. Away from the tip (iv), the Amplitude mode remains gapped and the Goldstone mode becomes quadratic with a vanishing sound velocity corresponding to a generic dynamical critical exponent $z = 2$. Figure taken with permission from [18] (with minor modifications)

phase boundary. This result has also been predicted using scaling arguments based on an effective action approach [13].

However, early Quantum Monte Carlo (QMC) calculations of the superfluid density [11] suggested that the universality class at the tip of the Mott lobe of the JCHM is different from the BHM. This controversy between analytical and numerical findings has been resolved in Ref. [20] in favour of the early analytical arguments in Refs. [12, 13]. In Ref. [20] extensive QMC simulations of the superfluid density and the compressibility were carried out on the two-dimensional square lattice by using much larger system sizes as compared to previous studies. Below we briefly summarise these results.

The finite-size scaling form of the superfluid density ρ_s is known from Ref. [69] and reads

$$\rho_s = L^{2-D-z}\tilde{\rho}_s[(J-J_c)L^{1/\nu}, \beta/L^z]. \tag{2.36}$$

Fixing the ratio $\alpha = \beta/L^z$, the quantity

$$X_{z\nu}(L) = L^{D-2+z}\rho_s[(J-J_c)L^{1/\nu}, \alpha] = \tilde{\rho}_s[(J-J_c)L^{1/\nu}, \alpha] \tag{2.37}$$

depends only on the distance from the critical point, i.e., $(J-J_c)L^{1/\nu}$. Thus plotting $X_{z\nu}(L)$ as a function of J for different system sizes L allows us to determine the critical value J_c, where curves for different L intersect. In Ref. [20] this quantity has been calculated via QMC simulations using world lines in the stochastic series expansion (SSE) representation (see Ref. [14] and references therein).

Figure 2.5a shows the rescaled superfluid density $\rho_s L$ as a function of the hopping strength J/g assuming $z = 1$ for system sizes ranging from 20×20 to 40×40. The intersect of the curves leads to the estimate of the critical hopping strength $J_c/g = 0.05241(1)$. Figure 2.5b shows $X_{z\nu}(L)$ as a function of $(J-J_c)L^{1/\nu}$. One observes a clear scaling collapse as expected from Eq. (2.37). Note, that here a correlation length exponent $\nu = 0.6715$ (as found numerically for the BHM [77]) has been assumed. We can thus conclude that universal scaling at the tip of the lobe is observed for a dynamical critical exponent $z = 1$ confirming unambiguously the prediction of the analytical calculations in Refs. [12, 13].

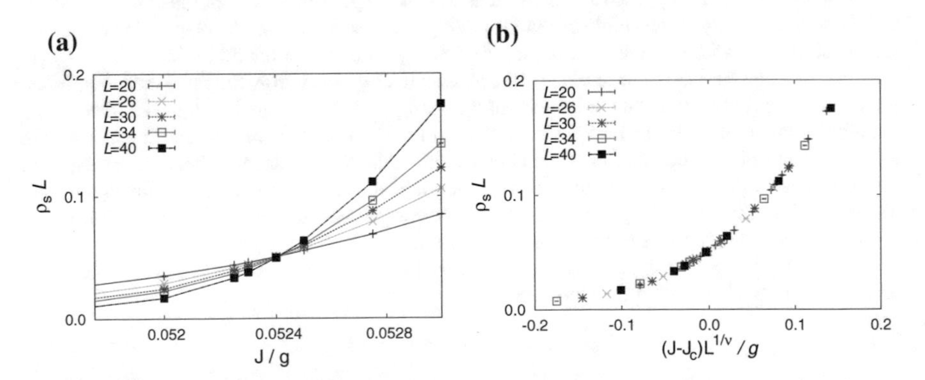

Fig. 2.5 (Color online) Scaling of the superfluid density ρ_s across the fixed-density transition at the tip of the lobe with $\mu/g = 0.185$ [11] using $L \times L$ square lattices and $\beta g = 2L$. The intersect of $L\rho_s$ for different lattice sizes L in a single point in panel **a** is evidence for a dynamical critical exponent $z = 1$, and defines the critical point at $J_c/g = 0.05242(1)$. **b** Scaling collapse using $J_c/g = 0.05242$ and the exact critical exponent $\nu = 0.6715$ [77]. Figure taken with permission from Ref. [20]

2.6 Relation to the Dicke Model

In this section we discuss a remarkable relation between the superfluid-Mott insulator transition of polaritons in coupled qubit-cavity arrays and the superradiance phase transition of the Tavis-Cummings or Dicke model [78–82]. In particular, we argue that a weak-coupling mean-field theory for the Dicke model predicts the existence of a single Mott lobe, where the universality class of the phase transition changes at the tip of the lobe just as for the JCHM [26].

The Dicke model describes the interaction of a single photonic mode with a number of N_s qubits and can be written as

$$H = \tilde{\omega}_c a_0^\dagger a_0 + \tilde{\omega}_x \sum_i \sigma_i^+ \sigma_i^- + \frac{g}{\sqrt{N_s}} \sum_i \left(\sigma_i^+ a_0 + \text{h.c.}\right) , \quad (2.38)$$

where we have also introduced a Dicke model chemical potential μ_D for polaritons according to the recipe in Eq. (2.3) leading to the definition $\tilde{\omega}_c = \omega_c - \mu_D$ and $\tilde{\omega}_x = \omega_x - \mu_D$. A weak-coupling mean-field theory for the Dicke model (at zero temperature and in a frame rotating at the cavity frequency ω_c) predicts a phase transition from a normal phase with $\psi = \langle a_0 \rangle = 0$ to a superradiant state with $\psi \neq 0$ at the critical coupling strength [79, 81]

$$g^2 = -\mu_D |\delta_D - \mu_D| . \quad (2.39)$$

However, for $\mu_D > \delta_D$ and negative detuning $\delta_D < 0$ we observe the appearance of a second normal phase corresponding to a Mott lobe with all two-level systems inverted, i.e., with one excitation per cavity ($n = 1$) [26]. The corresponding zero temperature phase diagram is shown in Fig. 2.6. The phase boundary of this lobe can be obtained from Eq. (2.39) as

$$\mu_D = \frac{1}{2} \left(\delta_D \pm \sqrt{\delta_D^2 - 4g^2} \right) . \quad (2.40)$$

with the tip of the lobe at $\delta_D = -2g$ and $\mu_D = -g$.

We now argue that this interesting result can be understood as a special limit of the JCHM: We first Fourier transform the photon operators in Eq. (2.1) to momentum space

$$a_i = \frac{1}{\sqrt{N_s}} \sum_{\mathbf{k}} a_{\mathbf{k}} e^{i \mathbf{k} \cdot \mathbf{r}_i} \quad (2.41)$$

such that the JCHM can be written as

$$H = \sum_{\mathbf{k}} \tilde{\omega}_{\mathbf{k}} a_{\mathbf{k}}^\dagger a_{\mathbf{k}} + \sum_i \tilde{\omega}_x \sigma_i^+ \sigma_i^- + \sum_{i\mathbf{k}} \frac{g_{i\mathbf{k}}}{\sqrt{N_s}} \left(\sigma_i^+ a_{\mathbf{k}} + \text{h.c.}\right) \quad (2.42)$$

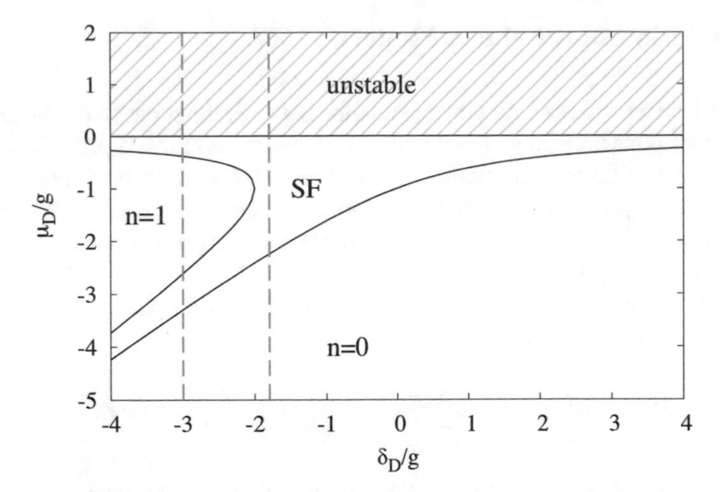

Fig. 2.6 Quantum phase diagram of the Dicke model and the JCHM at infinite bandwidth and infinite negative detuning for fixed μ_D, δ_D see text) showing the transition from the vacuum ($n = 0$) to a superfluid state as well as the existence of a single Mott lobe with $n = 1$. Figure taken with permission from Ref. [26]

where $\tilde{\omega}_{\mathbf{k}} = \tilde{\omega_c} - 2J\sum_{\alpha=1}^{D}\cos(k_\alpha)$ with $\tilde{\omega_c} = \omega_c - \mu$, $\tilde{\omega_x} = \omega_x - \mu$ and $g_{i\mathbf{k}} = ge^{i\mathbf{k}\cdot\mathbf{r}_i}$ (N_s denotes the number of lattice sites). This Hamiltonian represents a many-mode Dicke model, as studied in Refs. [83, 84]. The case studied in those works, however, considered a quadratic photon spectrum, equivalent to expanding the lattice dispersion for small $\mathbf{k}$ vectors yielding

$$\tilde{\omega}_{\mathbf{k}} = \tilde{\omega_c} - 2DJ + J\mathbf{k}^2 \equiv J\mathbf{k}^2 - \mu_D\,, \tag{2.43}$$

where we have defined a Dicke-model chemical potential $\mu_D = \mu + 2DJ - \omega_c$ (such that $\mu_D < 0$ is required for thermodynamic stability). Following Ref. [26], it is also useful to define a Dicke-model detuning $\delta_D = \delta + 2DJ$, which measures the detuning between the qubit (2LS) energy and the bottom of the photon band so that $\tilde{\omega_x} = \delta_D - \mu_D$. With this quadratic expansion the generalised Dicke model in (2.42) describes N_s localised two-level systems coherently coupled to a continuum of photonic modes with an effective photonic mass $1/2J$. Note, that if one neglects all finite momentum modes, Eq. (2.42) reduces to the single-mode Dicke model in Eq. (2.38) with $a_0 = a_{\mathbf{k}=\mathbf{0}}$.

Thus, if one considers the limit of infinite bandwidth $J \to \infty$ and infinite negative detuning $\delta \to -\infty$, the ground state of the JCHM at fixed chemical potential will mostly be composed of qubit excitations such that higher k modes can be neglected. In that case we expect that the long wavelength physics of the JCHM is similar to the Dicke model. Indeed, in Ref. [26] we have shown that if one takes the limit of infinite bandwidth and negative detuning such that μ_D and δ_D remain fixed, the phase diagram in Fig. 2.3 maps exactly to the one in Fig. 2.6. This is a remarkable connection since both phase diagrams have been derived using very different methods, i.e., a weak-coupling mean-field theory on one hand and a strong-coupling slave-boson theory on the other. Moreover, also the excitation spectra as calculated from both theories match exactly at the phase boundary of the Mott lobe [26]. In Fig. 2.3 one can see the reason for this success: the size of the lowest Mott lobe increases for large J and large negative detuning δ, while the size of all other lobes decreases. Thus, only one lobe survives in Fig. 2.6. All other modes are pushed towards $\mu_D = 0$ and vanish. This is in strong contrast to the Bose-Hubbard model, where a weak-coupling Bogoliubov-like theory describing weakly interacting atomic BEC's, fails to predict the existence of Mott lobes and gapped Higgs-like modes [85].

Acknowledgements We thank Martin Hohenadler, Markus Aichhorn, Jonathan Keeling and Lode Pollet for valuable discussions. This work was supported by a Ambizione award (S.S.) of the Swiss National Science Foundation.

References

1. A.D. Greentree, C. Tahan, J.H. Cole, L. Hollenberg, Nat. Phys. **2**, 856 (2006)
2. D. Angelakis, M. Santos, S. Bose, Phys. Rev. A **76**, 031805 (2007)
3. J. Cho, D.G. Angelakis, S. Bose, Phys. Rev. A **78**, 062338 (2008)
4. G. Zhu, S. Schmidt, J. Koch, New J. Phys. **15**, 115002 (2013)
5. M. Hartmann, F. Brandão, M. Plenio, Phys. Rev. Lett. **99**, 160501 (2007)
6. M. Hartmann, F. Brandão, M. Plenio, Nat. Phys. **2**, 849–855 (2006)
7. D. Rossini, R. Fazio, Phys. Rev. Lett. **99**, 186401 (2007)
8. D. Rossini, R. Fazio, G. Santoro, Europhys. Lett. **83**, 47011 (2008)
9. M. Aichhorn, M. Hohenadler, C. Tahan, P. Littlewood, Phys. Rev. Lett. **100**, 216401 (2008)
10. M.J. Hartmann, F.G.S.L. Brandao, M.B. Plenio, New J. Phys. **10**, 033011 (2008)
11. J. Zhao, A.W. Sandvik, K. Ueda (2008). arXiv:0806.3603
12. S. Schmidt, G. Blatter, Phys. Rev. Lett. **103**, 086403 (2009)
13. J. Koch, K.L. Hur, Phys. Rev. A **80**, 023811 (2009)
14. P. Pippan, H. Evertz, M. Hohenadler, Phys. Rev. A **80**, 033612 (2009)
15. P.A. Ivanov, S.S. Ivanov, N.V. Vitanov, A. Mering, M. Fleischhauer, K. Singer, Phys. Rev. A **80**, 060301 (2009)
16. M. Knap, E. Arrigoni, W. von der Linden, Phys. Rev. B **81**, 104303 (2010)
17. M. Knap, E. Arrigoni, W. von der Linden, Phys. Rev. B **82**, 045126 (2010)
18. S. Schmidt, G. Blatter, Phys. Rev. Lett. **104**, 216402 (2010)
19. A. Tomadin, V. Giovannetti, R. Fazio, D. Gerace, I. Carusotto, H. Türeci, A. Imamoglu, Phys. Rev. A **81**, 061801 (2010)
20. M. Hohenadler, M. Aichhorn, S. Schmidt, L. Pollet, Phys. Rev. A **84**, 041608(R) (2011)
21. K. Liu, L. Tan, C.H. Lv, W.M. Liu, Phys. Rev. A **83**, 063840 (2011)
22. M. Hohenadler, M. Aichhorn, L. Pollet, S. Schmidt, Phys. Rev. A **85**, 013810 (2012)
23. C. Nietner, A. Pelster, Phys. Rev. A **85**, 043831 (2012)
24. M. Schiró, M. Bordyuh, B. Öztop, H.E. Türeci, Phys. Rev. Lett. **109**, 053601 (2012)
25. F. Nissen, S. Schmidt, M. Biondi, G. Blatter, H.E. Türeci, J. Keeling, Phys. Rev. Lett. **108**, 233603 (2012)
26. S. Schmidt, G. Blatter, J. Keeling, J. Phys. B: At. Mol. Opt. Phys. **46**, 224020 (2013)
27. T. Grujic, S.R. Clark, D. Jaksch, D.G. Angelakis, Phys. Rev. A **87**, 053846 (2013)
28. T. Yuge, K. Kamide, M. Yamaguchi, T. Ogawa (2014). arXiv:1401.6229
29. D. Sarchi, I. Carusotto, M. Wouters, V. Savona, Phys. Rev. B **77**, 125324 (2008)

30. D. Gerace, H.E. Türeci, A. Imamoglu, V. Giovannetti, R. Fazio, Nat. Phys. **5**, 281–284 (2009)
31. S. Schmidt, D. Gerace, A. Houck, G. Blatter, H.E. Türeci, Phys. Rev. B **82**, 100507 (2010)
32. N. Schetakis, T. Grujic, S. Clark, D. Jaksch, D. Angelakis, J. Phys. B: At. Mol. Opt. Phys. **46**, 224025 (2013)
33. J. Raftery, D. Sadri, S. Schmidt, H.E. Türeci, A.A. Houck (2013). arXiv:1312.2963
34. M. Hartmann, Phys. Rev. Lett. **104**, 113601 (2010)
35. T. Grujic, S.R. Clark, D.G. Angelakis, D. Jaksch, New J. Phys. **14**, 103025 (2012)
36. A.L. Boité, G. Orso, C. Ciuti, Phys. Rev. Lett. **110**, 233601 (2013)
37. J. Jin, D. Rossini, R. Fazio, M. Leib, M.J. Hartmann, Phys. Rev. Lett. **110**, 163605 (2013)
38. I. Carusotto, D. Gerace, H.E. Türeci, S. De Liberato, C. Ciuti, A. Imamoğlu, Phys. Rev. Lett. **103**, 033601 (2009)
39. M. Kiffner, M. Hartmann, Phys. Rev. A **81**, 021806 (2010)
40. A.G. D'Souza, B.C. Sanders, D.L. Feder, Phys. Rev. A **88**, 063801 (2013)
41. J. Koch, A. Houck, K. Le Hur, S. Girvin, Phys. Rev. A **82**, 043811 (2010)
42. A. Nunnenkamp, J. Koch, S.M. Girvin, New J. Phys. **13**, 095008 (2011)
43. M. Hafezi, E.A. Demler, M.D. Lukin, J.M. Taylor, Nat. Phys. **7**, 907–912 (2011)
44. A. Kamal, J. Clarke, M.H. Devoret, Nat. Phys. **7**, 311–315 (2011)
45. R.O. Umucalilar, I. Carusotto, Phys. Rev. Lett. **108**, 206809 (2012)
46. M. Hafezi, S. Mittal, J. Fan, A. Migdall, J.M. Taylor, Nat. Photon **7**, 1001–1005 (2013)
47. C.E. Bardyn, S.D. Huber, O. Zilberberg (2013). arXiv:1312.6894
48. A. Petrescu, A.A. Houck, K. Le Hur, Phys. Rev. A **86**, 053804 (2012)
49. C.E. Bardyn, A. Imamoglu, Phys. Rev. Lett. **109**, 253606 (2012)
50. M.J. Hwang, M.S. Choi, Phys. Rev. B **87**, 125404 (2013)
51. B. Kumar, S. Jalal, Phys. Rev. A **88**, 011802 (2013)
52. A.A. Zvyagin, Phys. Rev. Lett. **110**, 217207 (2013)
53. J. Cho, D. Angelakis, S. Bose, Phys. Rev. Lett. **101**, 246809 (2008)
54. A. Hayward, A.M. Martin, A.D. Greentree, Phys. Rev. Lett. **108**, 223602 (2012)
55. M. Hafezi, M.D. Lukin, J.M. Taylor, New J. Phys. **15**, 063001 (2013)
56. R.O. Umucalilar, M. Wouters, I. Carusotto, Phys. Rev. A **89**, 023803 (2014)
57. M. Hartmann, F. Brandão, M. Plenio, Laser Photonics Rev. **2**, 527–556 (2008)
58. A. Tomadin, R. Fazio, J. Opt. Soc. Am. B **27**, 130–136 (2010)
59. A.A. Houck, H.E. Türeci, J. Koch, Nat. Phys. **8**, 292–299 (2012)
60. I. Carusotto, C. Ciuti, Rev. Mod. Phys. **85**, 299 (2013)
61. S. Schmidt, J. Koch, Annalen der Physik **525**, 395–412 (2013)
62. D.L. Underwood, W.E. Shanks, J. Koch, A.A. Houck, Phys. Rev. A **86**, 023837 (2012)
63. J. Klaers, J. Schmitt, F. Vewinger, M. Weitz, Nature **468**, 545–548 (2010)
64. P. Kirton, J. Keeling, Phys. Rev. Lett. **111**(Sep), 100404 (2013)
65. J. Schmitt, T. Damm, D. Dung, F. Vewinger, J. Klaers, M. Weitz, Phys. Rev. Lett. **112**(Jan), 030401 (2014)
66. J. Klaers, J. Schmitt, T. Damm, D. Dung, F. Vewinger, M. Weitz, Proc. SPIE **8600**, 8600L (2013)
67. H. Deng, H. Haug, Y. Yamamoto, Rev. Mod. Phys. **82**, 1489 (2010)
68. O. Astafiev, A.M. Zagoskin, A.A. Abdumalikov, Y.A. Pashkin, T. Yamamoto, K. Inomata, Y. Nakamura, J.S. Tsai, Science **327**, 840 (2010)
69. M.P.A. Fisher, P.B. Weichman, J. Watson, D.S. Fisher, G. Grinstein, Phys. Rev. B **40**, 546 (1989)
70. H. Kleinert, *Path Integrals in Quantum Mechanics, Statistics, Polymer Physics, and Financial Markets* (World Scientific, Singapore, 2006)
71. H. Kleinert, S. Schmidt, A. Pelster, Annalen der Physik **14**, 214 (2005)
72. F.E.A. dos Santos, A. Pelster, Phys. Rev. A **79**, 013614 (2009)
73. N. Teichmann, D. Hinrichs, M. Holthaus, A. Eckardt, Phys. Rev. B **79**, 100503 (2009)
74. W. Metzner, Phys. Rev. B **43**, 8549 (1991)
75. M. Ohliger, A. Pelster (2008). arXiv:0810.4399
76. S.D. Huber, E. Altman, H.P. Büchler, G. Blatter, Phys. Rev. B **75**, 085106 (2007)

77. B. Capogrosso-Sansone, S.G. Söyler, N. Prokof'ev, B. Svistunov, Phys. Rev. B **77**, 015602 (2008)
78. R. Dicke, Phys. Rev. **93**, 99–110 (1954)
79. K. Hepp, E.H. Lieb, Ann. Phys. **76**, 360–404 (1973)
80. Y.K. Wang, F.T. Hioe, Phys. Rev. A **7**, 831 (1973)
81. K. Hepp, E. Lieb, Phys. Rev. A **8**, 2517–2525 (1973)
82. M. Tavis, F.W. Cummings, Phys. Rev. **170**, 379 (1968)
83. J. Keeling, P.R. Eastham, M.H. Szymanska, P.B. Littlewood, Phys. Rev. Lett. **93**, 226403 (2004)
84. J. Keeling, P.R. Eastham, M.H. Szymanska, P.B. Littlewood, Phys. Rev. B **72**, 115320 (2005)
85. D. van Oosten, P. van der Straten, H.T.C. Stoof, Phys. Rev. A **63**, 053601 (2001)

Chapter 3
Out-of-Equilibrium Physics in Driven Dissipative Photonic Resonator Arrays

Changsuk Noh, Stephen R. Clark, Dieter Jaksch and Dimitris G. Angelakis

Abstract Coupled resonator arrays have been shown to exhibit interesting many-body physics including Mott and Fractional Hall states of photons. One of the main differences between these photonic quantum simulators and their cold atoms counterparts is in the dissipative nature of their photonic excitations. The natural equilibrium state is where there are no photons left in the cavity. Pumping the system with external drives is therefore necessary to compensate for the losses and realise non-trivial states. The external driving here can easily be tuned to be incoherent, coherent or fully quantum, opening the road for exploration of many body regimes beyond the reach of other approaches. In this chapter, we review some of the physics arising in driven dissipative coupled resonator arrays including photon fermionisation, crystallisation, as well as photonic quantum Hall physics out of equilibrium. We start by briefly describing possible experimental candidates to realise coupled resonator arrays along with the two theoretical models that capture their physics, the Jaynes-Cummings-Hubbard and Bose-Hubbard Hamiltonians. A brief review of the analytical and sophisticated numerical methods required to tackle these systems is included.

3.1 Introduction

One of the key problems in physics is to understand strongly correlated many-body systems. Fully solving a many-body Hamiltonian, analytically or numerically, is

C. Noh (✉) · S.R. Clark · D. Jaksch · D.G. Angelakis
Centre for Quantum Technologies, National University of Singapore,
2 Science Drive 3, Singapore 117542, Singapore
e-mail: undefying@gmail.com

D.G. Angelakis
School of Electrical and Computer Engineering, Technical University of Crete,
73100 Chania, Crete, Greece
e-mail: dimitris.angelakis@gmail.com

S.R. Clark · D. Jaksch
Clarendon Laboratory, University of Oxford, Parks Road, Oxford OX1 3PU, UK

D.G. Angelakis (ed.), *Quantum Simulations with Photons and Polaritons*,
Quantum Science and Technology, DOI 10.1007/978-3-319-52025-4_3

however a notoriously difficult problem. An increasingly plausible way around this issue is to use an experimentally accessible and well-controlled quantum system (e.g., cold atoms in optical lattices) to simulate the physics of another complex quantum system (e.g., electrons in a 2D lattice described by the Hubbard model) [1]. While cold atoms [2] and trapped ions [3] arose as strong candidates for quantum simulators [4, 5] of interacting fermions/bosons and quantum spin systems, respectively, photonic systems based on linear optics and Circuit QED architectures were also recently proposed as alternative candidates with their own advantages [6, 7]. One prominent example is the simulation of strongly interacting condensed matter models such as the Bose-Hubbard model [8] in coupled resonator arrays (CRAs) [7, 9–11].

The Bose-Hubbard (BH) model is a lattice model for a single bosonic species with nearest-neighbour hopping and an on-site interaction. In Ref. [8], it was shown that the competition between the two processes leads to a quantum phase transition between the superfluid and Mott-insulating phases, which has been demonstrated experimentally with ultracold atoms in an optical lattice [12, 13]. To realise a photonic superfluid-Mott phase transition in a CRA one needs to build up a Bose-Hubbard-like Hamiltonian. The hopping between the lattice sites is naturally provided by an optical coupling between resonators, whereas to realise the on-site interaction, one has to induce some kind of nonlinearity in each resonator. For this purpose, an (artificial) atom can be coupled to each resonator, or for the case of solid-state cavities an intrinsic Kerr-nonlinearity may be considered. In the latter case, the dynamics of the CRA is governed by the BH model, whereas in the former case the dynamics depends on the type of the doped atom. When the atom is a two-level system, for which the coupling to the cavity mode is of the Jaynes-Cummings type, the resulting CRA is now known as the Jaynes-Cummings-Hubbard (JCH) system.

Following an initial realisation that photons in CRAs undergo superfluid-Mott phase transition [14–16], the equilibrium phases of the JCH model has been extensively investigated [17] (see also chapters by Schmidt & Blatter and Tomadin & Fazio). However, for photons it is more natural to consider a driven dissipative scenario, where photons are continuously pumped into the system to counteract unavoidable losses. This has led to the study of CRAs in non-equilibrium driven dissipative regimes, where strongly correlated many-body behaviour can be observed. The aim of this chapter is to discuss efforts in this direction. We discuss in detail a few topical examples, mostly on driven dissipative systems, but also on transient effects in dissipative arrays. We begin by providing a short survey of experimental platforms that could realise CRAs and then discuss generic models that describe these systems.

Candidates for realising CRAs

Photonic crystals are defect semiconductors with holes periodically introduced to an otherwise uniform substrate. This induces a spatial dependence of refractive index, and thus gives rise to photonic band gaps within the infra-red to visible range [18]. A cavity is introduced to the system by creating an isolated photonic mode within the band gap, for example by removing one or more holes, that are strongly confined to the photonic crystal plane. Figure 3.1 illustrates such coupled photonic crystal cavities. Photon-photon interactions can be induced by growing a semiconductor

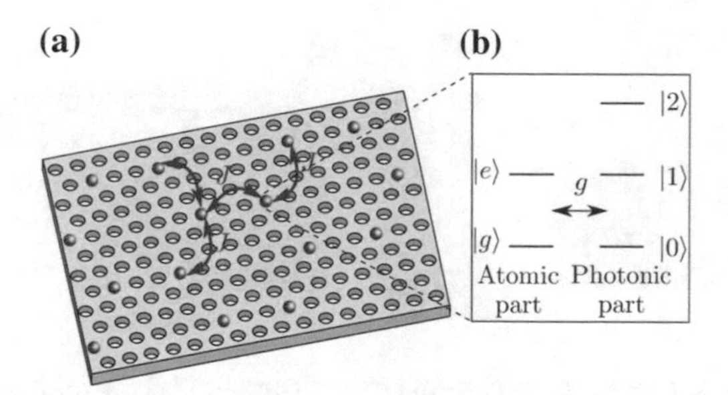

Fig. 3.1 Photonic crystal: **a** Periodic holes are introduced to modulate the refractive index of the material, giving rise to band gaps in the material. Defect cavities, represented by *red circles*, introduce isolated photonic states within the band gap. Photons can hop between neighbouring cavities, as represented by *red arrows*, due to modal overlap. **b** When a quantum dot is grown inside the defect cavity, the interaction between the quantised electronic levels of the quantum *dot* and the cavity modes can be described by the Jaynes-Cummings model

nanostructure called quantum dot in each cavity. Due to the tight confinement of electrons in a quantum dot, discrete energy levels form and when one of the transitions match the cavity frequency the resulting dynamics is well described by the Jaynes-Cummings Hamiltonian (see Sect. 3.2 for a quick introduction to the Jaynes-Cummings model). The advantages of photonics crystal cavities include low photon losses, and extremely small mode volumes on the order of a cubic wavelength λ^3.

In another promising semiconductor-based platform, a quantum well is sandwiched between two Bragg reflectors. The reflectors create an effective cavity, with a mode that interacts with the exciton in the quantum well. This creates the so-called exciton-polaritons which inherit nonlinearities from the electron-hole interaction. The interested reader is referred to chapters [Gerace] and [Kim] for more details on these devices.

Integrated waveguide arrays are another interesting candidate. In this set-up, the formation of resonators involve precise etching of concave mirror arrays into silicon [19] with Bragg-inscribed optical fibres placed above them (see Fig. 3.2a, b). CRAs can be realised by introducing laser-written waveguides in places of optical fibres [20], such that the photon tunneling is allowed by evanescent coupling between the waveguides (see Fig. 3.2c). Phase shifters can then be used to tune the coupling strength, while the nonlinearity can be introduced by placing an atom on each concave mirror cavity.

Superconducting circuits, our last candidate, is the most promising platform for realising CRAs [7, 21–24]. They enjoy low photon loss rates due to superconductivity and good scalability due to highly-accurate fabrication procedures available. Various circuit designs exist to create an artificial two-level atom, all of which however share the common feature of exploiting the intrinsic nonlinearity of one or more Josephson junctions. Such an artificial atom can be coupled to superconducting transmission line resonators to mimic the Jaynes-Cummings interaction in cavity QED, as shown in Fig. 3.3 [25]. The resulting system is often called the circuit QED (cQED) system

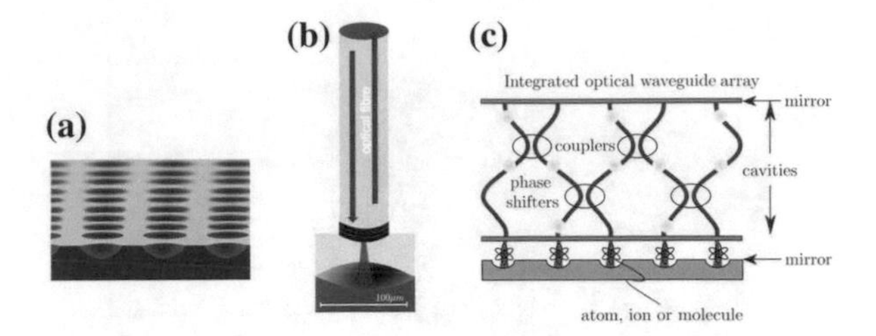

Fig. 3.2 Integrated optical waveguide: **a** An array of periodic concave mirrors etched into Si can be fabricated with high-precision [19], while **b** resonators may be formed by the addition of Bragg-inscribed fibres above them. **c** CRAs are realised by replacing the fibres with waveguides, inscribed with different refractive indices by a strong UV "writing" laser [20]. Photon tunneling between neighbouring waveguides/cavities is enhanced by optimising waveguide paths

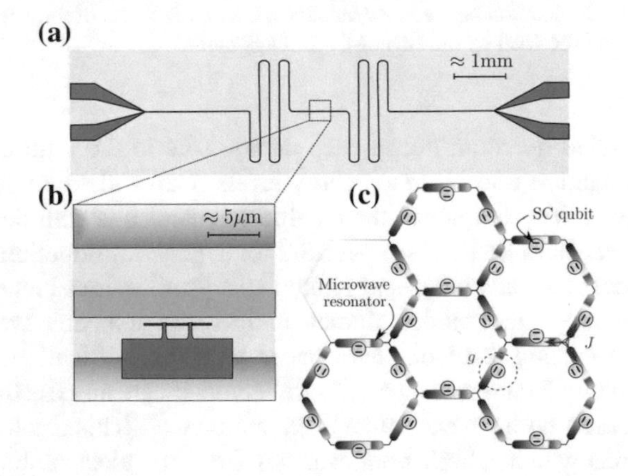

Fig. 3.3 Superconducting circuit QED: **a** A superconducting coplanar waveguide resonator fabricated on a silicon chip using optical lithography. *Green* tapering structures fanning out to the edges of the chip are input and output feed *lines*. The *center* conductor is separated from the lateral ground planes by a gap as shown in the zoom. **b** Effective 2-level system, a superconducting qubit, coupled to the coplanar waveguide resonator. A superconducting island connected by a pair of Josephson junctions to the superconducting reservoir is placed in the gap between the central conductor and the *bottom* ground plane. **c** Many stripline resonators fabricated into a JCH system

in analogy to cavity-QED. A comparison between the various realisations of CRAs using currently available parameters are summarised in Table 3.1. Note that a large cooperativity (the ratio between the interaction strength g and the loss rate γ) is required in most proposals for equilibrium simulations.

Table 3.1 Comparison of typical CRA parameters for different architectures. Note that the values quoted for integrated optics platform are not taken from actual experiments, but are realistic parameters predicted in Ref. [26]

Parameter	Symbol	Photonic crystals	Integrated optics	Superconducting
Resonance frequency	$\omega_c/2\pi$	325 THz	380 THz	10 GHz
JC parameter	$g/2\pi$, g/ω_c	20 GHz	33 MHz	200 MHz
Cavity decay rate	$\gamma/2\pi$	1 GHz	10 MHz	100 kHz
Atom decay rate	$\kappa/2\pi$	8 GHz	10 MHz	2 kHz
Cooperativity	$g/\gamma\kappa$	2.5	10	$\gg$1
Resonator coupling	$J/2\pi$	100 GHz	2 GHz	10 MHz

3.2 Modelling CRAs

There are two models often used to describe CRAs, depending on whether the nonlinearity is assumed to be intrinsic to the resonator, or is explicitly introduced by coupling to each resonator a two-level atom. When losses are neglected, the former case is described by the BH model (from here on, we assume $\hbar = 1$)

$$\hat{H}_{\mathrm{BH}} = -J \sum_{\langle i,j \rangle} \left(\hat{a}_i^\dagger \hat{a}_j + \mathrm{h.c.} \right) + \frac{U}{2} \sum_i \hat{a}_i^\dagger \hat{a}_i^\dagger \hat{a}_i \hat{a}_i + \omega_c \sum_i \hat{a}_i^\dagger \hat{a}_i, \tag{3.1}$$

whereas the latter is described by the JCH model:

$$\begin{aligned} \hat{H}_{\mathrm{JCH}} = &-J \sum_{\langle i,j \rangle} \left(\hat{a}_i^\dagger \hat{a}_j + \mathrm{h.c.} \right) + g \sum_i \left(\hat{a}_i^\dagger \hat{\sigma}_i^- + \mathrm{h.c.} \right) \\ &+ \omega_c \sum_i \left(\hat{a}_i^\dagger \hat{a}_i + \hat{\sigma}_i^+ \hat{\sigma}_i^- \right) - \Delta \sum_i \hat{\sigma}_i^+ \hat{\sigma}_i^- . \end{aligned} \tag{3.2}$$

Here $\hat{a}$ is the photon annihilation operator, $\hat{\sigma}_-$ is the atomic lowering operator, J is the photon hopping rate between neighbouring cavities, U is the intrinsic nonlinearity of the resonator, g is the coupling strength between the resonator and the two-level system, ω_c is the natural frequency of the resonator, and Δ is the detuning between the resonator and the two-level system. When $J = 0$, $\hat{H}_{JCH}$ becomes the Jaynes-Cummings model, which describes the physics of a two-level atom coupled to a resonator mode when the coupling strength g is much smaller than the atomic and resonator frequencies [27]. The Kerr-nonlinearity in the BH model also comes from some sort of interaction with matter, but in the regime where only the overall effects of the material on the resonator mode are considered. These effects are usually weak, but can be enhanced in some systems as reviewed in [10].

With losses included, the Hamiltonian description is no longer sufficient. However, it is possible to model such open systems by coupling the system of interest to a bath of harmonic oscillators. Assuming weak system-bath coupling and a few general properties of the bath, one can obtain an equation of motion for the reduced density matrix of the system alone [27]. The resulting quantum master equation (QME) can be written as

$$d\rho/dt = -i\left[\hat{H}_{\alpha}, \rho\right] + \sum_{j} \frac{\gamma_j}{2}\left(2\hat{a}_j\rho\hat{a}_j^{\dagger} - \hat{a}_j^{\dagger}\hat{a}_j\rho - \rho\hat{a}_j^{\dagger}\hat{a}_j\right), \tag{3.3}$$

where ρ is the density operator and $\alpha \in \{\text{JCH, BH}\}$. The second term on the right hand side is a Lindblad noise term we will call $\mathcal{L}_{\text{loss}}$, which accounts for the dissipation in the cavities where γ_j is the loss rate of cavity j. For the JCH systems, an additional source of dissipation, due to decay of the excited atomic state, can also be taken into account by adding another Lindblad noise term

$$\mathcal{L}_{\text{decay}}\{\rho\} = \sum_{j} \frac{\kappa_j}{2}\left(2\hat{\sigma}_j^{-}\rho\hat{\sigma}_j^{+} - \hat{\sigma}_j^{+}\hat{\sigma}_j^{-}\rho - \rho\hat{\sigma}_j^{+}\hat{\sigma}_j^{-}\right), \tag{3.4}$$

where κ_j is the decay rate for the jth two-level atom. To compensate for the losses, experimentalists usually drive one or more resonators with lasers. Such driving can be modelled by adding the term

$$\hat{H}_{\text{drive}} = \sum_{j} \Omega_j(t)\hat{a}_j^{\dagger} + \Omega_j^{*}(t)\hat{a}_j, \tag{3.5}$$

to the Hamiltonian, where $\Omega_j(t) = \Omega_j e^{-i\omega_L t}$ is the Rabi frequency of the driving laser. Incoherent driving can also be added in the form of a Lindblad noise operator.[1] For example, the cavity array driven by a thermal reservoir field with average photon number $\bar{n}$ is described by

$$\begin{aligned} d\rho/dt = -i\left[\hat{H}_{\alpha}, \rho\right] &+ (\bar{n}+1)\sum_{j} \frac{\gamma_j}{2}\left(2\hat{a}_j\rho\hat{a}_j^{\dagger} - \hat{a}_j^{\dagger}\hat{a}_j\rho - \rho\hat{a}_j^{\dagger}\hat{a}_j\right) \\ &+ \bar{n}\sum_{j} \frac{\gamma_j}{2}\left(2\hat{a}_j^{\dagger}\rho\hat{a}_j - \hat{a}_j\hat{a}_j^{\dagger}\rho - \rho\hat{a}_j\hat{a}_j^{\dagger}\right). \end{aligned} \tag{3.6}$$

3.3 Computing the Properties of CRAs

Given the QME description of a CRA, in most cases we will concentrate on the properties of the non-equilibrium steady state (NESS), found by solving $d\rho/dt = 0$, in analogy to the ground state properties of the equilibrium systems. We will also

[1] One way to see this is to derive the rate equation for the diagonal density operator elements from the master equation.

be interested in the transient open system dynamics given the system is prepared in some initial state. For either task we are faced with integrating the QME for a BH or JCH system—a technically challenging problem due to the exponential growth in the Hilbert space dimension with the number of sites (e.g. resonators or resonator + atoms) considered. Developing numerical algorithms to aid in studying driven dissipative strongly-correlated many-body quantum systems is one of the most difficult problems in computational physics. However, the last decade has seen impressive progress in the development of computational representations of the state of quantum lattice systems as tensor networks [28–30], in particular in 1D via so-called matrix product states (MPS) [31, 32].

The key feature of MPS is that they provide a compact and efficient ansatz for describing weakly-entangled pure quantum states well suited to open chains of sites [32]. This has been used with much success to describe the low-lying excited states and zero-temperature unitary dynamics of cold-atom systems, e.g. the $-i\left[H'_{\alpha}, \rho\right]$ term in the QME [33–37]. These methods have been readily adapted [38–40] to describe dissipative systems with Lindblad noise terms which are local incoherent processes, like $\mathcal{L}_{\text{loss}}$ and $\mathcal{L}_{\text{decay}}$ relevant to CRA. This is done by unravelling the QME into quantum trajectories where many evolutions of pure states, described by an MPS, are made and stochastically interrupted by quantum jumps [41–45]. By averaging over many trajectories then recovers the properties of the full density matrix ρ [46]. In the appendix we describe in more detail this very powerful method.

3.4 Non-equilibrium Many-Body Phases of Photons in CRAs

In this section, we will review in detail a selection of many-body phenomena arising in non-equilibrium scenarios. To set the background, we start with a comparison between the physics of JC and Kerr-nonlinear resonator, followed by a description of strong photon-bunching phenomenon in a driven dissipative two-site (dimer) case. Next, a transient phenomenon called localisation-delocalisation of photons in a JCH dimer is presented, which shows a nice example of how CRAs are implemented in a circuit-QED platform. Continuing our review with driven dissipative phenomena, we discuss fermionisation and crystallisation of photons in 1D CRAs as well as the NESS phases and quantum Hall-like physics in 2D arrays. Last subsection provides a brief survey of other interesting driven dissipative studies that we do not have enough space to cover in detail.

3.4.1 Jaynes-Cummings Versus Kerr-Nonlinear Resonator

In the limit of zero atom-photon detuning, the eigenstates of the Jaynes-Cummings Hamiltonian, i.e., the $J = 0$ and $\Delta = 0$ case of (3.2), are the polariton (dressed

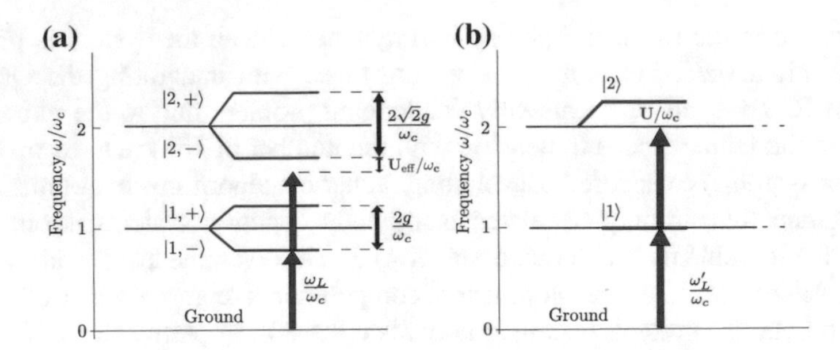

Fig. 3.4 Low-lying eigenstates of the **a** Jaynes-Cumming model and the **b** Bose-Hubbard model. Effective photon-photon interaction is provided by the anharmonicity in the spectrum of a given polariton branch. Taken from Grujic et al. [47]

photon-atom) states $|m, \pm\rangle = \frac{1}{\sqrt{2}}(|G, m\rangle \pm |E, m-1\rangle)$ having eigenenergies $\mathcal{E}_{m\pm} = \omega_c m \pm g\sqrt{m}$ (see Fig. 3.4a). Here, G and E refer to the ground and excited states of the atom, while m is the number of photons in the resonator. Imagine driving the lowest '− polariton' branch $|1, -\rangle$ by setting $\omega_L = \omega_c - g$. Because of the nonlinearity induced by the atom, the transition to the excited state $|2, -\rangle$ is detuned by the amount $\mathcal{E}_{2-} - 2\mathcal{E}_{1-} = g(2 - \sqrt{2})$. Compare this to the single-site BH system—a Kerr-nonlinear resonator—where the detuning is simply U as depicted in Fig. 3.4b. When photons are restricted to the lowest manifolds, we may therefore take the effective on-site interaction strength of the JC system as $U_{eff} = g(2 - \sqrt{2})$ for zero atom-photon detuning. For non-zero detuning the expression becomes

$$\frac{U_{eff}}{g} = \frac{\Delta}{2g} + 2\sqrt{\left(\frac{\Delta}{2g}\right)^2 + 1} - \sqrt{\left(\frac{\Delta}{2g}\right)^2 + 2}. \tag{3.7}$$

The 'repulsion' between photons induced by the nonlinearity can be quantified by the (equal-time) second-order intensity correlation function

$$g^{(2)} = \langle \hat{a}^\dagger \hat{a}^\dagger \hat{a} \hat{a} \rangle / \langle \hat{a}^\dagger \hat{a} \rangle^2, \tag{3.8}$$

as shown in Fig. 3.5. Note the dips in $g^{(2)}$ when the laser field (with laser-cavity detuning $\Delta_c = \omega_L - \omega_c$) is resonant with the eigenstates of each model. In particular, the dip is most prominent and falls below 1 when the lowest excited state is driven (the situation depicted in Fig. 3.4), signifying that the presence of a photon in a resonator prevents another one from entering the it. This phenomenon is called photon-blockade or photon anti-bunching. Stronger the repulsion, stronger the blockade or anti-bunching effect, i.e., $g \to 0$.

Despite this similarity between the JC and Kerr resonators, there are qualitative differences which become important when modelling CRAs as first investigated by Grujic et al. [47]. Let we look at the single resonator case for simplicity. First,

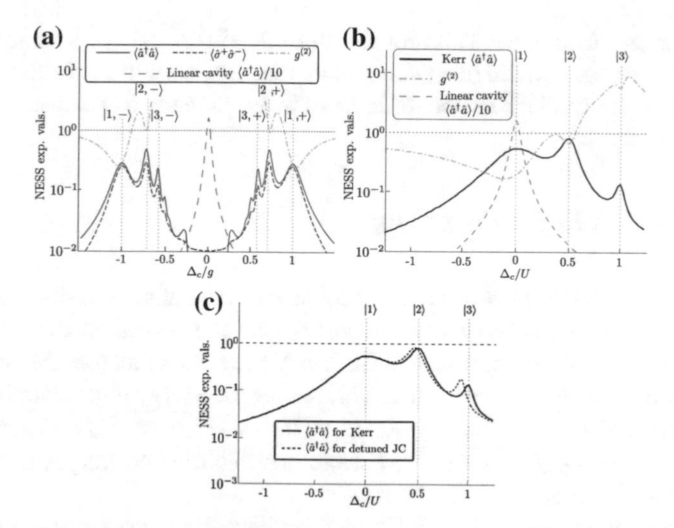

Fig. 3.5 Steady state values of photonic/atomic population (*blue solid/red dashed curves*) and the second-order intensity correlation function $g^{(2)}$ (*green dot-dashed curve*) as functions of the driving laser-cavity detuning. **a** Jaynes-Cummings resonator with zero atom-photon detuning. The Lorentzian profile of the empty driven dissipative resonator is shown for comparison (*blue dashed curve*). **b** Kerr-nonlinear resonator. *Vertical dashed lines* mark expected resonances in each model, corresponding to the labeled states. Parameters are: $\Omega/\gamma = 2$, $g/\gamma = 20$ (JC), and $U/\gamma \approx 11.7$ (Kerr, as calculated from Eq. (3.7)). **c** Comparison of the steady-state photon numbers for a Kerr resonator with parameters as used in (**b**) and a JC resonator with $\Delta/g = -10$ and $g/\gamma \approx 1.6 \times 10^4$. Taken from Grujic et al. [47]

broad similarities between the two models are apparent upon comparing the left wing of Fig. 3.5a (corresponding to the lower ($-$) polariton branch) to Fig. 3.5b: there are peaks at the eigen-modes of the resonators accompanied by the dips in $g^{(2)}$. However, the JC model displays richer structure, as indicated by the presence of two symmetric wings (corresponding to the two species of polaritons) and shift in the lowest excitation state with respect to the linear case (due to the atom-photon interaction).

It is then natural to ask if the two models can be made equivalent at all, so that the physics of the JC resonator can be described by an effective Kerr model. Indeed this is possible as shown in Fig. 3.5c. To achieve good agreement, one needs to take the photonic limit of the JC model in which the photonic component of the polaritons are dominant. For the—polaritons this requires setting a large negative detuning $\Delta \ll -g$. However, this means reduced effective atom-photon interaction and g must be increased in order to compensate for it. In Fig. 3.5c, $g/\gamma \approx 1.6 \times 10^4$ was required to achieve good agreement for $\Delta/g = -10$.

The above example shows that the JC and Kerr resonators are qualitatively different systems unless care is taken to ensure that they are in the right parameter regimes. The same conclusion holds for an array of resonators, meaning that, within realistic regimes of interactions, a phenomenon found in one type of array is not guaranteed to be reproduced in another. This provides flexibility in choosing CRAs and enriches physics that can be found in them. Later in this section, we will see further examples that highlight the similarities and differences between the two systems.

3.4.2 Photon Super-Bunching

Let us start with the simplest system of 2 coupled Kerr-nonlinear cavities and investigate the effects of an interplay between the hopping and on-site interaction. Here we focus on the small-sized equivalent of the 'Mott' state, meaning that the driving field will be set on two-photon resonance with the lowest-energy eigenstate in the two-photon manifold $|E_-^{(2)}\rangle \propto |2,0\rangle + |0,2\rangle + (U + \sqrt{U^2+16J^2})/2\sqrt{2}J|1,1\rangle$. That is, $\omega_L = \omega_c + (U - \sqrt{U^2+16J^2})/4$. Note that this state converges to the 'Mott' state $|1,1\rangle$ as $U \to \infty$.

Studying this set-up, Grujic and coworkers have discovered that strong photon bunching ($g^{(2)} \gg 1$) can be observed even when U is significantly stronger than J [48]. This is depicted in Fig. 3.6, which plots the local $g^{(2)}$ of the emitted photons as a function of J and U. Note that the diagram is broadly divided into three regions, defined by the Poissonian ($g^{(2)} \approx 1$), anti-bunched ($g^{(2)} < 1$), and bunched ($g^{(2)} > 1$) statistics of the emitted photons. The crossover points from anti-bunching to bunching are depicted by the black line, showing that there is a critical coupling strength, J_c, below which no bunching can be observed. This is further illustrated in Fig. 3.6b, where the auto-correlations along the two white lines in (a) are drawn for various values of the driving strength. In both the linear ($U \ll \gamma$) and the hardcore ($U \gg \gamma$) limit, expected behaviours are observed: in the linear regime, the steady state is a

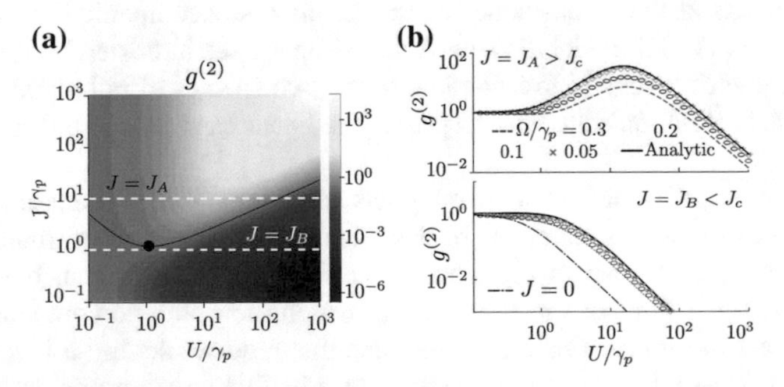

Fig. 3.6 Photon super-bunching due to photon tunneling in the presence of the Kerr nonlinearity. **a** Emitted photon statistics $g^{(2)}$ as a function of the photon tunneling strength J and Kerr nonlinearity U. Above some critical threshold J_c indicated by the *black curve*, photon emission is bunched ($g^{(2)} > 1$). Otherwise, photon emission is anti-bunched. **b** Emitted photon statistics $g^{(2)}$ as a function of Kerr nonlinearity U above and below the critical tunneling strength J_c. Super-bunching can be observed for tunneling strengths larger than J_c. We note that the definition of U used in the figure is two times larger than the definition used in Eq. (3.1). Taken from Grujic et al. [48]

coherent state, inheriting the Poissonian statistics from the driving laser, while in the hardcore regime, no more than a single photon per resonator can be injected, resulting in completely anti-bunched light. The behaviour of photons in between these extreme limits, however, depends strongly on the actual values of U and J. For $J > J_c$, there can be strong bunching ($g^{(2)} \gg 1$) even with significant repulsion between photons $U > J$. Similar behaviour was found in the JCH dimer also [48].

The origin of strong bunching can be traced to the relative enhancement of the two-photon sector in the NESS, relative to the single-photon sector, as the nonlinearity is increased. In the linear case the NESS is a coherent state, meaning that the probability of having two-photons in the second cavity is equal to the square of the probability to have a single photon, giving rise to the observed coherence. Initially with increasing U/J, this distribution is modified such that the probability of having two photons is enhanced, i.e., it is larger than the square of the probability of having a single photon, because the driving field is detuned from the single-particle manifold. This explains the initial rise in $g^{(2)}$. However, this trend cannot continue indefinitely because the $|2,0\rangle + |0,2\rangle$ portion in the driven two-photon eigenstate decreases with increasing U/J. At some point the $|1,1\rangle$ part of the two-photon eigenstate starts to dominate and the system exhibits antibunching. The interplay between the enhanced total two-photon sector and the suppressed local two-photon sector with increase in U/J gives rise to the observed behaviour in $g^{(2)}$.

3.4.3 Localisation-Delocalisation Transition of Photons

Instead of continuously driving the system, let us now prepare the system in a certain state and observe its dynamics. The steady state of the system will be the trivial vacuum state, but we can still observe interesting physics in the transient dynamics. In particular, strong atom-photon interaction gives rise to a dramatic transition from a localised phase to a delocalised phase as first discovered by Schmidt et al. [49]. In this set-up, photons are initially localised within a single site of a dimer. Without atom-photon coupling, the initially localised photons hop into the unoccupied cavity. However, as the atom-photon interaction is increased, there is a sharp transition to the localised regime where photons are trapped in the initial cavity. Similar delocalisation-localisation transition, or photon *freezing*, was first discovered in the BH-type systems and is known as self-trapping [50–55]. The result has been generalised to larger arrays by Schetakis et al. [56].

The system is represented schematically in Fig. 3.7. There are M resonators with the left-side resonators initially occupied by N_0 photons per resonator with the atoms in the ground state. The transition between localised and delocalised phases is nicely captured by the photon imbalance defined as

$$Z(t) = \frac{\sum_{j=1}^{M/2} \langle \hat{a}_j^\dagger(t)\hat{a}_j(t)\rangle - \sum_{j=M/2+1}^{M} \langle \hat{a}_j^\dagger(t)\hat{a}_j(t)\rangle}{\sum_{j=1}^{M} \langle \hat{a}_j^\dagger(t)\hat{a}_j(t)\rangle}, \tag{3.9}$$

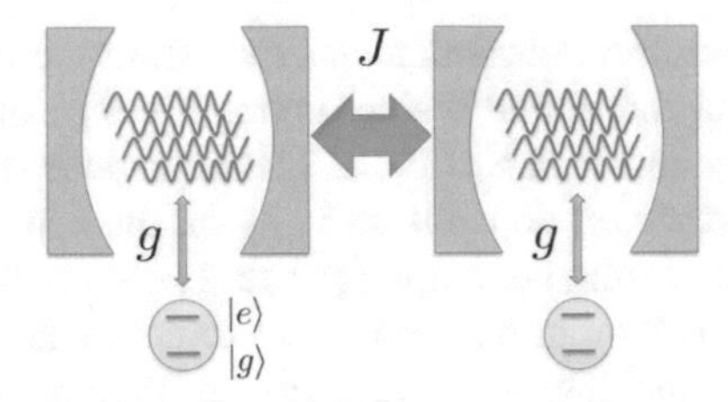

Fig. 3.7 Schematic representation of the JCH system exhibiting 'freezing' of photons. The left cavity of a dimer is initially pumped to contain 4 photons per resonator. The tunneling ratio is J and the atom-photon coupling strength is g

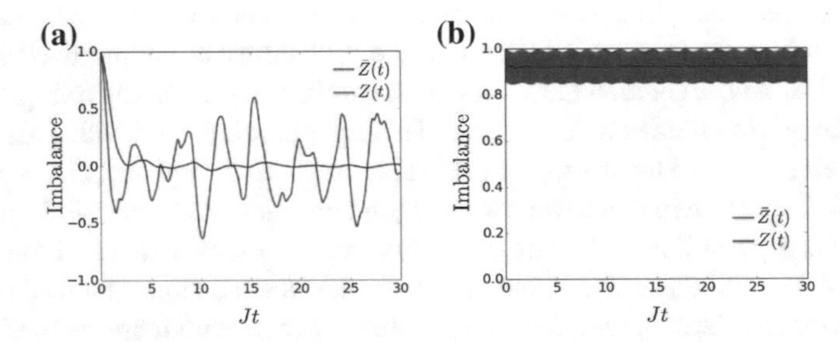

Fig. 3.8 Photon number imbalance (*blue curves*) and the time-averaged imbalance (*red curves*) for the lossless dimer with 7 initial photons. **a** *Below* the critical coupling strength, $g = 0.3g_c^{cl}$, showing delocalised dynamics. **b** *Above* the critical coupling strength, $g = 2.0g_c^{cl}$, showing localised dynamics

or rather the time-averaged version of it, which we will denote as $\bar{Z}$. Using this definition, the localised regime is characterised by $|\bar{Z}| \approx 1$ and the delocalised regime by $\bar{Z} \approx 0$.

Figure 3.8 provides a succinct account of the delocalised (left hand side) and localised (right hand side) dynamics for the lossless dimer with 7 initial photons. It shows the imbalance Z and the time average of it ($\bar{Z}(t)$) for two values of atom-photon coupling strength: One well below the critical coupling strength, $g = 0.3g_c^{cl}$ (left hand side), and the other is well above it, $g = 2.0g_c^{cl}$ (right hand side), where g_c^{cl} is the critical coupling strength which in the semiclassical approximation can be derived as $g_c^{cl} \approx 2.8\sqrt{N_0}J$ [49]. Below the critical coupling strength, photons are free to travel back and forth between the cavities and the average imbalance $\bar{Z}$ goes to zero with time. Above the critical strength, however, photons are self-trapped and cannot move to the right half of the array (rapid oscillations in the photon number is due to an energy exchange with the atom). This self-trapping behaviour stems from the fact that, for the symmetric system, the two states with opposite imbalance 1 and -1 are approximate eigenstates of the system. For the dissipationless dimer, the time scale of the oscillation between the states of opposite imbalance can be found within the degenerate perturbation theory as $1/(c_{N_0}J(J/g)^{N_0-1})$ with c_{N_0} a constant that depends on N_0 [49]. In the strongly nonlinear regime, the time scale diverges with increasing N_0, i.e., the photons are effectively trapped in the initial site.

Detailed calculation shows that the localisation transition is shifted to smaller g values and the transition is smoothened in the quantum case, i.e. $g_c^{qu} < g_c^{cl}$. With dissipation, the imbalance goes to zero trivially as the photons leave the system, but within the decay time, the localisation behaviour can be clearly observed. Interestingly, the dissipation can help one to achieve the transition from the delocalised regime to the localised regime. This follows from the fact that the critical coupling strength g_c^{qu} increases with the number of photons in the system (as described by the semiclassical approximate formula $g_c^{cl} \propto \sqrt{N_0 J}$). Consider an experimental situation in which g is fixed and less than g_c^{qu}. As the photons leave the system, the total photon number decreases and consequently g_c^{qu} decreases along with it. This means that g_c^{qu} will eventually dip below g, at which point the system enters the localised regime.

Such dissipation-induced transition has been observed experimentally in a superconducting circuit platform by Rafery et al. [57]. In this experiment, the atom-photon interaction is initially turned off by detuning the superconducting qubits out of resonance, and the resulting linear system is driven by a coherent pulse. Coherent oscillations of photons automatically prepare the required initial state with imbalance 1 at a certain time, at which point the interaction is ramped up by bringing the qubits into resonance. Figure 3.9 depicts the experimentally measured phase diagram, in which the homodyne signal of the photons from the initially unoccupied site is shown as a function of the initial mean photon number N_i and time. Up to about $N_i \approx 20$, the atom-photon interaction strength g is greater than the critical strength g_c^{qu} and the photons are localised. Increasing the initial mean photon number beyond $N_i > 20$, g becomes smaller than g_c^{qu} and the photons start to oscillate between the two sites. However, as the system loses photons, the localised regime is recovered as the total number of photons in the system dips below some critical number.

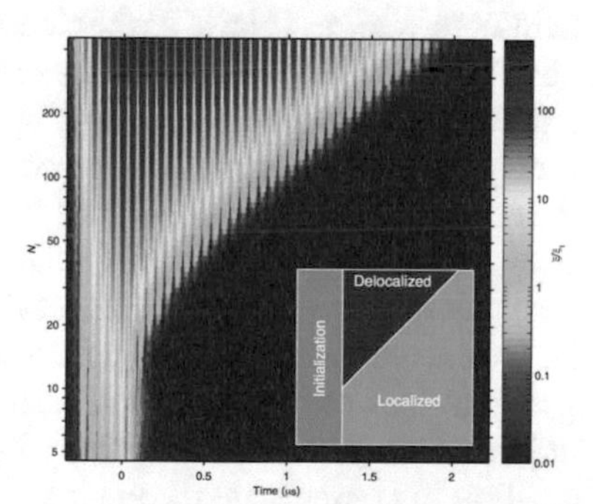

Fig. 3.9 Phase diagram of the dissipation-induced delocalisation-localisation transition for the two-site system. Homodyne signal from the initially unoccupied site as a function of initial photon number (N_i) and time. For large enough N_i the system starts in the delocalised regime, but as time progresses and photons are lost from the cavity, the system enters the localised regime. Taken from Raftery et al. [57]

3.4.4 Fermionisation of Hardcore Photons

In Sect. 3.4.2, we have seen that the local second-order intensity correlation function reveals anti-bunching behaviour in the strongly interacting limit ($U/J \gg 1$). The physical picture is that two-photons cannot occupy the same site because of strong repulsion between them. This is reminiscent of the fermionic behaviour, where the Pauli exclusion principle prohibits two fermions from occupying the same state. In fact, in 1 dimensional systems in equilibrium, strong interactions between bosons are known to induce fermionic behaviour [58]. As the on-site nonlinearity U in the BH model approaches infinity, double occupancy of any resonator is completely suppressed and the bosonic wave function of the system can be written in terms of the wave function of free fermions [59]. That such fermionisation can be observed in driven dissipative CRAs was first noticed by Carusotto et al. [60]. By calculating the spectrum and intensity-intensity correlation functions of the output light in the NESS, it was shown that the strongly correlated many-body nature of the photons can be readily observed. Here we follow the work by Grujic et al. [47], which investigated an analogous phenomenon in the JCH system. Readers are refered to Ch. [Gerace] or the original article [60] for further details on the BH system.

Consider a homogeneously driven 3-site cyclic array of nonlinear resonators. The eigenstates of this system in the hardcore regime can be determined by the aforementioned bose-fermi mapping. These eigenstates are conveniently labelled by quasi-momenta (of the Bloch modes), where the lowest eigenstates are found to be the one-particle state at zero momentum $|q = 0\rangle$, and the two-particle state $|q, -q\rangle$ with equal and opposite momenta $q = \pi/3$ [60].

The first evidence of the presence of fermionised photons can be observed in the steady-state excitation number as shown in Fig. 3.10a. For the BH model, we clearly see the two lowest-lying fermionic energy eigenstates $|q = 0\rangle$ and $|\pi/3, -\pi/3\rangle$ at $\Delta_c = -2J, -J$ in the hard-core regime (black dashed curve). Also shown is an effect of finite U/J: the peak is red-shifted towards lower energy (black solid curve). The JCH system shows the almost identical behaviour in the 'photonic regime' of polaritons, where the atom-photon detuning is large compared to the coupling strength g such that the photonic contribution to polaritons dominates over the atomic contribution. Photon numbers in the hardcore regime of the JCH model are calculated by truncating to at most one polariton excitation in each cavity.

From these calculations we conclude that the (JC) polaritons are also fermionised in the strongly interacting regime. However, achieving strongly interacting regime for the photonic polaritons requires a very large value of coupling strength ($g/\gamma \approx 2 \times 10^5$ has been used in Fig. 3.10a) to compensate for the large atom-photon detuning. So, the question arises: Can we observe the same phenomenon for a more modest value of g? Fig. 3.10b displays the total excitation number for $g/\gamma \approx 800$, showing that similar physics can be observed by reducing the values of Δ and g. It is interesting to note that the quantitative agreement to the fermionised limit is better achieved for $\Delta = g$.

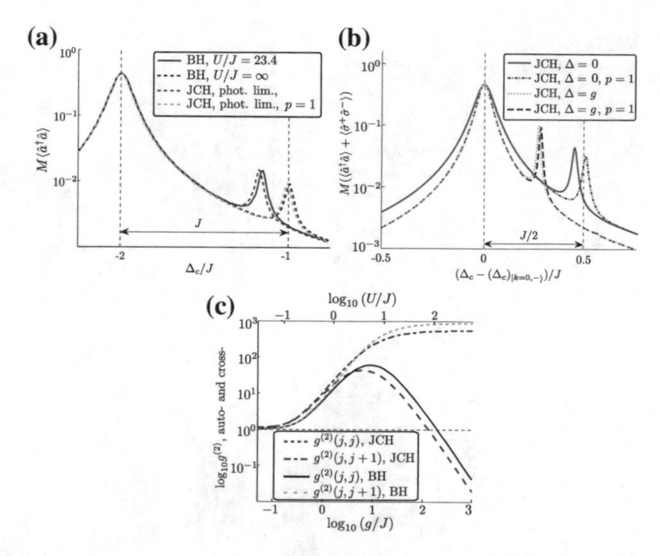

Fig. 3.10 **a** Steady state photon number as a function of laser-cavity detuning for both the BH and JCH systems. One particle peaks are located on the *left* ($\Delta_c = -2J$) and two-particle peaks on the *right*. Strongly nonlinear cases are shown by the *solid blue* (BH) and the *dotted red* (JCH) *curves*, whereas the strictly hard-core limit is shown by the *dotted black* (BH) and the *dotted green* (JCH) *curves*. Parameters are $M = 3$ cavities, $J/\gamma = 20$, $\Omega/\gamma = 0.5$, $\Delta/g = -10$, and $g/\gamma \approx 2 \times 10^5$. **b** Analogous photon number spectra for the JCH model with a smaller atom-photon detuning Δ. $g/\gamma \approx 800$ as calculated from Eq. (3.7). **c** The auto- and cross-correlations measured at the two-particle peak as a function of nonlinearity in both the BH and JCH ($\Delta = 2J$) models. Values of U are determined from Eq. (3.7). Taken from Grujic et al. [47]

Further evidence of fermionisation can be obtained by inspecting the second-order intensity correlations between photons. For this purpose, the auto-correlations $g^{(2)}_{j,j}$ and the cross-correlations $g^{(2)}_{j,j+1}$ are depicted in Fig. 3.10c. The driving field is tuned at the two-particle resonance. We have already seen the behaviour of the auto-correlations in Sect. 3.4.2. In the weakly nonlinear regime, the resonance peaks corresponding to different values of N (the total number of photons in the array) overlap, and the photons inherit the Poissonian nature of the pump field [60]. With increasing nonlinearity, the two-photon peak initially gives rise to strong photon bunching, before fermionisation takes place at which point the photons become strongly anti-bunched. Fermionisation of photons is further corroborated by the cross-correlations, which increases with the nonlinearity, meaning that the photons are preferentially located next to each other.

3.4.5 Polariton Crystallisation

In the above scenario we assumed homogeneous driving, i.e., all sites were driven with the driving field of the same intensity and phase. Let us now see an example of what happens when the system is driven inhomogeneously. The case where

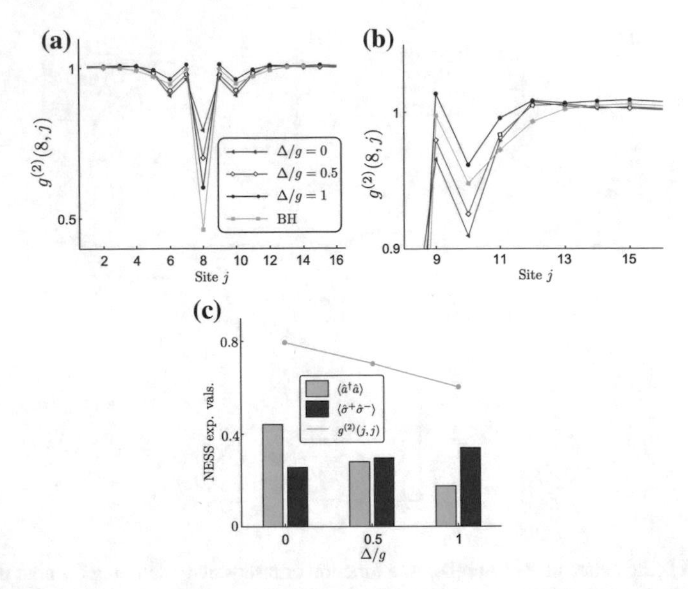

Fig. 3.11 **a** Steady state intensity correlations for a cyclic 16-site JCH system, driven by lasers exciting the $\pi/2$ momentum mode. Connecting *lines* are drawn to guide the eye. Parameters are $J/\gamma = 2, g/\gamma = 10$, and $|\Omega_k|/\gamma = 2$. Also shown (*solid green*) are the intenstiy correlations for the BH system with a 'matched' Kerr nonlinearity $U/\gamma \approx 6$. **b** Closer view of the correlation function. **c** The atomic and photonic population in each resonator in the steady state for three different values of atom-resonator detuning, as well as the on-site intensity correlations. Taken from Grujic et al. [47]

the nearest-neighbour driving fields have a phase difference of $\pi/2$ ($\Omega_k = \Omega \exp(ik\pi/2)$) has been studied by Hartmann [61]. This system exhibits an intriguing phenomenon of photon crystallisation, which will be briefly reviewed here. We follow the presentation in [47], which generalised the BH system in the original proposal to the JCH system. Interested readers are recommended to read the original paper [61] for an in-depth analysis of the effects of the inhomogeneous driving.

The system consists of a cyclic ring of 16 cavities, with the above mentioned driving field arrangement. The latter creates a flow of polaritons around the ring, with momentum $\pi/2$, when the laser frequency is set to the lowest energy transition. The interplay between the flow and on-site interaction gives rise to the phenomenon of polariton crystallisation: there is a larger probability of finding photons in neighbouring cavities than in cavities further apart. This is evident in the second-order correlation function as shown in Fig. 3.11a, b: the on-site anti-bunching is accompanied by nearest-neighbour intensity correlations that are stronger than correlations between more distant cavities. The interpretation is that the polaritons form dimers across two neighbouring sites which then move together around the array.

Both the BH and JCH systems clearly yield similar qualitative results, although there are qualitative differences in the small detuning limit considered here. As noted in the previous subsection, only in the 'photonic regime'—where the photon-atom detuning is very large and g even larger—would one expect a quantitative match.

Furthermore, the atom-photon detuning provides an additional knob in the JCH system that is absent in the BH system: it controls the composition of the polaritons. The photonic and atomic contributions to the polaritons are depicted in Fig. 3.11c for three different values of atom-cavity detuning. In this regime, the atomic contribution increases with Δ, which explains the enhanced on-site anti-bunching with increasing Δ.

3.4.6 Phases in 2D Arrays

So far, we have limited our attention to 1D arrays, for which the steady-state or time-evolution could be studied numerically using either exact diagonalisation methods or the MPS formalism. Now we would like to discuss the physics of 2D arrays, where a whole new range of physics are waiting to be explored. However, there is a slight problem. In 2D, the application of tensor network ansatz does not work very well due to an unfavourable scaling of entanglement and are therefore difficult to investigate numerically. For this reason studies up to now have either focused on small systems or employed variants of mean-field analysis. Here, we give a brief survey of recent studies of 2D CRAs based on the latter.

Initially, the dynamics of interacting photons after an initial preparation in the Mott state were studied by Tomadin and coworkers [62]. To solve the master equation, the authors employed the cluster mean-field approach, which reduces the whole system to a cluster of cavities plus mean-fields representing the rest of the system it is interacting with. For a single-site-cluster case, this amounts to making the mixed state equivalent of the usual mean-field approximation $\hat{a}_i^\dagger \hat{a}_j \rightarrow \langle \hat{a}_i^\dagger \rangle \hat{a}_j + \hat{a}_i^\dagger \langle \hat{a}_j \rangle$. In the typical quench scenario, the dynamics of interacting photons show characteristic differences depending on whether the system parameters are in the Mott or superfluid regime. Such signatures are also observed in the dissipative cases and are manifest in the coherence of the cavity emission. In the Mott-regime the equal-time second-order intensity correlation function stays zero, while in the superfluid-regime it gains a finite value within the photonic lifetime.

A mean-field phase diagram of the steady-state density matrix in the driven dissipative scenario has also been explored. Firstly, the phase diagram of the driven dissipative 2D BH array was mapped out by Le Boité et al. [63, 64], who found regions of unstable, mono-stable, and bi-stable phases. Despite Mott-lobe-like structures characterised by one stable solution with photon-antibunching, the phase diagram is very different from its equilibrium counterpart, showing rich many-body physics arising in the non-equilibrium scenario. Secondly, Jin and co-workers have investigated the phases of a 2D BH array, but with an extra component of nearest-neighbour cross-Kerr interaction ($V \sum_{i,j} \hat{n}_i \hat{n}_j$) [65, 66]. Due to this extra term, the NESS phases were shown to be classified into a uniform phase and a checkerboard phase. In the uniform phase, two neighbouring cavities have equal densities whereas in the checkerboard (or staggered) phase they have a finite density-difference. Furthermore, the NESS can show non-trivial oscillatory behaviour, hinting at a non-equilibrium analog of the supersolid phase.

Clearly, the validity of these mean-field results should be confirmed with approaches that take into account longer-ranged quantum correlations across the lattice. Numerical methods that go beyond the cluster mean-field approaches are under development. These include a hybrid real-space renormalisation group approach called 'Corner space renormalisation method' [67] and a self-consistent projection operator theory that derives an exact equation of motion for the reduced density matrices of an arbitrary sub-lattice [68].

3.4.7 *Quantum Hall Physics with Light*

It is possible to engineer artificial gauge fields for photons in a 2D CRA and thereby achieve photonic quantum Hall systems. There are several proposals to prepare and probe (integer or fractional) photonic quantum Hall states, but we will focus on recent developments on driven dissipative signatures in photonic quantum Hall systems and refer the interested reader to Ch. [Hafezi].

In a lattice system, gauge fields such as magnetic fields can be included by the so-called Peierls substitution:

$$J \sum_{\langle i,j \rangle} \hat{a}_i^\dagger \hat{a}_j + \text{h.c.} \rightarrow J \sum_{\langle i,j \rangle} e^{i\phi_{i,j}} \hat{a}_i^\dagger \hat{a}_j + \text{h.c.} \tag{3.10}$$

Essentially, a particle traveling around any loop accumulates a phase factor corresponding to the sum of $\phi_{i,j}$ over the loop. It is possible to engineer such site-dependent hopping amplitudes in two-dimensional CRAs, to create an artificial magnetic field. Then, in analogy to the electronic systems, the presence of the on-site interaction gives rise to a photonic fractional Hall effect. The latter can be checked by performing measurements on the output photons and calculating the overlap with the bosonic Laughlin state [69], or can be inferred from the second-order intensity correlation function of the collective mode [70]. Here, we review the work by Umucalılar and Carusotto [69].

Consider a 4 by 4 lattice in the hard-core limit $U/J \rightarrow \infty$, which is weakly pumped such that only up to 2 photons are allowed in the system at any given time. The effective magnetic field strength B is set such that the number of flux quanta, the total magnetic flux divided by the flux quantum, is 4. This choice gives the filling fraction 1/2 when there are two photons in the system. In a conventional fractional quantum Hall system, the ground state is described by the bosonic analogue of a generalised Laughlin wave function [71]. However, the steady-state of the proposed driven dissipative set-up surely does not coincide with this state because there are 0 and 1 photon manifolds. Interestingly, though, the two-photon sub-manifold of the steady-state density matrix exhibits a strong overlap with the generalised Laughlin wave function. This has been shown in [69], where the two-photon amplitude $\psi_{ij} \equiv \text{Tr}[\rho_{ss}\hat{a}_i\hat{a}_j]$ has been calculated numerically and compared with the bosonic Laughlin state. Note that ψ_{ij} can be measured experimentally by performing multiple

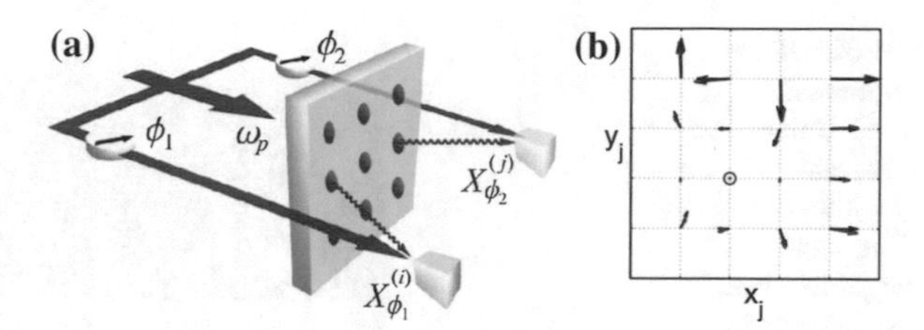

Fig. 3.12 **a** Schematic illustration of the experimental set-up to measure the two-photon amplitude ψ_{ij}. **b** Comparison of the normalised two-photon amplitude for a fixed $i = i_0$ (*red thick arrows*) and the generalised Laughlin function (*black thin arrows*). The reference site i_0 is marked by a *circle*. Reprinted with permission from Umucalılar and Carusotto [69]. Copyright (2012) by the American Physical Society

homodyne measurements as schematically illustrated in Fig. 3.12a. An example of a numerically calculated ψ_{ij} is shown in Fig. 3.12b, along with the corresponding Laughlin wave function. The system is driven at a two-photon resonance and weak dissipation $\gamma/J = 0.03$ and weak driving $\Omega/\gamma = 0.1$ are assumed. An overlap of up to 98.9% can be achieved for $\gamma/J = 0.002$, while for a more realistic loss rate $\gamma/J = 0.05$, the overlap value is 90.0%. These high overlap values show that despite the losses and driving, one can prepare strongly-correlated photonic states which are analogues of the Laughlin state of fractional quantum Hall systems.

A driven dissipative analogue of the integer quantum Hall effect has also been proposed in a linear 2D CRA, where an experimentally viable scheme to measure the global Chern number (and even the local Berry curvature) was devised [72]. In this proposal, a photonic analogue of the integer quantum Hall system is investigated, in which there is a constant force driving a current. Such a force can be modelled by a position dependent energy term $H_f = F \sum_{m,n} n\hat{a}^{\dagger}_{m,n}\hat{a}_{m,n}$ (force in -y direction). Recall that in integer quantum Hall systems the driven current induces a Hall current in the x-direction, whose conductivity is quantised. This quantisation is robust against various imperfections due to its topological nature, and is captured by the Chern number associated with each Bloch band. Remarkably, it is possible to measure the Chern number in a driven dissipative set-up as proposed by Ozawa and Carusotto [72]. Alternative methods, based on tuning of twisted boundary conditions, to experimentally measure the topological invariants in driven-dissipative settings were proposed by Hafezi [73] and Bardyn and coworkers [74].

To understand the method by Ozawa and Carusotto, consider a linear 2D array (of size 41 × 41), with the central site driven by a coherent field of amplitude Ω. Then, because the system is linear, we can replace the operators $\hat{a}_{m,n}$ by complex amplitudes $a_{m,n}e^{-i\omega_0 t}$. When the driving force is absent, i.e., $F = 0$, the injected photons either disperse through the lattice (Fig. 3.13a, b) or stay localised within the central site (Fig. 3.13c), depending on whether the driving frequency lies within a Bloch band or a band gap. With the driving force turned on ($F = 0.1J$), however, the photons clearly 'travel' towards the left as shown in Fig. 3.13d. It is possible to show that the displacement in the center of mass $\langle x\rangle \equiv \sum_{m,n} m|a_{m,n}|^2/\sum_{m,n}|a_{m,n}|^2$ is related to the Berry curvature $\Omega_B(k)$ by the formula [72]

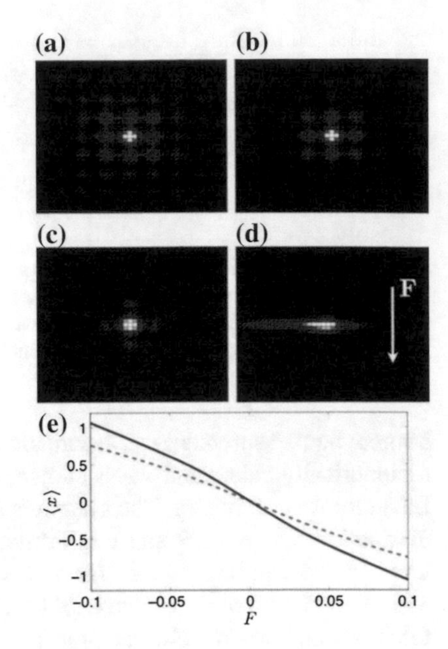

Fig. 3.13 **a–d** Distributions of the photon amplitudes $|a_{m,n}|$ on a 41×41 lattice. **a** $F = 0, \gamma = 0.01J$, and a pump frequency within the lowest band. **b** The same as in (**a**) but with larger loss rate $\gamma = 0.02J$. **c** The same as in (**a**) but the pump frequency lies within a band gap. **d** The same as in (**a**) but with $F = 0.1J$. **e** Displacement $\langle x \rangle$ as a function of F/J, for a pump frequency within the lowest band. The *solid blue line* is for $\gamma/J = 0.05$ and the *dashed green line* is for $\gamma/J = 0.08$. Reprinted with permission from Ozawa and Carusotto [72]. Copyright (2014) by the American Physical Society

$$\langle x \rangle = F \frac{\int_{\text{MBZ}} \gamma \Omega_B(k) n^2(k)}{\int_{\text{MBZ}} n(k)}. \tag{3.11}$$

Here, MBZ stands for the magnetic Brillouin zone; $n(k) = [(\omega_L - E(k))^2 + \gamma^2]^{-1}$, where ω_L is the driving frequency; $E(k)$ is the energy dispersion of the corresponding band. In the large-loss limit where γ is bigger than the width of the Bloch band under consideration but smaller than the band gap, the above formula reduces to

$$\langle x \rangle \approx \frac{qCF}{2\pi\gamma}, \tag{3.12}$$

where C is the Chern number. On the other hand, if the loss rate is very small compared to the width of the Bloch band, the formula reduces to

$$\langle x \rangle \approx \frac{\bar{\Omega}_B(\omega_L)F}{2\gamma}, \tag{3.13}$$

where $\bar{\Omega}_B(\omega_L)$ is the averaged Berry curvature on the $E(k) = \omega_L$ curve. Examples of the linear dependence of $\langle x \rangle$ on F is shown in Fig. 3.13e.

Therefore, depending on the value of the loss rate, one can either directly measure the Chern number or the local Berry curvature. For the lowest Bloch band, which has $C = -1$, the theoretically calculated value using the above method was shown to be as close as -0.97, demonstrating that the signature of the integer quantum Hall effect can be directly measured in the driven dissipative set-up. Furthermore, the ability to measure the Berry curvature allows one to detect nontrivial properties of the photonic honeycomb lattice in the absence of an external magnetic field. In the latter, there are two bands near the Dirac points whose degeneracy is lifted when two neighbouring sites have different energies. Even though the total Chern number vanishes in the absence of a synthetic magnetic field, the two bands are known to possess nontrivial Berry curvatures. Because the Chern number vanishes, one needs to resolve the Berry curvature in order to detect the system's nontrivial properties, which can be done using the above method.

3.4.8 Other Works

There are a number of interesting works on driven dissipative CRAs that we could not cover due to spatial limitations. Here we provide a brief survey of some of these, so that the interested readers can refer to original articles.

First of all, there are many works that investigate properties of photons emitted from a dimer. Liew and Savona discovered the phenomena of 'unconventional photon blockade', where photon blockade is achieved in the weakly Kerr-nonlinear regime in which anti-bunching would otherwise be absent [75]. This was shown to result from an interference of excitation pathways [76] and was generalised to a coupled resonator with second-order nonlinearity [77]. The fluorescence spectrum and the spectrum of the second-order intensity correlation function of the JC dimer were calculated in [78], both for coherent and incoherent driving fields, while photon statistics in a BH dimer were investigated with emphasis on photon anti-bunching in [79, 80]. Scaling of this anti-bunching behaviour with increasing array size was studied within a mean-field theory [81]. In a slightly different setting, transport of photons in a BH array of up to 60 sites was studied using an MPS method adapted to mixed states [82]. Here, only the first cavity is driven and the underlying many-body states were shown to be visible in the transmission spectrum. Transport of quantum light in a BH dimer was also studied, showing that more detailed information about the underlying correlated states is revealed when one uses two photons of different energies instead of a coherent field as the input [83].

Other works look at connections with well-known condensed matter systems. A quantum optical analogue of the Josephson interferometer in a three-site set-up was proposed by Gerace and co-workers [84]. There, in analogy to the Josephson junction, two end cavities are driven by coherent fields with different phases, while the middle cavity contains single-photon nonlinearity. The interplay between tunneling and interaction gives rise to rich physics in the steady-state, details of which can be found in Ch. [Gerace]. A driven dissipative realisation of the Kitaev chain [85] was proposed by Bardyn and İmamoğlu [86], showing that Majorana zero modes can be created and detected in a CRA. To obtain the so-called p-wave pairing term, the authors propose to use parametric driving in the strongly-interacting regime. Photons are then effectively spin-1/2 particles, allowing one to use Jordan-Wigner transformation to achieve the Kitaev chain.

3.5 Summary and Outlook

In this chapter, we have reviewed out-of-equilibrium many-body physics in coupled resonator arrays. A short survey of possible experimental platforms was provided, along with two theoretical models (Jaynes-Cummings-Hubbard and Bose-Hubbard) to describe them as well as the master equation description of driven dissipative CRAs. Then we reviewed a range of many-body non-equilibrium phenomena, starting from those arising in a two-site set-up and finishing with phenomena in 2D arrays. A brief survey of many interesting works that could not be covered in detail was then given.

The study of driven dissipative CRAs has only recently begun and there are many exciting avenues to be explored. Experimentally, it is still quite challenging to build an array of resonators with large nonlinearity, although splendid progress is being made in the field of circuit-QED as we have seen in Sect. 3.4.3. Controlling the fluctuations of local parameters is another limitation. Theoretically, one can look at ways to enhance the effects of nonlinearity, as in unconventional quantum blockade, to help ease the requirement of strong nonlinearity. More pressing is the development of theoretical tools to study driven dissipative 2D arrays. As we have alluded to earlier, 1D systems can be simulated efficiently using the MPS formalism, but 2D system are waiting for further developments. There is already progress in this direction as we briefly mentioned at the end of the last section, and more will surely follow. With the on-going developments in both the experimental and theoretical fronts, many more exciting discoveries will surely follow in this nascent field.

Acknowledgements C. Noh and D.G. Angelakis would like to acknowledge the financial support provided by the National Research Foundation and Ministry of Education Singapore (partly through the Tier 3 Grant "Random numbers from quantum processes"), and travel support by the EU IP-SIQS. S.R. Clark and D. Jaksch acknowledge support from the European Research Council under the European Union's Seventh Framework Programme (FP7/2007–2013)/ERC Grant Agreement no. 319286 Q-MAC.

Appendix: Matrix Product States and Quantum Trajectories

In this Appendix we provide some more details about MPS methods and quantum trajectories approach. First we specify the problem to be solved in general.

Quantum Master Equation

It is known that the most general form of a Markovian time-local master equation that preserves the trace, Hermiticity, and positivity of the system's density matrix ρ must be of so-called Lindblad form [44, 46]:

$$\frac{d\rho}{dt} = -i[\hat{H}, \rho] + \sum_{\alpha} \left(\hat{L}_\alpha \rho \hat{L}_\alpha^\dagger - \frac{1}{2} \left\{ \hat{L}_\alpha^\dagger \hat{L}_\alpha, \rho \right\} \right), \tag{3.14}$$

where the Lindblad operators $\hat{L}_\alpha = \sqrt{\gamma_\alpha} A_\alpha$ and $\hat{H}$ is the Hamiltonian describing the unitary part of the system's dynamics. The γ_α are identified as the characteristic rates at which 'jump operators' A_α act on the system. In the case of CRAs the jump operators are the bosonic mode operators $\hat{a}_j$ or atomic ladder operator $\hat{\sigma}_j^-$ which act locally on a site j. The locality of the jump operators and the nearest-neighbour nature of the JCH and BH Hamiltonians in 1D are crucial to the applicability of MPS methods to solve this problem.

Matrix Product States

Generally we consider an L site chain composed of subsystems with d-dimensional Hilbert space spanned by states enumerated as $|j\rangle$, with $j \in \{1, \ldots, d\}$. The Hilbert space for the total system is then spanned by the product basis $|\mathbf{j}\rangle = |j_1, \ldots, j_L\rangle \equiv |j_1\rangle \otimes \cdots \otimes |j_L\rangle$. An arbitrary pure state of the system is then

$$|\Psi\rangle = \sum_{\mathbf{j}} c_{j_1 j_2 \ldots j_L} |\mathbf{j}\rangle , \tag{3.15}$$

which is described by d^L complex amplitudes $c_{j_1 j_2 \ldots j_L}$. To circumvent this intractable description the MPS ansatz parameterises the amplitudes of a state in terms of a product of matrices indexed by the local configurations as

$$|\Psi\rangle = \sum_{\mathbf{j}} \left(\bar{L}^{\mathrm{T}} \mathbf{A}^{j_1} \mathbf{A}^{j_2} \ldots \mathbf{A}^{j_L} \bar{R} \right) |\mathbf{j}\rangle . \tag{3.16}$$

Here the $\mathbf{A}^{j_m}$ are a set of d matrices (one for each local state $|j_m\rangle$) of dimension $\chi \times \chi$, while $\bar{L}$ and $\bar{R}$ are boundary vectors which collapse the product into a scalar complex amplitude. This ansatz has only $Ld\chi^2$ complex numbers, so if χ is fixed and small it provides a compact class of states [32].

A special feature of MPS is that the dimension of the matrices χ is directly related to the entanglement of the system when it is bipartitioned. If a state is weakly entangled then it will possess an accurate MPS representation with a small fixed χ. Given an MPS, reduced density matrices for subsets of sites, norms and expectation values of observables can be efficiently computed, essentially via products of the state's constituent matrices [31, 32]. These features become especially useful for 1D systems which are described by a Hamiltonian $\hat{H} = \sum_m \hat{h}_{m,m+1}$ composed of at most nearest-neighbour terms. Many such systems are now known to possess weakly entangled ground states and low-lying excited states. Thus an MPS description of them is both natural and effective [32].

To find ground states and excited states of $\hat{H}$ the highly successful density matrix renormalisation group (DMRG) method [32–34, 37] was developed which is essentially a variational minimiser over the MPS ansatz. For our purposes here however the extension of DMRG to time-evolution via the time evolving block decimation (TEBD) algorithm is most relevant [35, 36]. Here we sketch some of the key points to this method. Our task is to apply unitary time-evolution described by the propagator $\hat{U}(\delta t) = \exp(-i\hat{H}\delta t)$ to an initial MPS for a small time step δt.

First, we break up the exponential using a Suzuki-Trotter decomposition [87] such as

$$\hat{U}(\delta t) = \left[\prod_{\text{odd } m} \exp(-i\hat{h}_{m,m+1}\delta t) \right] \left[\prod_{\text{even } m} \exp(-i\hat{h}_{m,m+1}\delta t) \right] + \mathcal{O}(\delta t^2), \quad (3.17)$$

where we have utilised that all the odd Hamiltonian terms $\hat{h}_{m,m+1}$ commute among themselves, and similarly for the even terms. Higher order Trotter decompositions have a similar form. What we are left with is a product of unitary operators (or gates) $\exp(-i\hat{h}_{m,m+1}\delta t)$ acting on nearest-neighbouring pairs of sites which have to be applied to the MPS initial state in sequence.

Second, upon applying a gate $\exp(-i\hat{h}_{m,m+1}\delta t)$ to an MPS the matrices $\mathbf{A}^{j_m}$ and $\mathbf{A}^{j_{m+1}}$ for the two sites m and $m+1$ get merged together into a joint matrix $\mathbf{B}^{j_m,j_{m+1}}$. To bring the state back into MPS form the matrix $\mathbf{B}^{j_m,j_{m+1}}$ needs to be factorised. This is achieved by using a singular value decomposition (SVD) which breaks $\mathbf{B}^{j_m,j_{m+1}} \mapsto \tilde{\mathbf{A}}^{j_m}\tilde{\mathbf{A}}^{j_{m+1}}$ yielding new updated matrices describing the evolved MPS [35, 36]. During this operation the dimensions of the new $\tilde{\mathbf{A}}$ matrices can grow and so after repeated applications of gates the MPS can eventually become intractably large. However, the singular values outputted by the SVD provide a quantitative means of truncating the dimension down, essentially compressing the state and thus ensuring that the MPS remains tractable. This truncation will only be accurate if the evolution does not generate too much entanglement. Once all the gates in the decomposition are applied we have an MPS approximation for $|\Psi(\delta t)\rangle = \hat{U}(\delta t)\,|\Psi\rangle$, and the procedure can be repeated to evolve further in time [32].

Quantum Trajectory Algorithm

Rather than simulating the dynamics of an open system by evolving its density matrix ρ directly, an alternative approach is to propagate stochastic realisations of individual system state vectors $\{|\Psi_i(t)\rangle\}$ [41–43]. These 'quantum trajectories' are piecewise deterministic processes, interrupted randomly by 'quantum jumps' due to the system's interaction with the environment. An approximation to the density matrix may be obtained by averaging the contributions $\Pi_i = \{|\Psi_i(t)\rangle\langle\Psi_i(t)|\}$ of many trajectories [45]. The details of the 'unravelling' of the Lindblad master equation into stochastic wave functions are described in Ref. [46]. Here we give a practical recipe on how to generate quantum trajectories.

For a set of Lindblad operators $\{\hat{L}_\alpha\}$, we construct a non-unitary effective Hamiltonian $\hat{H}_{\text{eff}}$:

$$\hat{H}_{\text{eff}} = \hat{H} - \frac{i}{2}\sum_\alpha \hat{L}_\alpha^\dagger \hat{L}_\alpha. \quad (3.18)$$

Evolution under the Schrödinger equation with this effective Hamiltonian leads to a decay in the norm of the wave-function as a consequence of interactions with the environment. Trajectories are then generated as follows.

1. Begin the simulation by initialising the system in a state $|\Psi\rangle = |\Psi(0)\rangle$.
2. Draw a randomly chosen number $r \in [0, 1]$.
3. Evolve the system under the action of the non-Hermitian Hamiltonian of Eq. (3.18). This will involve computing $|\Psi(t)\rangle = \exp(-i\hat{H}_{\text{eff}}t)\,|\Psi(0)\rangle$. Due to the decaying norm of the state, physical expectation values are calculated during this evolution using the normalised form of the state $|\Psi(t)\rangle/||\,|\Psi(t)\rangle||$.
4. Continue the evolution until the norm falls below the randomly chosen number, i.e. when the condition $||\,|\Psi(t_{\text{jump}})\rangle|| < r$ is first met. One of the possible quantum jumps now occurs.
5. The jump operator to be applied is chosen by first generating the normalised probability distribution

$$P_\alpha = \frac{||\hat{L}_\alpha \Psi(t_{\text{jump}})\rangle||^2}{\sum_\beta ||\hat{L}_\beta \Psi(t_{\text{jump}})\rangle||^2}, \tag{3.19}$$

then randomly selecting a jump operator index j from this weighted distribution. The normalised state after the application of the chosen operator is

$$\frac{\hat{L}_j|\Psi(t_{\text{jump}})\rangle}{||\hat{L}_j|\Psi(t_{\text{jump}})\rangle||}. \tag{3.20}$$

We continue looping over steps 2–5 until the end of our simulation window. Crucially all the steps in this algorithm, such as the time-evolution in step 3, the computation of norms in step 4 and the application of a jump operator in step 5, can be implemented efficiently on MPS so long as the Hamiltonian and Lindblad operators act locally or on nearest-neighbour sites.

References

1. R.P. Feynman, Simulating physics with computers. Int. J. Theor. Phys. **21**, 467–488 (1982)
2. I. Bloch, J. Dalibard, S. Nascimbéne, Quantum simulations with ultra cold quantum gases. Nat. Phys. **8**, 267–276 (2012)
3. R. Blatt, C.F. Roos, Quantum simulations with trapped ions. Nat. Phys. **8**, 277–284 (2012)
4. J.I. Cirac, P. Zoller, Goals and opportunities in quantum simulation. Nat. Phys. **8**, 264–266 (2012)
5. T.H. Johnson, S.R. Clark, D. Jaksch, What is a quantum simulator? EPJ Quantum Technol. **1**, 10 (2014)
6. A. Aspuru-Guzik, P. Walther, Photonic quantum simulators. Nat. Phys. **8**, 285–291 (2012)
7. A.A. Houck, H.E. Türeci, J. Koch, On-chip quantum simulation with superconducting circuits. Nat. Phys. **8**, 292–299 (2012)
8. M.P.A. Fisher, P.B. Weichman, G. Grinstein, D.S. Fisher, Boson localization and the superfluid-insulator transition. Phys. Rev. B **40**(1), 546–570 (1989)
9. A. Tomadin, R. Fazio, Many-body phenomena in QED-cavity arrays. J. Opt. Soc. Am. B **27**(6), A130–A136 (2010)
10. I. Carusotto, C. Ciuti, Quantum fluids of light. Rev. Mod. Phys. **85**, 299–366 (2013)

11. S. Schmidt, J. Koch, Circuit QED lattices: towards quantum simulation with superconducting circuits. Ann. Phys. **525**, 395–412 (2013)
12. D. Jaksch, C. Bruder, J.I. Cirac, C.W. Gardiner, P. Zoller, Cold Bosonic atoms in optical lattices. Phys. Rev. Lett. **81**, 3108–3111 (1998)
13. M. Greiner, O. Mandel, T. Esslinger, T.W. Hänsch, I. Bloch, Quantum phase transition from a superfluid to a Mott insulator in a gas of ultracold atoms. Nature **415**, 39–44 (2002)
14. M.J. Hartmann, F.G.S.L. Brandão, M.B. Plenio, Strongly interacting polaritons in coupled arrays of cavities. Nat. Phys. **2**, 849–855 (2006)
15. A.D. Greentree, C. Tahan, J.H. Cole, L.C.L. Hollenberg, Quantum phase transitions of light. Nat. Phys. **2**, 856–861 (2006)
16. D.G. Angelakis, M.F. Santos, S. Bose, Photon-blockade-induced Mott transitions and XY spin models in coupled cavity arrays. Phys. Rev. A **76**, 031805 (2007)
17. M.J. Hartmann, F.G.S.L. Brandao, M.B. Plenio, Quantum many-body phenomena in coupled cavity arrays. Laser Photon Rev. **2**, 527–556 (2008)
18. J.D. Joannopoulos, S.G. Johnson, J.N. Winn, *Photonic Crystals: Molding the Flow of Light* (Princeton University Press, 2008)
19. M. Trupke, E.A. Hinds, S. Eriksson, E.A. Curtis, Z. Moktadir, E. Kukharenka, M. Kraft, Microfabricated high-finesse optical cavity with open access and small volume. App. Phys. Lett. **87**, 211106 (2005)
20. G. Lepert, M. Trupke, E.A. Hinds, H. Rogers, J.C. Gates, P.G.R. Smith, Elementary array of Fabry-Pérot waveguide resonators with tunable coupling. App. Phys. Lett. **103**, 111112 (2013)
21. A. Blais, R.S. Huang, A. Wallraff, S.M. Girvin, R.J. Schoelkopf, Cavity quantum electrodynamics for superconducting electrical circuits: an architecture for quantum computation. Phys. Rev. A **69**, 062320 (2004)
22. O. Astafiev, A.M. Zagoskin, A.A. Abdumalikov, Y.A. Pashkin, T. Yamamoto, K. Inomata, Y. Nakamura, J.S. Tsai, Resonance fluorescence of a single artificial atom. Science **327**, 840 (2010)
23. L. Frunzio, A. Wallraff, D. Schuster, J. Majer, R. Schoelkopf, Fabrication and characterization of superconducting circuit qed devices for quantum computation. Appl. Supercond. **15**, 860–863 (2005)
24. M.H. Devoret, S. Girvin, R. Schoelkopf, Circuit-qed: how strong can the coupling between a Josephson junction atom and a transmission line resonator be? Ann. Phys. **16**, 767–779 (2007)
25. A. Wallraff, D.I. Schuster, A. Blais, L. Frunzio, R.S. Huang, J. Majer, S. Kumar, S.M. Girvin, R.J. Schoelkopf, Strong coupling of a single photon to a superconducting qubit using circuit quantum electrodynamics. Nature **431**, 162–167 (2004)
26. G. Lepert, M. Trupke, M.J. Hartmann, M.B. Plenio, E.A. Hinds, Arrays of waveguide-coupled optical cavities that interact strongly with atoms. New J. Phys. **13**, 113002 (2011)
27. D.F. Walls, G.J. Milburn, *Quantum Optics* (Springer, Berlin, 1994)
28. F. Verstraete, V. Murg, J.I. Cirac, Matrix product states, projected entangled pair states, and variational renormalization group methods for quantum spin systems. Adv. Phys. **57**, 143 (2008)
29. J.I. Cirac, F. Verstraete, Renormalization and tensor product states in spin chains and lattices. J. Phys. A Math. Theor. **42**, 504004 (2009)
30. S. Al-Assam, S.R. Clark, D. Jaksch, Tensor Network Theory (TNT) Library. http://www.ccpforge.cse.rl.ac.uk/gf/project/tntlibrary/
31. D. Perez-Garcia, F. Verstraete, M.M. Wolf, J.I. Cirac, Matrix product state representations. Quantum Inf. Comput. **7**, 401 (2007)
32. U. Schollwöck, The density-matrix renormalization group in the age of matrix product states. Ann. Phys. **326**, 96 (2011)
33. S.R. White, Density matrix formulation for quantum renormalization groups. Phys. Rev. Lett. **69**, 2863 (1992)
34. S.R. White, Density-matrix algorithms for quantum renormalization groups. Phys. Rev. B **48**, 10345 (1993)
35. G. Vidal, Efficient classical simulation of slightly entangled quantum computations. Phys. Rev. Lett. **91**, 147902 (2003)

36. G. Vidal, Efficient simulation of one-dimensional quantum many-body systems. Phys. Rev. Lett. **93**, 40502 (2004)
37. U. Schollwöck, The density-matrix renormalization group. Rev. Mod. Phys. **77**, 259 (2005)
38. M. Zwolak, G. Vidal, Mixed-state dynamics in one-dimensional quantum lattice systems: a time-dependent superoperator renormalization algorithm. Phys. Rev. Lett. **93**, 207205 (2004)
39. F. Verstraete, J.J. Garcia-Ripoll, J.I. Cirac, Matrix product density operators: simulation of finite-temperature and dissipative systems. Phys. Rev. Lett. **93**, 207204 (2004)
40. A.J. Daley, J.M. Taylor, S. Diehl, M. Baranov, P. Zoller, Atomic three- body loss as a dynamical three-body interaction. Phys. Rev. Lett. **102**, 40402 (2009)
41. C.W. Gardiner, A.S. Parkins, P. Zoller, Wave-function quantum stochastic differential equations and quantum-jump simulation methods. Phys. Rev. A **46**, 4363 (1992)
42. R. Dum, A.S. Parkins, P. Zoller, C.W. Gardiner, Monte Carlo simulation of master equations in quantum optics for vacuum, thermal, and squeezed reservoirs. Phys. Rev. A **46**, 4382 (1992)
43. J. Dalibard, Y. Castin, K. Molmer, Wave-function approach to dissipative processes in quantum optics. Phys. Rev. Lett. **68**, 580 (1992)
44. H.J. Carmichael, *An Open Systems Approach to Quantum Optics* (Springer, 1993)
45. M.B. Plenio, P.L. Knight, The quantum-jump approach to dissipative dynamics in quantum optics. Rev. Mod. Phys. **70**, 101 (1998)
46. H.P. Breuer, F. Petruccione, *The Theory of Open Quantum Systems* (Oxford University Press, 2002)
47. T. Grujic, S.R. Clark, D. Jaksch, D.G. Angelakis, Non-equilibrium many-body effects in driven nonlinear resonator arrays. New J. Phys. **14**, 103025 (2012)
48. T. Grujic, S.R. Clark, D. Jaksch, D.G. Angelakis, Repulsively induced photon super-bunching in driven resonator arrays. Phys. Rev. A **87**, 053846 (2013)
49. S. Schmidt, D. Gerace, A.A. Houck, G. Blatter, H.E. Türeci, Nonequilibrium delocalization-localization transition of photons in circuit quantum electrodynamics. Phys. Rev. B **82**, 100507(R) (2010)
50. S.M. Jensen, The nonlinear coherent coupler. IEEE J. Quantum Electron. **QE-18**, 1580 (1982)
51. J.C. Eilbeck, P.S. Lomdahl, A.C. Scott, The discrete self-trapping equation. Phys. D **16**, 318 (1985)
52. A. Smerzi, S. Fantoni, S. Giovanazzi, S.R. Shenoy, Quantum coherent atomic tunenlling between two trapped Bose-Einstein condensates. Phys. Rev. Lett. **79**, 4950 (1997)
53. M. Albiez, R. Gati, J. Fölling, S. Hunsmann, M. Cristiani, M.K. Oberthaler, Direct observation of tunenlling and nonlinear self-trapping in a single bossing Josephson junction. Phys. Rev. Lett. **95**, 010402 (2005)
54. S. Levy, E. Lahoud, I. Shomroni, J. Steinhauer, The ac and dc Josephson effects in a Bose-Einstein condensate. Nature **449**, 579–583 (2006)
55. D. Sarchi, I. Carusotto, M. Wouters, V. Savona, Coherent dynamics and parametric instabilities of microcavity polarities in double-well systems. Phys. Rev. B **77**, 125324 (2008)
56. N. Schetakis, T. Grujic, S.R. Clark, D. Jaksch, D.G. Angelakis, Frozen photons in Jaynes Cummings arrays. J. Phys. B **46**, 224025 (2013)
57. J. Raftery, D. Sadri, S. Schmidt, H.E. Türeci, A.A. Houck, Observation of a dissipation-induced classical to quantum transition. Phys. Rev. X **4**, 031043 (2014)
58. M.A. Cazalilla, R. Citro, T. Giamarchi, E. Orignac, M. Rigol, One dimensional bosons: from condensed matter systems to ultracold gases. Rev. Mod. Phys. **83**, 1405–1466 (2011)
59. M. Girardeau, Relationship between systems of impenetrable bosons and fermions in one dimension. J. Math. Phys. **1**, 516 (1960)
60. I. Carusotto, D. Gerace, H.E. Tureci, S. De Liberato, C. Ciuti, A. Imamoğlu, Fermionized photons in an array of driven dissipative nonlinear cavities. Phys. Rev. Lett. **103**, 033601 (2009)
61. M.J. Hartmann, Polariton crystallization in driven arrays of lossy nonlinear resonators. Phys. Rev. Lett. **104**, 113601 (2010)
62. A. Tomadin, V. Giovannetti, R. Fazio, D. Gerace, I. Carusotto, H.E. Türeci, A. Imamoglu, Signatures of the superfluid-insulator phase transition in laser-driven dissipative nonlinear cavity arrays. Phys. Rev. A **81**, 061801(R) (2010)

63. A. Le Boité, G. Orso, C. Ciuti, Steady-state phases and tunenlling-induced instabilities in the driven dissipative Bose-Hubbard model. Phys. Rev. Lett. **110**, 233601 (2013)
64. A. Le Boité, G. Orso, C. Ciuti, Bose-Hubbard model: relation between driven dissipative steady states and equilibrium quantum phases. Phys. Rev. A **90**, 063821 (2014)
65. J. Jin, D. Rossini, R. Fazio, M. Leib, M.J. Hartmann, Photon solid phases in driven arrays of nonlinearly coupled cavities. Phys. Rev. Lett. **110**, 163605 (2013)
66. J. Jin, D. Rossini, M. Leib, M.J. Hartmann, R. Fazio, Steady-state phase diagram of a driven QED-cavity array with cross-Kerr nonlinearities. Phys. Rev. A **90**, 023827 (2014)
67. S. Finazzi, A. Le Boité, F. Storme, A. Baksic, C. Ciuti, Corner space renormalization method for driven dissipative 2D correlated systems. arXiv:1502.05651 (2015)
68. P. Degenfeld-Schonburg, M.J. Hartmann, Self-consistent projection operator theory for quantum many-body systems. Phys. Rev. B **89**, 245108 (2014)
69. R.O. Umucalılar, I. Carusotto, Fractional quantum Hall states of photons in an array of dissipative coupled cavities. Phys. Rev. Lett. **108**, 206809 (2012)
70. M. Hafezi, M.D. Lukin, J.M. Taylor, Non-equilibrium fractional quantum Hall state of light. New. J. Phys. **15**, 063001 (2013)
71. F.D.M. Haldane, E.H. Rezayi, Periodic Laughlin-Jastrow wave functions for the fractional quantized Hall effect. Phys. Rev. B **31**, 2529 (1985)
72. T. Ozawa, I. Carusotto, Anomalous and quantum Hall effects in lossy photonic lattices. Phys. Rev. Lett. **112**, 133902 (2014)
73. M. Hafezi, Measuring topological invariants in small photonic systems. Phys. Rev. Lett. **112**, 210405 (2014)
74. C.-E. Bardyn, S.D. Huber, O. Zilberberg, Measuring topological invariants in small photonic lattices. New. J. Phys. **16**, 123013 (2014)
75. T.C.H. Liew, V. Savona, Single photons from coupled quantum modes. Phys. Rev. Lett. **104**, 183601 (2010)
76. M. Bamba, A. Imamoglu, I. Carusotto, C. Ciuti, Origin of strong photon antibunching in weakly nonlinear photonic molecules. Phys. Rev. A **83**, 021802(R) (2011)
77. D. Gerace, V. Savona, Unconventional photon blockade in doubly resonant microcavities with second-order nonlinearity. Phys. Rev. A **89**, 031803(R) (2014)
78. M. Knap, E. Arrigoni, W. von der Linden, Emission characteristics of laser-driven dissipative coupled-cavity systems. Phys. Rev. A **83**, 023821 (2011)
79. M. Leib, M.J. Hartmann, Bose-Hubbard dynamics of polaritons in a chain of circuit quantum electrodynamics cavities. New. J. Phys. **12**, 093031 (2010)
80. S. Ferretti, L.C. Andreani, H.E. Türeci, D. Gerace, Photon correlations in a two-site nonlinear cavity system under coherent drive and dissipation. Phys. Rev. A **82**, 013841 (2010)
81. F. Nissen, S. Schmidt, M. Biondi, G. Blatter, H.E. Türeci, J. Keeling, Nonequilibrium dynamics of coupled qubit-cavity arrays. Phys. Rev. Lett. **108**, 233603 (2012)
82. A. Biella, L. Mazza, I. Carusotto, D. Rossini, R. Fazio, Photon transport in a dissipative chain of nonlinear cavities. Phys. Rev. A **91**, 053815 (2015)
83. C. Lee, C. Noh, N. Schetakis, D. Angelakis, Few-photon transport in nonlinear cavity arrays: probing signatures of strongly correlated states. Phys. Rev. A **92**, 063817 (2015)
84. D. Gerace, H.E. Türeci, A. Imamoglu, V. Giovannetti, R. Fazio, The quantum-optical Josephson interferometer. Nat. Phys. **5**, 281–284 (2009)
85. A.Y. Kitaev, Unpaired Majorana fermions in quantum wires. Phys. Usp. **44**, 131 (2001)
86. C.-E. Bardyn, A. İmamoğlu, Mjorana-like modes of light in a one-dimensional array of nonlinear cavities. Phys. Rev. Lett. **109**, 253606 (2012)
87. M. Suzuki, Fractal decomposition of exponential operators with applications to many-body theories and monte carlo simulations. Phys. Lett. A **146**, 319 (1990)

Chapter 4
Topological Physics with Photons

Mohammad Hafezi and Jacob Taylor

Abstract In this chapter, we review the recent progress in the implementation of synthetic gauge fields for photons and the investigation of topological features such as quantum Hall physics in photonic systems. We first discuss the implementation of magnetic-like Hamiltonians in coupled resonator systems and provide a pedagogical connection between the transfer matrix approach and the couple mode theory to evaluate the system Hamiltonian. Furthermore, we discuss the investigation of non-equilibrium fractional quantum Hall physics in photonic systems. In particular, we show that driven strongly interacting photons exhibit interesting many-body behaviors which can be probed using the conventional optical measurement techniques. Finally, we present a scheme to implement three-body interaction in circuit-QED architecture to realize some of the fractional quantum Hall parent Hamiltonians with Pfaffian wavefunction as ground state.

4.1 Introduction

In physics, symmetries play a crucial role in determining the behavior of a system. Translation invariance leads to momentum conservation, while time reversal invariance leads to energy conservation. However, in large systems, symmetries of the system, which are present in the microscopic Hamiltonian, may not be stable to interactions between bodies, leading to the concept of spontaneous symmetry breaking, and the natural emergence of different phases of matter. However, in the early 1980s, it was realized that some physical observables are dictated by the topology of the system, rather than the symmetries of the Hamiltonian [1].

The hallmark example of such topological features is the quantum Hall regime of two dimensional electron gas systems. This simplest case has provided the greatest illumination, in which semi-classical understandings of single particle physics, in the

M. Hafezi · J. Taylor (✉)
Joint Quantum Institute, College Park, USA
e-mail: hafezi@umd.edu

J. Taylor
e-mail: jmtaylor@umd.edu

D.G. Angelakis (ed.), *Quantum Simulations with Photons and Polaritons*,
Quantum Science and Technology, DOI 10.1007/978-3-319-52025-4_4

form the skipping orbits and edge states, provides useful qualitative and even quantitative predictions of the transverse conductance in such a system. More generally, non-interacting theories of particles in lattices have a natural method of examining topological features, simply by computing topological invariants of the system at an energy band over the exact solutions of the Bloch wave vectors. This is physically motivated by the connection to electronic systems, where a Fermi sea fills all available states up to a specified energy, and it is the properties near that energy that determine the low-frequency response of the system.

An analogous behavior can also been seen in bosonic systems. Specifically, working with photons in the non-interacting regime, a single-particle 'shell' of fixed energy can be explored in a photonic system simply by looking at the monochromatic response of the system, i.e., examining its transmission spectrum.

In both cases, the crucial theoretical object to compute is the Berry curvature. Mathematically, if we define single-particle solutions to the (classical) wave equation for light $u_k(r) = \exp(-ikr)\psi_k(r)$ in a lattice system with translation invariance (such that k is a good 'quantum' number), we can define the Berry potential

$$\boldsymbol{A} = -\mathrm{Im}\langle u_k|\boldsymbol{\nabla}_k|u_k\rangle$$

Unfortunately, while measuring the phase of the single-particle solutions is in principle feasible, in practice is leaves much to be desired. Instead, there is another topological feature which may be easier to observe: the existence of transport (extended solutions of the wave equation) near the edges of the system when the 'bulk' of the system, has only localized solutions. These edge states are the hallmark experimental signature of non-trivial topological order, and arise due to the boundary conditions—in this case, the edge of the disk is different than a sphere or a torus.

In this chapter we describe an approach for generating and testing topological features in photonic (light-based) systems. Our approach relies heavily upon dramatic successes in understanding related physics in spin Hall systems in condensed matter physics, and in approaches to generating 'synthetic' magnetic-like effects (synthetic gauge fields) in atomic systems. Many of these ideas are available in the literature for further reading; here we focus on the pedagogical aspects of these approaches.

4.2 Gauge Field and Topological Features in Noninteracting Photonic System

We begin by examining the simplest case for a topological system: charged particles in a magnetic field as in the integer quantum Hall effect. Recall that the generator of translations for a charged particle in a magnetic field is fundamentally different than that of a particle without the field. Specifically, the transformation of the momentum from $\boldsymbol{p}$ to $\boldsymbol{p} - \frac{e}{c}\boldsymbol{A}$ takes an operator $U = \exp(-i\boldsymbol{a}\cdot\boldsymbol{p}/\hbar)$ to one with a non-trivial additional phase $\phi_{ij} = \frac{e}{c\hbar}\int_{r_i}^{r_j}\boldsymbol{A}(r)\cdot d\boldsymbol{r}$. Thus, for the particle translating about a closed loop, the total phase picked up becomes $\phi = \frac{e}{\hbar c} \oint \boldsymbol{A} \cdot d\boldsymbol{r}$, i.e., the flux enclosed in the loop for $\nabla \times \boldsymbol{A} \neq 0$.

The role the magnetic field is then simply to generate terms in translation that act as a phase in hopping. More generally, it was realized in atomic systems that any added terms to the Hamiltonian that play an analogous role to ϕ_{ij}, i.e., to have a non-trivial phase when the particle hops from lattice site to lattice site, can produce behavior analogous to the phase a charged particle accrues while moving in a real magnetic field [2]. Such effective fields are denoted *synthetic* gauge fields, as they do not explicitly have a gauge symmetry attached to them, but merely mimic the effects of a real, classical gauge field such as the vector potential in electromagnetism. The primary benefit of such synthetic gauge field ideas is the wide variety of possible approaches that implement such fields.

Similarly, for particles confined to a lattice, we replace the momentum terms by the usual Bloch-band approximation, $p^2 \to -J\sum_{\langle ij\rangle} a_i^\dagger a_j$ where the sum is over nearest neighbors and $a_i^\dagger$ creates a particle at lattice site i. Using our analogy of translations, we see immediately that $a_i^\dagger a_j$ is an approximation to the translation $U_{r_{ij}}$; an effective field arises when we have a nontrivial phase in hopping (Peierls's substitution). Specifically,

$$-J\sum_{\langle ij\rangle} a_i^\dagger a_j e^{-i\phi_{ij}}$$

where $\phi_{ij} \sim \int_{r_i}^{r_j} \boldsymbol{A} \cdot d\boldsymbol{l}$ is the effective phase in tunneling, and now $\boldsymbol{A}$ is the effective gauge potential which can be obtained by inverting this relationship.

4.2.1 Magneto-Optic Effects and MW Experiment

In direct analogy to quantum Hall physics for electronic systems, photonic systems can also exhibit similar physics and have edge states. More specifically, in the presence of strong Faraday effect, it was theoretically predicted [3], and later experimentally demonstrated [4], can lead to generation of edge state. However, this approach relied upon magneto-optical effects, which are extremely weak at optical frequencies. Thus, studying the effect of gauge fields—such as the magnetic field—on photons was limited to the microwave domain. Fortunately, there are alternative ways to simulate the effect of gauge field for photons. In the following we provide a detailed discussion on a scheme which relies on differential optical path lengths in an array of coupled ring resonators. In Sect. 4.2.3, we briefly discuss other approaches to synthesize gauge fields photonic systems.

We note that there have been parallel efforts in the ultracold atoms community to synthesize magnetic field for neutral particles [2, 5–8] and various techniques to detect the resulting topological invariants [9–15].

4.2.2 Synthetic Gauge Field in Coupled Resonators

In this section, we show that the coupled resonators can simulate a magnetic-like Hamiltonian based on differential optical path lengths [16–18]. In particular, a non-zero magnetic field can be synthesized using an imbalance between the optical paths that connect resonators. In order to show this, we first review the equivalence between the transfer matrix method and the coupled mode theory for two simple systems: a single resonator and a three-resonator system coupled to two probing waveguides. For the three-ring system, we show that the asymmetric coupling of the middle resonator can be described by a non-zero magnetic-phase in the effective Hamiltonian.

We begin by the simplest example: a ring resonator coupled to two waveguides, as shown in Fig. 4.1a. In photonics, such system is known as the add/drop filter. We assume that the resonator has a single optical mode. Therefore, using the coupled mode theory (input-output formalism [19]), we can write the dynamics of the field inside the resonator as:

$$\frac{d\mathcal{E}}{dt} = (-\kappa_{\text{in}} - 2\kappa_{\text{ex}})\mathcal{E} - \sqrt{2\kappa_{\text{ex}}}\mathcal{E}_{\text{in}}e^{-i\omega t}, \tag{4.1}$$

where κ_{ex} is the coupling rate between the probing-waveguide and the resonator and κ_{in} is the field decay rate to undesired modes. We assumed that a monochromatic field with the amplitude $\mathcal{E}_{\text{in}}$ and the detuning ω drives the system, as shown in Fig. 4.1a. The output field in the drop port is related to the field inside the resonator by: $\mathcal{E}^{\text{out}} = \sqrt{2\kappa_{\text{ex}}}\mathcal{E}$. Consequently, by solving the dynamics in the Fourier domain, we obtain the transmission in the drop port as:

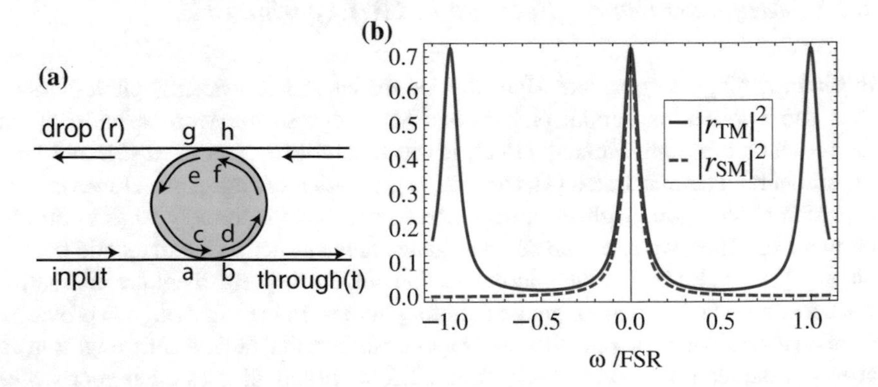

Fig. 4.1 **a** A resonator connected to two probing waveguides **b** The transfer matrix formalism and the coupled mode theory yield the same results, for frequencies around the resonance. The parameters are $\epsilon = 0.5$, $\alpha' L = 0.05$

$$r_{\mathrm{SM}} = \frac{\sqrt{2\kappa_{\mathrm{ex}}}\mathcal{E}}{\mathcal{E}_{\mathrm{in}}e^{-i\omega t}} = \frac{2\kappa_{\mathrm{ex}}}{i\omega - 2\kappa_{\mathrm{ex}} - \kappa_{\mathrm{in}}}. \tag{4.2}$$

Now, we evaluate the transmission properties using the transfer matrix formalism. The waveguide-resonator coupling regions can be described as:

$$M_{\mathrm{coupl}} = \frac{1}{t}\begin{pmatrix} -r^2 + t^2 & r \\ -r & 1 \end{pmatrix}, \begin{pmatrix} d \\ c \end{pmatrix} = M_{\mathrm{coupl}} \begin{pmatrix} a \\ b \end{pmatrix}, \begin{pmatrix} g \\ h \end{pmatrix} = M_{\mathrm{coupl}} \begin{pmatrix} f \\ e \end{pmatrix},$$

where t and r are the transmission and reflection coefficients of the coupling regions. Furthermore, we assume that the propagation constant is $\beta = 2\pi n/\lambda$, where n is the index of refraction and λ is the wavelength, and the absorption constant is α'. Therefore, the free propagation inside the resonators is given by:

$$M_{\mathrm{prop}} = \begin{pmatrix} e^{i\beta \mathrm{L}/2 - \alpha' \mathrm{L}/2} & 0 \\ 0 & e^{-i\beta \mathrm{L}/2 + \alpha' L/2} \end{pmatrix}, \begin{pmatrix} f \\ e \end{pmatrix} = M_{\mathrm{prop}} \begin{pmatrix} d \\ c \end{pmatrix}. \tag{4.3}$$

We are interested in a limit where the coupling loss can be ignored (i.e. $|t|^2 + |r|^2 = 1$) and the junctions are highly reflective. In other words: $r \to \sqrt{1-\epsilon^2}, t \to i\epsilon$, where $\epsilon \ll 1$. The regime of interest is near the resonant frequency of the resonator, and is much smaller than the free spectral range (FSR), so we consider $\beta L \ll 1$. Since the propagation loss over a typical distance in these systems is not large, we take $\alpha' L \ll 1$. The input field is only present at one port, as shown in Fig. 4.1a, so we can replace: $a = 1, h = 0$. Keeping terms to the total 2nd order in $\epsilon^2, \beta L, \alpha' L$, both in the numerator and the denominator, we find that the field in the drop channel can be simplified as:

$$r_{\mathrm{SM}} = \frac{\epsilon^2}{i\beta L - \alpha' L - \epsilon^2}. \tag{4.4}$$

Now, if we use the following substitutions:

$$\epsilon^2 \to \frac{4\pi\kappa_{\mathrm{ex}}}{\mathrm{FSR}}, \quad \alpha' \mathrm{L} \to \frac{2\pi\kappa_{\mathrm{in}}}{\mathrm{FSR}}, \quad \beta \mathrm{L} \to 2\pi\frac{\omega}{\mathrm{FSR}}, \tag{4.5}$$

we see that Eqs. (4.4) and (4.2) are equivalent. One can also compare these two approaches, using the exact expressions, as shown in Fig. 4.1b. Both formalism agree with each other around the resonant frequency.

In the following, we show that the dynamics of two ring resonators that are coupled through a middle off-resonant ring can be effectively written as two resonators coupled with a "hopping phase". We evaluate transport properties in both cases and show that they are identical, around the resonant frequency of the side resonators.

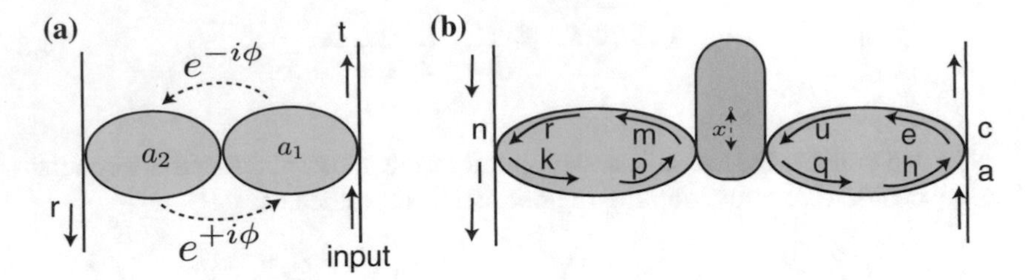

Fig. 4.2 **a** Two ring resonator coupled with a hopping phase (ϕ) **b** Two ring resonators coupled through an off-resonant middle ring. These two systems are equivalent around the resonant frequency of side resonators

First, we derive the transmission and reflection coefficients of two rings coupled with a hopping phase. The Hamiltonian describing such system can be written as:

$$H = -J\hat{a}_2^\dagger \hat{a}_1 e^{-i\phi} - J\hat{a}_1^\dagger \hat{a}_2 e^{+i\phi} \tag{4.6}$$

where J is the tunneling rate and ϕ is the hopping phase phase, as depicted in Fig. 4.2a. Note that by hopping phase we mean a Hamiltonian of the kind written in Eq. (4.6). This should not be confused with optical non-reciprocity which requires an external field, cf. Refs. [20, 21]. Similar to the single resonator case, using coupled mode theory, we can write the dynamics of the field inside the resonators as:

$$\frac{d}{dt}\begin{pmatrix} a_1 \\ a_2 \end{pmatrix} = \begin{pmatrix} -\kappa_{\text{in}} - \kappa_{\text{ex}} & iJe^{+I\phi} \\ iJe^{-I\phi} & -\kappa_{\text{in}} - \kappa_{\text{ex}} \end{pmatrix}\begin{pmatrix} a_1 \\ a_2 \end{pmatrix} - \sqrt{2\kappa_{\text{ex}}}\begin{pmatrix} \mathcal{E}_{\text{in}} \\ 0 \end{pmatrix}. \tag{4.7}$$

The output field of the resonators is given by: $a_2^{\text{out}} = \sqrt{2\kappa_{\text{ex}}}a_2$, $a_1^{\text{out}} = 1 + \sqrt{2\kappa_{\text{ex}}}a_1$. Consequently, the transmission and reflection coefficients, as defined in Fig. 4.2, are given by:

$$\begin{aligned} r_{\text{SM}} &= \frac{\sqrt{2\kappa_{\text{ex}}}a_2}{\mathcal{E}_{\text{in}}} = -\frac{2ie^{-i\phi}J\kappa_{\text{ex}}}{J^2 + (i\omega - \kappa_{\text{ex}} - \kappa_{\text{in}})^2} \\ t_{\text{SM}} &= \frac{\sqrt{2\kappa_{\text{ex}}}a_1 + 1}{\mathcal{E}_{\text{in}}} = 1 + \frac{2\kappa_{\text{ex}}(+i\omega - \kappa_{\text{ex}} - \kappa_{\text{in}})}{J^2 + (+i\omega - \kappa_{\text{ex}} - \kappa_{\text{in}})^2}. \end{aligned} \tag{4.8}$$

Now, we consider two ring resonators that are coupled through a middle off-resonant ring, as shown in Fig. 4.1b. We use the transfer matrix formalism to derive the transmission of the system. The transfer matrix for the waveguide-resonator and resonator-resonator coupling regions, respectively, are given by:

$$M_1 = \frac{1}{t_1}\begin{pmatrix} -r_1^2 + t_1^2 & r_1 \\ -r_1 & 1 \end{pmatrix}$$
$$M_2 = \frac{1}{t_2}\begin{pmatrix} -r_2^2 + t_2^2 & r_2 \\ -r_2 & 1 \end{pmatrix}.$$

where t_i and r_i are the transmission and reflection coefficients of the coupling regions. Therefore, the waveguide-resonator couplings can be written as:

$$\begin{pmatrix} e \\ h \end{pmatrix} = M_1 \begin{pmatrix} a \\ c \end{pmatrix} \ , \ \begin{pmatrix} l \\ s \end{pmatrix} = M_1 \begin{pmatrix} r \\ k \end{pmatrix}. \tag{4.9}$$

We assume that the length of the side resonators (the middle resonator) is L $(L + \eta)$, so that the middle resonator remains off-resonant with the other two. Furthermore, the middle resonator is shifted vertically by x to induce the hopping phase, as we show. Therefore, the free propagation inside the side resonators is described by Eq. (4.3). The propagation inside the middle rings is given by:

$$\begin{pmatrix} m \\ p \end{pmatrix} = M_2 \begin{pmatrix} e^{i\,\beta \mathrm{L}/2 + i\beta\eta - i2\beta x - \alpha' \mathrm{L}/2} & 0 \\ 0 & e^{-i\,\beta \mathrm{L}/2 -\, i\beta\eta - i2\beta x + \alpha' \mathrm{L}/2} \end{pmatrix} M_2 \begin{pmatrix} u \\ q \end{pmatrix}. \tag{4.10}$$

Again, keeping terms to the total 2nd order in ϵ_i^2, βL, $\alpha' L$, we find that the field in the drop channel can be simplified as:

$$r_{\mathrm{TM}} = \frac{2e^{-i\beta x}\epsilon_1^2\epsilon_2^2}{2\left(2\alpha'\mathrm{L} - 2i\beta\mathrm{L} + \epsilon_1^2\right)\epsilon_2^2\cos(\beta\eta) - i\left(\left(2\alpha'\mathrm{L} - 2i\beta\mathrm{L} + \epsilon_1^2\right)^2 + \epsilon_2^4\right)\sin(\beta\eta)}. \tag{4.11}$$

To compare this expression with the one obtained from the single-mode approximation (Eq. 4.8), we use the following substitutions:

$$\epsilon_1{}^2 \rightarrow \frac{4\pi\kappa_{\mathrm{ex}}}{\mathrm{FSR}} \ , \ \epsilon_2{}^2 \rightarrow \frac{4\pi J}{\mathrm{FSR}} \ , \ \alpha'\mathrm{L} \rightarrow \frac{2\pi\kappa_{\mathrm{in}}}{\mathrm{FSR}} \ , \ \beta\mathrm{L} \rightarrow 2\pi\frac{\omega}{\mathrm{FSR}} \ , \ \beta x \rightarrow \phi \tag{4.12}$$

and we have

$$r_{\mathrm{TM}} = \frac{2e^{-i\phi}J\kappa_{\mathrm{ex}}}{2J(\kappa_{\mathrm{ex}} + \kappa_{\mathrm{in}} - i\omega)\cos(\beta\eta) - i\left(J^2 + (\kappa_{\mathrm{ex}} + \kappa_{\mathrm{in}} - i\omega)^2\right)\sin(\beta\eta)}. \tag{4.13}$$

In the special case where $\beta\eta = 3\pi/2$, this expression reduces to Eq. (4.8). This means that if the middle ring is precisely anti-resonant with the side rings, the three rings can be effectively described by two resonators coupled with a hopping phase. If we have $\beta\eta = \pi/2$, the two models are again the same, only the sign of the tunneling is reversed ($J \rightarrow -J$). When the middle resonator deviates from the anti-resonant condition ($\beta\eta = \pi/2, 3\pi/2, \ldots$), the system can still be effectively described by

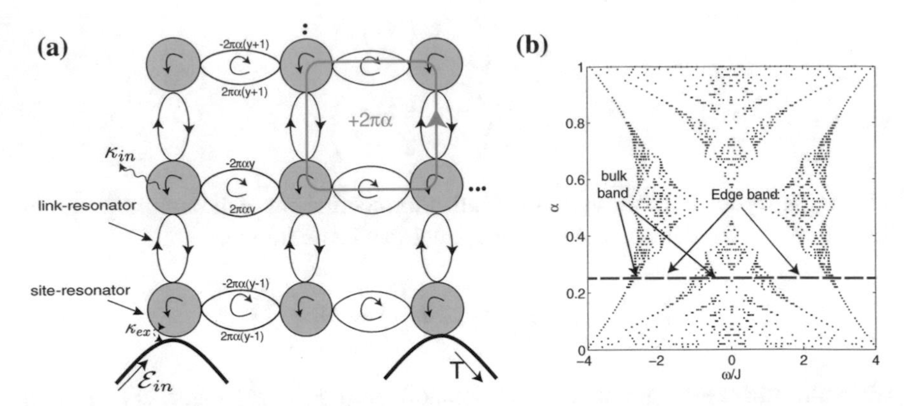

Fig. 4.3 **a** A 2D lattice of coupled resonators which be described by a magnetic tight-binding model (Eq. 4.14) **b** Hofstadter butterfly spectrum. Each point represents a transmission greater than 0.005, for a 10×10 lattice with torus boundary condition and coupling $\kappa_{\text{ex}}/J = 0.02$. The *red line* is a guide for the eye to show the spectrum at the specific magnetic field ($\alpha = 0.25$). The bulk and edge band are highlighted

two resonators with a hopping phase, however, the effective tunneling is $J_{\text{eff}} \rightarrow J/\sin(\beta\eta)$ and system is shifted in frequency by $\omega \rightarrow \omega - J\cot(\beta\eta)$.

Now, we consider a two-dimensional system of coupled resonators, where two sets sets of resonators play the role of sites and links of a lattice, as shown in Fig. 4.3. We arrange the system so that the phase imbalance in only present in the row link-resonator and increases by the row number. The dynamics of the system near the resonance of the site resonators is described by the following Hamiltonian

$$H_{\text{mag}} = -J \sum_{x,y} \hat{a}^{\dagger}_{x+1,y}\hat{a}_{x,y}e^{i2\pi\alpha y} + \hat{a}^{\dagger}_{x,y}\hat{a}_{x+1,y}e^{-i2\pi\alpha y} + \hat{a}^{\dagger}_{x,y+1}\hat{a}_{x,y} + \hat{a}^{\dagger}_{x,y}\hat{a}_{x,y+1}, \tag{4.14}$$

where $a^{\dagger}_{x,y}$ is the creation operator at site (x, y), J is the effective tunneling rate between resonators and α characterizes the phase imbalance. In particular, a photon hopping around a plaquette, in the counter-clockwise direction, acquires the phase $2\pi\alpha$, in direct analogy to Aharanov-bohm phase. Therefore, α is the effective magnetic flux per plaquette and the total magnetic flux is $N_{\phi} = \alpha N_x N_y$.

In this discussion, we assumed only counter-clockwise photons in the site- (link-) resonators. The opposite circulating photon experience the opposite magnetic field. Therefore, the system is equivalent to two copies of integer quantum Hall systems with opposite magnetic fields, in direct analogy to a special case of the quantum spin Hall physics in electronic systems [22].

The Hamiltonian in Eq. 4.14 is identical to that of an electron on a lattice with a uniform magnetic field and the spectrum of the system is known as the Hofstadter butterfly, when $N_x, N_y \rightarrow \infty$. The spectrum of such photonic system can be probed using transmission spectroscopy. By applying the input-output formalism, we can evaluate different transport coefficients. In this section, the dynamics is linear, therefore, the "quantum" input-output formalism reduces to the "classical" coupled mode theory, by replacing the field operator with the corresponding expectation value, i.e., $\langle \hat{a}_{x,y} \rangle = a_{x,y}$. The field dynamics of the resonators is given by:

$$\frac{d\hat{a}_{x,y}}{dt} = i[H, \hat{a}_{x,y}] + \left[-\kappa_{\text{in}} - \kappa_{\text{ex}}(\delta_{x,x_{\text{in}}}\delta_{y,y_{\text{in}}} + \delta_{x,x_{\text{out}}}\delta_{y,y_{\text{out}}})\right]\hat{a}_{x,y} \\ - \sqrt{2\kappa_{\text{ex}}}(\delta_{x,x_{\text{in}}}\delta_{y,y_{\text{in}}})\mathcal{E}_{\text{in}}e^{-i\omega t} \quad (4.15)$$

where κ_{ex} is the coupling rate between the probing-waveguide and the site resonators and κ_{in} is the field decay rate to undesired modes. "in" ("out") represents the resonators to which the input (output) probing waveguides are connected. We assume a monochromatic input field at the left-bottom corner resonator, with amplitude $\mathcal{E}_{\text{in}}$ and detuned by ω from the resonance, as shown in Fig. 4.3a. Going to the Fourier domain and evaluating the steady-state solution, we can obtain the transmission in the output channel as $T = |a_{x_{\text{out}},y_{\text{out}}}/\mathcal{E}_{\text{in}}|^2$.

Since an infinite system can be simulated by a finite system with periodic boundary conditions, we study a 10×10 lattice with torus boundary condition. Figure 4.3b, shows the transmission profile when the mangetic flux varies from 0 to 2π. We observe that a finite version of the Hofstadter's fractal appears.

In a finite lattice, there exist states between magnetic bulk bands which are known as 'edge states'. In direct analogy to quantum Hall physics, such quasi-one dimensional states localized at the perimeter of the system which carry current. In particular, for certain frequency bands, the field in resonators located in the bulk (away from the edges) undergoes destructive interference and, therefore, the light intensity is nonzero only at the edges. This is illustrated in Fig. 4.4a. For each edge state, there is a corresponding edge state with an opposite chirality. More specifically, the forward- and backward-propagating edge states take different paths, and consequently, they have different resonances at detunings, equal in magnitude and opposite in sign.

Again in direct analogy to integer quantum Hall edge state, such edge states are immune to disorders in the form of random potential. In particular, when an impurity is located on the edge—the resonator is detuned ($U\hat{a}^{\dagger}_{x,y}\hat{a}_{x,y}$)—the edge state routes around it, as shown in Fig. 4.4b for the test case of a single disordered site. More precisely, scattering which would reverse the current is prevented because the backward going edge state has a different energy, as discussed above, preventing elastic scattering.

Transport through edge states requires the photon to traverse the perimeter of the system, leading to a time delay proportional to the transverse resonators. Since such transport is robust against frequency mismatch among resonators, such two-dimensional systems provide a robust alternative to conventional CROW in photonic delay lines [16].

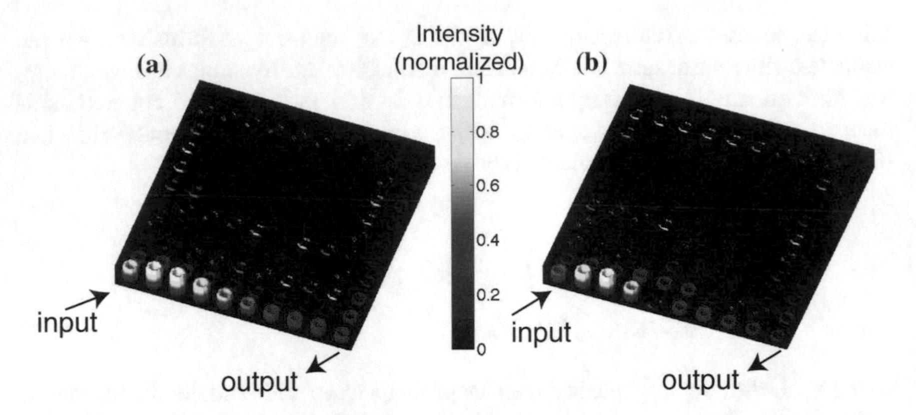

Fig. 4.4 Light intensity for a 10×10 lattice **a** in the absence and **b** in the presence of disorder. For (**b**) the resonator (x,'y)=(5, 1) is detuned by $U = 20J$. For both figures, the parameters are: $(\kappa_{\text{ex}}, \kappa_{\text{in}})/J = 0.2, 0.1$, $\alpha = .25$) and the system is excited at $\omega = 2J$

While the implementation of gauge fields has been achieved, the measurement of the expected topological orders remains elusive due to the inapplicability of the conventional Hall conductance measurements to photonic systems. How can one directly measure the integer topological invariants, e.g., the winding number of the edge states or the Chern number of the bulk state in a photonic system? In other words, how do the integer values manifest themselves in an optical realization of quantum Hall Hamiltonians. Recently, there have been a number of proposals to achieve this end by benefiting from individual site addressability to manipulate synthetic gauge fields in photonic systems [23–25]. In particular, following Laughlin/Halperin's argument [26, 27], it is suggested that the spectral shift and edge state transfer can be experimentally observed using standard transmission spectroscopy [24].

4.2.3 Other Approaches: Helical Waveguides, Polarization Scheme, Modulation

A wide variety of additional approaches have been considered, and are detailed elsewhere in this book. We briefly highlight that synthetic gauge fields can be implemented by the application of strong magnetic field [3, 4], polarization control [28], opto-mechanically [20], via bi-anisotropic metamaterials [29]. Other examples include: emulation of edge states as localized state at the two ends of a one-dimentional array [30], topological states in photonic quasi-crystals [31], strain induced magnetic field [32]. At the same time, there have been efforts to implement synthetic gauge fields in circuit-QED systems [33, 34]. In particular, breaking time-reversal symmetry using biased circulators [35, 36], qubit-assisted tunneling [37].

Focusing on time- or space-dependent modulation scheme [21, 38], a repeating structural along the direction of propagation of a waveguide, or a periodically varying two dimensional system, can implement another version of a topological system: a Floquet insulator [39, 40]. Here we highlight such Floquet approaches with a simple example, from Refs. [21, 41]. Consider two adjacent optical resonances with annihilation operators a_L and a_R at frequencies ω_L and ω_R. Furthermore, imagine the two resonators are coupled parametrically (via, e.g., a Josephson parametric coupler), with a coupling

$$V = -J\left(\cos(\nu t + \phi)a_L^\dagger a_R + \text{H.c.}\right).$$

Working in the interaction picture ($a_L \to e^{-i\omega_L t}a_L$, and similarly for R), we see that the cosine can be split into a forward and backward rotating phase term:

$$V_I = -\frac{J}{2}\left(e^{-i\nu t - i\phi}e^{i(\omega_L-\omega_R)t}a_L^\dagger a_R + e^{i\nu t + i\phi}e^{i(\omega_L-\omega_R)t}a_L^\dagger a_R\right) + \text{H.c.}$$

When $|\omega_L - \omega_R| \approx \nu$, i.e., the pump in the parametric coupler matches the frequency difference, only the first term contributes, while the second term rapidly averages to zero in a rotating wave approximation. Consequently, a photon in the left resonator can 'hop' to the right resonator, picking up a phase $e^{-i\phi}$ which is set by the phase of the pump field. Thus, in an array of such resonators, varying ϕ with index number in the array can implement a synthetic gauge field.

4.3 Interacting Photons and Many-Body Regime

In this section, we show that the above system can be extended to investigate strongly interacting photonic states. In particular, by inducing strong photon-photon interaction, we show that certain fractional quantum Hall states can be generated, and their signature can be probed using the correlation function measurement.

4.3.1 Strong Kerr Nonlinearity and Laughlin State of Light

Photon-photon interaction can be mediated by coupling emitters (e.g., atoms) to each resonator, as shown in Fig. 4.5. In the strong coupling regime, where the photon blockade is observed [42–44], this interaction can be represented as an on-site interaction in a Bose-Hubbard model for a coupled array of resonators [45–47]. As shown in Refs. [48–51], in the presence of a synthetic magnetic field, such interacting system of photons can be described by a Hamiltonian which is identical to the fractional quantum Hall Hamiltonian on a lattice:

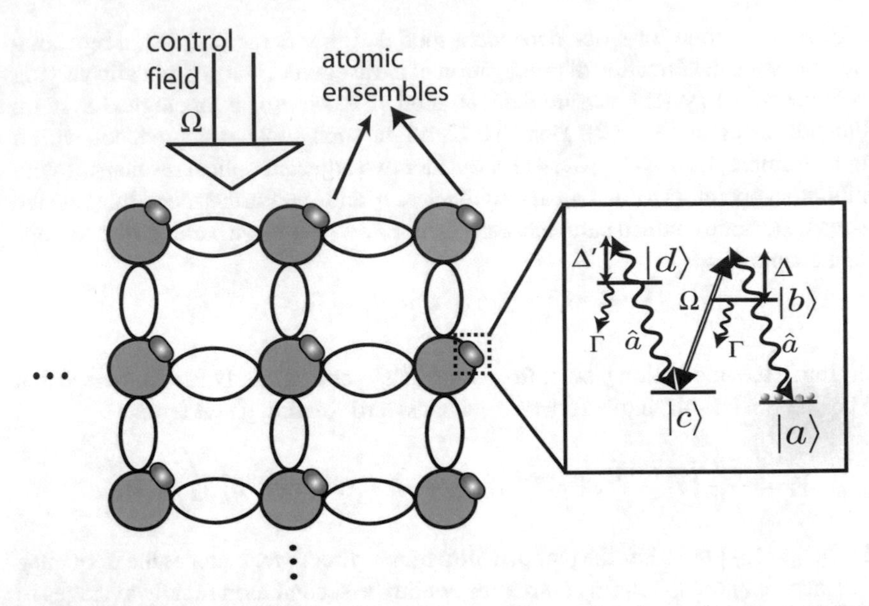

Fig. 4.5 Atomic ensembles are coupled to resonators to mediate interaction. inset: Mediating interacting using N-level atomic ensemble. A control field couples the internal levels, and provides on-site interaction for photons

$$\begin{aligned} H_{\text{mag}} = & -J \sum_{x,y} \hat{a}^{\dagger}_{x+1,y}\hat{a}_{x,y}e^{i2\pi\alpha y} + \hat{a}^{\dagger}_{x,y}\hat{a}_{x+1,y}e^{-i2\pi\alpha y} \\ & + \hat{a}^{\dagger}_{x,y+1}\hat{a}_{x,y} + \hat{a}^{\dagger}_{x,y}\hat{a}_{x,y+1} + U\hat{a}^{\dagger}_{x,y}\hat{a}_{x,y}(\hat{a}^{\dagger}_{x,y}\hat{a}_{x,y} - 1). \end{aligned} \tag{4.16}$$

Such Hamiltonian can be implemented by adding strong optical nonlinearity in the scheme presented in the previous section. As an example case, one can use an ensemble of N-level atoms to mediate onsite two-body interaction of the Kerr-type (Fig. 4.5) [46], which still preserves the propagation direction (clockwise or counterclockwise) used in Ref. [16]. In this approach, the optical cavity and ensemble enter into a slow-light regime, where the excitations are dark state polaritons [52] $\hat{\Psi}_{x,y} \propto \Omega\hat{a}_{x,y} - g\sqrt{N}\hat{S}_{x,y}$, where Ω is the pump field, g is the vacuum Rabi coupling, N is the number of ensemble atoms, and $\hat{S}_{x,y}$ is the spin-wave operator describing coherence between two atomic states $|a\rangle$ and $|c\rangle$ (from Fig. 4.5 inset). These bosonic excitations lead to an overall increase of dynamical timescales by $\eta = c/v_g \gg 1$, the ratio between the speed of light and group velocity for the dark state polariton, but they can also interact via a self-Kerr interaction with state $|d\rangle$ [53]. Coupling atoms to the photonic system introduces loss which can be reduced by detuning the cavity resonance from the emitter transitions ($\Delta, \Delta' \gg \Gamma$). As discussed in Ref. [54], to observe any many-body effect and to have a finite gap, the effective interaction between photons ($U \simeq g^2/\Delta'$) should be at least comparable to the tunneling rate J. These conditions can be satisfied for systems with a large

Purcell factor ($g^2/\kappa\Gamma \gg 1$). The same criterion applies to implementation of such scheme in the microwave domain.

Following Refs. [54, 55], the ground state of Eq. (4.16) are the fractional Quantum Hall states when the mangetic field is dilute ($\alpha < 0.4$). In particular, when the filling factor ($\nu = N_{\text{ph}}/N_{\phi}$), which is the ratio between the number of photons (N_{ph}) and the total magnetic field $N_{\phi} = \alpha N_x N_y$, is one half, the ground state of the system can be faithfully described by Laughlin wavefunction ($\alpha < 0.25$). For numerical simulation, one has to consider a torus boundary condition to mimic the effect of an infinite system. One a torus the Laughlin wavefunction is written using the Jacobi theta function. The detailed discussion of the overlap calculation can be found in Ref. [54]. The remarkable overlap with the Laughlin state in the photonic case is discussed in Refs. [48–51].

One can prepare a Laughlin state by adiabatically melting a Mott-insulator of photons, similar to the atomic method discussed in Ref. [55]. However, this requires both preparation of N_{ph} Fock states and photon lifetimes long enough to allow for the melting to be adiabatic. Moreover, the direct experimental verification of the Laughlin overlap is a difficult task which requires number post-selection (N_{ph}) and state tomography in a Hilbert space with dimension $\binom{N_x N_y}{N_{\text{ph}}}$. Therefore, it is important to find an implementation and a detection scheme that is more relevant for photonic system.

More generally most of the studies in many-body photonic system is inspired by analogies to electronic systems which are mainly focused on ground state properties. However, a photonic system is naturally an open driven system. Therefore, the most relevant approach to understand and manipulate many-photon states involves understanding the non-equilibrium dynamics in such systems [56–61]. For example, in a one-dimensional system strong interaction between photons leads to their fermionization, which can be probed in the output correlation functions of an externally driven system, both in a discrete array [56] and in the continuum limit [58].

For the fractional quantum Hall system, Ref. [51] demonstrates that by weakly driving the system, a few photon Laughlin state can be prepared and experimentally-relevant observables such as the correlation function of the zero-mode can show certain signatures of the Laughlin state. Since the system is driven and lossy, it may seem that the master equation formalism is required to fully described the system. However, as shown in Ref. [51], when the system is weakly driven, one can resort to the stochastic wavefuction approach. More precisely, if we are interested in evaluating the correlation function of any order, finding the steady-state of the effective Hamiltonian is sufficient and the effect of the "quantum jumps" can be ignored. Such simplification allows one to explore larger systems, compared to the master equation approach, and to avoid finite size effects, which usually undermines the numerical results.

When all the resonators are driven with a laser field, the input field consists of a Poisson distribution of photons. When photons are injected at the frequency corresponding to the Laughlin state at the N_{ph}-photon manifold, photons reconfigure themselves and form a wave function which corresponds to the Laughlin state. The

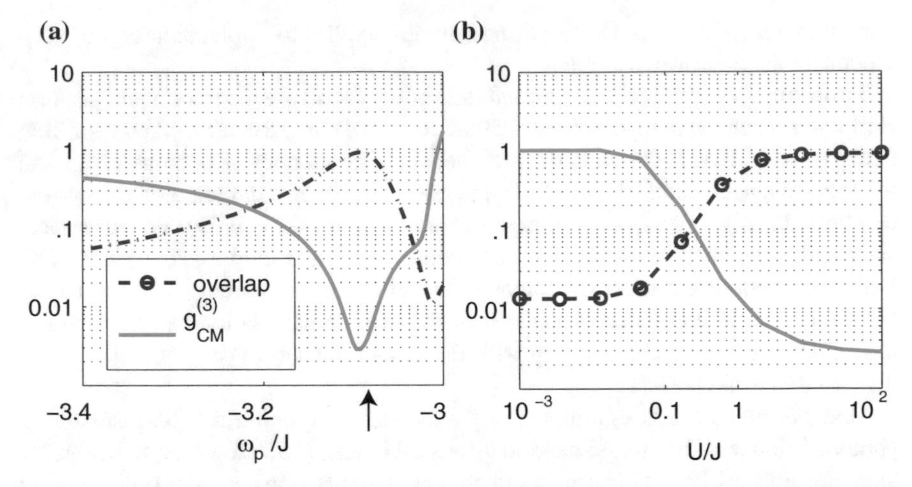

Fig. 4.6 Overlap with the Laughlin wave function ($\nu = 1/2$), and the correlation function of the zero mode ($g^{(3)}_{CM}$) are shown as a function of: **a** the pump frequency for hard-core bosons **b** the interaction strength for $\Delta = -3.095J$, as shown by an *arrow* on (**a**). The overlap with the Laughlin function is evaluated for $N_{\rm ph} = 3$ manifold. The total magnetic flux is $N_\phi = 6$. The simulations are performed for a 6×6 lattice, torus boundary condition, and the maximum number of photon is 3. $\kappa = 0.01J$, $\beta = 0.01$. All calculated quantities are dimensionless

remarkable overlap of this photonic state with the Laughlin wave function in the $N_{\rm ph}$-photon manifold is shown in Fig. 4.6a. Note the frequency required to be resonant with the Laughlin state is at the vicinity of the free photon state (Hofstadter's spectrum). We can relax the hard-core constraint and investigate the same observables. In the weak interaction limit, the system approaches the classical response, as shown in Fig. 4.6b. In the absence of interaction, using transport measurements—varying the pump frequency and measuring reflection/transmission– one recovers the Hofstadter's butterfly spectrum as in Fig. 4.3b, but regardless of the pump frequency, the correlation function remains equal to one.

To summarize, the driven strongly interacting photons exhibits interesting many-body behaviors and the emergence of FQH states of photons can be probed by using the conventional optical measurement techniques. One of the remaining question is to investigate other many-body signatures of these states such as their topological properties and fractional statistics.

4.3.2 *Three-Body Interaction and Pfaffian States of Light*

One particularly curious class of interacting topological systems are those corresponding to parent Hamiltonians, where theoretical understanding of fractional quantum Hall states can be inverted to construct a Hamiltonian who has the desired state as its ground state. Notably, states with anyonic quasiparticle excitation spectra correspond to particularly simple classes of parent Hamiltonians [62]. For the case of the Pfaffian, the Hamiltonian in the continuum limit involves just an uniform external magnetic field and a short-range three-particle interaction [41]. In the case of the Read-Rezayi state [63]—universal for quantum computing—it is similarly simple, but with a four-particle short range interaction [64].

Crucially, in both cases two-body interactions are detrimental for creating the appropriate ground state and low energy quasiparticle spectrum. This necessitates some degree of creativity in developing devices with appropriate three- or four-body interactions without having a corresponding low-energy two-body scattering term. Potential approaches using resonances in scattering such as Feshbach resonances for cold atomic systems have suggested that resonant behavior can dramatically alter the short-range nature of interaction, and have lead to suggestions for developing appropriate parent Hamiltonians using cold atoms.

An analogous approach is also available for superconducting systems, where the nature excitations of the system are bosonic, i.e., in the microwave photonic regime. Restricting the discussion to a lattice of resonators, one seeks resonators with the property that there is an on-site, three-particle interaction but no two-particle interaction. This mimics the appropriate low-energy behavior necessary to recover the parent Hamiltonian of choice for, e.g., the Pfaffian state. Towards that end, we can examine the simple case of a superconducting qubit with an inductive shunt. Writing

$$H = E_C \frac{n^2}{2} + E_L \frac{(\phi - \phi_x)^2}{2} - E_J \cos(\phi)$$

with $[\phi, n] = i$ as appropriate canonical flux and number variables, we can setup a scenario near $\phi_x = \pi/4$ where the potential (E_L and E_J terms taken together) is an asymmetric double-well. The excitations of the lower energy well can be thought of as various photon number states of the resonator, consistent with an harmonic oscillator approximation to the well. However, the first bound state of the higher well can cause a resonance, which dramatically shifts the position of an energy eigenstate (Fock state) that sits near the bound state (Fig. 4.7a). This leads to an effective n-body interaction on the site, where n is the nth energy eigenstate above the ground state for the lower well.

Suppression of the $n - 1$ and smaller interactions, as well as maintaining nearly bosonic coupling between resonators, is beyond the scope of this chapter but is considered in Ref. [41]. There, several of us realized that stronger interactions are possible by further reducing the barrier, where the residual effect of the resonance persists long after the nominal bound state has been removed. Combining such a circuit QED approach to nonlinear elements with periodic variation of couplers many enable a complete generation of the Pfaffian parent Hamiltonian, as shown numerically in Ref. [41].

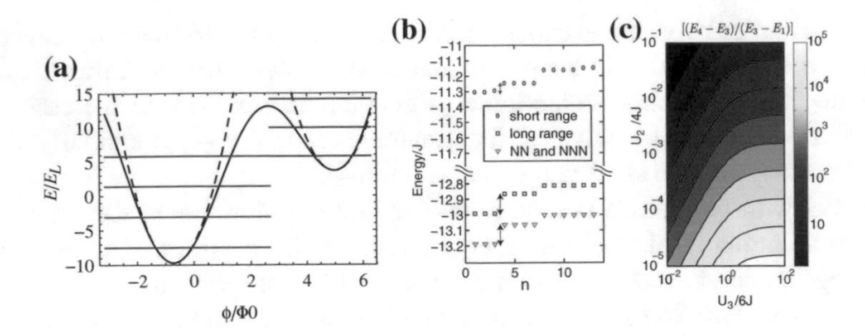

Fig. 4.7 **a** Anharmonic potential of a hybrid qubit $V(\phi)$ at finite external flux. Nominal solutions of the *left well* and *right well* can be energetically aligned by appropriate choice of E_C and ϕ_x, leading to a resonance enhancement of nonlinearity near the third excited state. **b** Lowest energy many-body eigenenergies for several different effective magnetic fields, demonstrating the many-body gap between the three-fold degenerate ground Pfaffian state and excited states for a system with bosons hopping on a lattice, from Ref. [41]. **c** Many-body gap as a function of two-particle U_2 and three particle U_3 interactions, indicating the benefit of reduced two-body interaction to improve the many-body gap for Pfaffian state preparation

4.4 Outlook

As we briefly discussed in this review, there are promising platforms to investigate synthetic gauge fields in the optical domain. The advantage in the optical domain is the relative ease of implementing the synthetic gauge fields and the possibility of preforming the experiments on the non-interacting models at room temperature. However, the challenge in the optical domain remains to be the weakness of nonlinearity and the difficulties in reaching photon-blockade on each resonators without inducing any inhomogeneity.

On the other hand, there has been recent investigation in the microwave domain to implement synthetic gauge field [35, 37] and explore many-body effects [65]. The advantage in the microwave domain is the presence of strong nonlinearity provided by the Josephson junctions. However, the challenges are the operation at very low temperature which requires dilution fridges and the presence of inhomogeneity in fabricated arrays of qubits.

In both cases, due to the open nature of photonic systems, preparation and detection should be performed in a driven regime. This arises many interesting theoretical questions to investigate many-body features such as incompressibility, fractional statistics in non-equilibrium systems. For example, new effective field theories should be developed to treat excitations in such bosonic systems. An intriguing bright future awaits for theoretical and experimental investigation of many-body physics, both in the optical and the microwave domains.

References

1. W.X. Gang, *Quantum Field Theory of Many-Body Systems: from the Origin of Sound to an Origin of Light and Electrons* (Oxford University Press, USA, 2007)
2. J. Dalibard, F. Gerbier, G. Juzeliūnas, P. Öhberg, Colloquium: artificial gauge potentials for neutral atoms. Rev. Mod. Phys. **83**(4), 1523–1543 (2011)
3. F.D.M. Haldane, S. Raghu, Possible realization of directional optical waveguides in photonic crystals with broken time-reversal symmetry. Phys. Rev. Lett. **100**(1), 13904 (2008)
4. Z. Wang, Y. Chong, J.D. Joannopoulos, M. Soljacic, Supplementary: observation of unidirectional backscattering-immune topological electromagnetic states. Nature **461**(7265), 772–775 (2009)
5. V. Galitski, I.B. Spielman, Spin-orbit coupling in quantum gases. Nature (2013)
6. Y.J. Lin, R.L. Compton, K. Jimenez-Gracia, J.V. Porto, I.B. Spielman, Synthetic magnetic fields for ultracold neutral atoms. Nature **462**(7273), 628 (2009)
7. J. Struck, C. Ölschläger, M. Weinberg, P. Hauke, J. Simonet, A. Eckardt, M. Lewenstein, K. Sengstock, P. Windpassinger, Tunable gauge potential for neutral and spinless particles in driven optical lattices. Phys. Rev. Lett. **108**(22), 225304 (2012)
8. M. Aidelsburger, M. Atala, S. Nascimbène, S. Trotzky, Y.A. Chen, I. Bloch, Experimental realization of strong effective magnetic fields in an optical lattice. Phys. Rev. Lett. **107**(25), 255301 (2011)
9. H.M. Price, N.R. Cooper. Mapping the berry curvature from semiclassical dynamics in optical lattices. Phys. Rev. A (2012)
10. E. Alba, X. Fernandez-Gonzalvo, J. Mur-Petit, J.K. Pachos, J.J. García-Ripoll, Seeing topological order in time-of-flight measurements. Phys. Rev. Lett. **107**(23), 235301 (2011)
11. D.A. Abanin, T. Kitagawa, I. Bloch, E. Demler, Interferometric approach to measuring band topology in 2D optical lattices. Phys. Rev. Lett. **110**, 165304 (2013)
12. N. Goldman, J. Beugnon, F. Gerbier, Detecting chiral edge states in the hofstadter optical lattice. Phys. Rev. Lett. **108**(25), 255303 (2012)
13. X.J. Liu, K.T. Law, T.K. Ng, P.A. Lee, Detecting topological phases in cold atoms. Phys. Rev. Lett. **111**, 120402 (2013)
14. A. Dauphin, N. Goldman, Extracting the chern number from the dynamics of a fermi gas: implementing a quantum hall bar for cold atoms. Phys. Rev. Lett. **111**(13), 135302 (2013)
15. Mittal et al., Nat. Photonics **10**, 180–183 (2016)
16. M. Hafezi, E.A. Demler, M.D. Lukin, J.M. Taylor, Robust optical delay lines with topological protection. Nat. Phys. **7**(11), 907–912 (2011)
17. M. Hafezi, S. Mittal, J. Fan, A. Migdall, J.M. Taylor, *Imaging topological edge states in silicon photonics* (Nat, Photon, 2013)
18. G.Q. Liang, Y.D. Chong. Optical resonator analog of a two-dimensional topological insulator. Phys. Rev. Lett., **110**(20) (2013)
19. C. Gardiner, M. Collett, Input and output in damped quantum systems: quantum stochastic differential equations and the master equation. Phys. Rev. A **31**(6), 3761–3774 (1985)
20. M. Hafezi, P. Rabl, Optomechanically induced non-reciprocity in microring resonators. Opt. Express **20**(7), 7672–7684 (2012)
21. K. Fang, Z. Yu, S. Fan, Realizing effective magnetic field for photons by controlling the phase of dynamic modulation. Nat. Photon **6**(11), 782–787 (2012)
22. B.A. Bernevig, S.-C. Zhang, Quantum spin Hall effect. Phys. Rev. Lett. **96**(10), 106802 (2006)
23. T. Ozawa, I. Carusotto, Anomalous and quantum hall effects in lossy photonic lattices. Phys. Rev. Lett. **112**(13), 133902 (2014)
24. M. Hafezi, Measuring topological invariants in photonic systems. Phys. Rev. Lett. **112**(21), 210405 (2014)
25. A.V. Poshakinskiy, A.N. Poddubny, L. Pilozzi, E.L. Ivchenko, Radiative topological states in resonant photonic crystals (2013)
26. R. Laughlin, Quantized Hall conductivity in two dimensions. Phys. Rev. B **23**(10), 5632–5633 (1981)

27. B. Halperin, Quantized Hall conductance, current-carrying edge states, and the existence of extended states in a two-dimensional disordered potential. Phys. Rev. B **25**(4), 2185–2190 (1982)
28. R.O. Umucalilar, I. Carusotto, Artificial gauge field for photons in coupled cavity arrays. Phys. Rev. A **84**, 043804 (2011)
29. A.B. Khanikaev, S.H. Mousavi, W.K. Tse, M. Kargarian, Photonic topological insulators. Nat. Mater. (2012)
30. Y. Kraus, Y. Lahini, Z. Ringel, M. Verbin, O. Zilberberg, Topological states and adiabatic pumping in quasicrystals. Phys. Rev. Lett. **109**(10), 106402 (2012)
31. M. Verbin, O. Zilberberg, Y.E. Kraus, Y. Lahini, Y. Silberberg, Observation of topological phase transitions in photonic quasicrystals. Phys. Rev. Lett. **110**(7), 076403 (2013)
32. M.C. Rechtsman, J.M. Zeuner, A. Tünnermann, S. Nolte, Strain-induced pseudomagnetic field and photonic Landau levels in dielectric structures. Nat. Photon (2012)
33. R.J. Schoelkopf, S.M. Girvin, Wiring up quantum systems. Nature **451**(7179), 664–669 (2008)
34. M.H. Devoret, R.J. Schoelkopf, Superconducting circuits for quantum information: an outlook. Science **339**(6124), 1169–1174 (2013)
35. J. Koch, A.A Houck, K. Le Hur, S.M. Girvin, Time-reversal symmetry breaking in circuit-QED based photon lattices. Phys. Rev. A **82**, 043811 (2010)
36. A. Petrescu, A.A. Houck, K. Le Hur, Anomalous Hall effects of light and chiral edge modes on the Kagome lattice. Phys. Rev. A **86**, 053804 (2012)
37. E. Kapit, Quantum Simulation architecture for lattice bosons in arbitrary, tunable external gauge fields. Phys. Rev. A **87**, 062336 (2013)
38. M.C. Rechtsman, J.M. Zeuner, Y. Plotnik, Y. Lumer, D. Podolsky, F. Dreisow, S. Nolte, M. Segev, A. Szameit, Photonic floquet topological insulators. Nature **496**(7444), 196–200 (2013)
39. T. Kitagawa, M.S. Rudner, E. Berg, E. Demler, Exploring topological phases with quantum walks. Phys. Rev. A **82**(3), 033429 (2010)
40. N.H. Lindner, G. Refael, V. Galitski, Floquet topological insulator in semiconductor quantum wells. Nat. Phys. **7**(6), 490–495 (2011)
41. M. Hafezi, P. Adhikari, J.M. Taylor, Engineering three-body interaction and Pfaffian states in circuit QED systems. Phys. Rev. B **90**, 060503 (R) (2014)
42. K.M. Birnbaum, A. Boca, R. Miller, A.D. Boozer, T.E. Northup, H.J. Kimble, Photon blockade in an optical cavity with one trapped atom. Nature **436**(7047), 87–90 (2005)
43. D. Englund, A. Faraon, I. Fushman, N. Stoltz, P. Petroff, J. Vuckovic, Controlling cavity reflectivity with a single quantum dot. Nature **450**(7171), 857–861 (2007)
44. K. Srinivasan, O. Painter, Linear and nonlinear optical spectroscopy of a strongly coupled microdisk-quantum dot system. Nature **450**(7171), 862–866 (2007)
45. A.D. Greentree, C. Tahan, J.H. Cole, L.C.L. Hollenberg. Quantum phase transitions of light. Nat. Phys. **2**(12), 856–861 (2006)
46. M.J. Hartmann, F.G.S.L. Brandao, M.B. Plenio, Strongly interacting polaritons in coupled arrays of cavities. Nat. Phys. **2**(12), 849–855 (2006)
47. D.G. Angelakis, M.F. Santos, S. Bose, Photon-blockade-induced Mott transitions and XY spin models in coupled cavity arrays. Phys. Rev. A **76**(3), 31805 (2007)
48. J. Cho, D. Angelakis, S. Bose, Fractional quantum Hall state in coupled cavities. Phys. Rev. Lett. **101**(24), 246809 (2008)
49. R.O. Umucalilar, I. Carusotto, Fractional quantum Hall states of photons in an array of dissipative coupled cavities. Phys. Rev. Lett. **108**, 206809 (2012)
50. A.L.C. Hayward, A.M. Martin, A.D. Greentree, Fractional quantum Hall physics in Jaynes-Cummings-Hubbard lattices. Phys. Rev. Lett. **108**(22), 223602 (2012)
51. M. Hafezi, M.D. Lukin, J.M Taylor. Non-equilibrium fractional quantum Hall state of light. New J. Phys. **15**, 063001 (2013)
52. M. Fleischhauer, M. Lukin, Dark-state polaritons in electromagnetically induced transparency. Phys. Rev. Lett. **84**(22), 5094–5097 (2000)
53. A. Andre, M. Bajcsy, A.S. Zibrov, M.D. Lukin, Nonlinear optics with stationary pulses of light. Phys. Rev. Lett. **94**(6), 063902 (2005)

54. M. Hafezi, A.S. Sorensen, E. Demler, M.D. Lukin, Fractional quantum Hall effect in optical lattices. Phys. Rev. A **76**(2), 023613 (2007)
55. A. Sørensen, E. Demler, M. Lukin, Fractional quantum Hall states of atoms in optical lattices. Phys. Rev. Lett. **94**(8), 086803 (2005)
56. I. Carusotto, D. Gerace, H. Tureci, S. De Liberato, C. Ciuti, A. Imamoğlu, Fermionized photons in an array of driven dissipative nonlinear cavities. Phys. Rev. Lett. **103**(3), 033601 (2009)
57. A. Tomadin, V. Giovannetti, R. Fazio, D. Gerace, I. Carusotto, H.E. Tureci, A. Imamoglu, Signatures of the superfluid-insulator phase transition in laser-driven dissipative nonlinear cavity arrays. Phys. Rev. A **81**(6), 061801 (2010)
58. M. Hafezi, D.E. Chang, V. Gritsev, E.A. Demler, M.D. Lukin, Photonic quantum transport in a nonlinear optical fiber. EPL (Europhys. Lett.) **94**, 54006 (2011)
59. A. Nunnenkamp, J. Koch, S.M. Girvin, Synthetic gauge fields and homodyne transmission in Jaynes-Cummings lattices. New J. Phys. **13**, 095008 (2011)
60. F. Nissen, S. Schmidt, M. Biondi, G. Blatter, H.E. Tureci, J. Keeling, Nonequilibrium dynamics of coupled qubit-cavity arrays. Phys. Rev. Lett. **108**(23), 233603 (2012)
61. M. Schiró, M. Bordyuh, B. Öztop, H. Tureci, Phase transition of light in cavity QED lattices. Phys. Rev. Lett. **109**, 053601 (2012)
62. M. Greiter, Mapping of Parent Hamiltonians: from Abelian and Non-Abelian Quantum Hall States to Exact Models of Critical Spin Chains, vol. 244 (Springer, Heidelberg, 2011)
63. N. Read, E. Rezayi, Beyond paired quantum Hall states: parafermions and incompressible states in the first excited Landau level. Phys. Rev. B **59**(12), 8084–8092 (1999)
64. C. Nayak, S.H. Simon, A. Stern, M. Freedman, S.D. Sarma, Non-Abelian anyons and topological quantum computation. Rev. Mod. Phys. **80**(3), 1083 (2008)
65. A.A. Houck, H.E. Türeci, J. Koch, On-chip quantum simulation with superconducting circuits. Nat. Phys. **8**(4), 292–299 (2012)

Chapter 5
Exciton-Polariton Quantum Simulators

Na Young Kim and Yoshihisa Yamamoto

Abstract A quantum simulator is a purposeful quantum machine that can address complex quantum problems in a controllable setting and an efficient manner. This chapter introduces a solid-state quantum simulator platform based on exciton-polaritons, which are hybrid light-matter quantum quasi-particles. We describe the physical realization of an exciton-polariton quantum simulator in semiconductor materials (hardware) and discuss a class of problems, which the exciton-polariton quantum simulators can address well (software). A current status of the experimental progress in building the quantum machine is reviewed, and potential applications are considered.

5.1 Introduction

We live in the twenty-first century—the heart of the Information Age. A transition to the Information Age was greatly accelerated in the late 1990s by the rapid development of the technologies in digital computation, and short- and long-distance communications. Now computers are indispensable appliances in modern society, influencing all aspects of our lifestyle and activities. Classical computer hardware has been sophisticated as well as functionalized in terms of size, integrability and speed, pacing up with the advancement of electronics. Figure 5.1 collects images of landmark machines throughout the computer history: a primitive Babbage automatic

N.Y. Kim (✉)
Department of Electrical and Computer Engineering, Institute for Quantum Computing, University of Waterloo, 200 University Ave. West, Waterloo, ON N2L 3G1, Canada
e-mail: nayoungstanford@gmail.com

Y. Yamamoto
Edward L. Ginzton Laboratory, Stanford University, 348 Via Pueblo Mall, Stanford, CA 94305, USA

Y. Yamamoto
National Institute of Informatics, 2-1-2 Hitotsubashi, Chiyoda-ku, Tokyo 101-8430, Japan

D.G. Angelakis (ed.), *Quantum Simulations with Photons and Polaritons*, Quantum Science and Technology, DOI 10.1007/978-3-319-52025-4_5

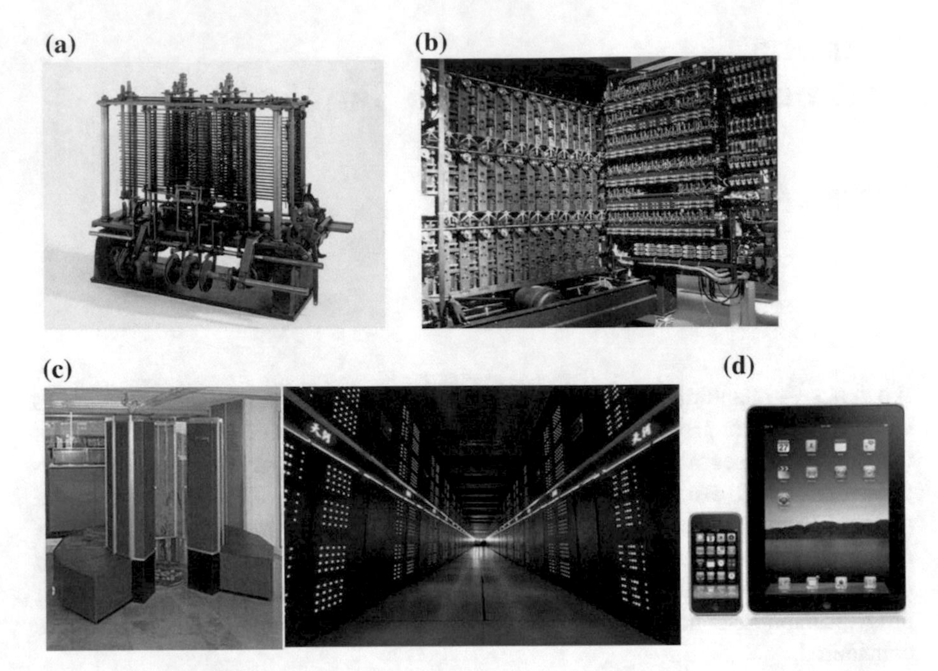

Fig. 5.1 Historical computing machines. **a** The first fully automatic Babbage analytic calculator in the 19th century. **b** An image of the Colossus, the first electronic digital Turing computer. **c** Photos of the supercomputers, the first generation CRAY-1 built in 1970s (*left*), and Milkyway-2, the most powerful supercomputer in 2013. **d** A multi-functional smart phone and a tablet computer

calculator, which could only compute simple arithmetics (Fig. 5.1a), a first digital Turing machine (Fig. 5.1b), which was programmable, and cutting-edge supercomputers equipped with multiple processors (Fig. 5.1c) to palm-sized personal computers, which allow an individual to multitask actively and productively (Fig. 5.1d). On top of the state-of-the-art computer platforms, myriads of software programs have also been designed and implemented successfully.

Notwithstanding that contemporary computers are impressively powerful and continuously improving, there still exist countless physical problems associated with many degrees of freedom, which demand unprecedented levels of performance in speed, flexibility and required resources beyond what the currently available computers can offer. Computer scientists theorized the computational complexity in regard to inherent difficulties of problem nature. Mechanically, this complexity may be quantified by computation time and resources for obtaining the right solution of a given problem. Quantum many-body problems belong to the category of the most complex class, that is often considered to be intractable by classical computers. This perplexing fact ignites the need of novel computing machines for tackling such effortful challenges.

In 1982, Richard P. Feynman conceived a simple but brilliant idea, building a machine governed by quantum mechanical rules [1]. Hence, this quantum machine should be able to solve any quantum many-body problems [2] by nature in an efficient and economical way. This approach is dubbed quantum simulation. This seed of idea is evolved to a universal quantum simulator by Lloyd [3], which mimics a quantum system of interest through coherent operations on the simulating system. Last decade, quantum simulation became a fast-growing theme in quantum science and technology both theoretically and experimentally. Several review articles have been published on the progress of quantum simulation research activities [4–6].

5.1.1 Digital and Analog Quantum Simulators

The ultimate goal of the engineered quantum machine is to reach a universally acceptable explanation of unanswered phenomena [4–6]. Often, physicists set up toy models, anticipating to explain the key features of the target phenomenon. Similarly, a quantum simulator aims a designated toy model, and it is prepared in two distinct ways: one is to build a bottom-up system with local addressing and individual manipulation, and the other is to directly map its time evolution of the system of interest onto the controlled time evolution of the simulating one [4–6]. The former approach is named *digital quantum simulator* (DQS). In the DQS, quantum problems are tackled by arranging a large number of qubits, a unit of quantum information or a state vector, followed by executing coherent unitary operations to manipulate these qubits. The universal DQS is an omnipotent quantum computer that can solve any quantum many-body problems efficiently [7]. The advantage of this scheme is the ability to supervise and identify errors during operations, which can be fixed by error-correction algorithms. A reliable and accurate solution can thusly be obtained. However, even simple prime number factorization problems require tremendous resources by the very nature, which is a major challenge to be overcome towards the implementation of a large-scale DQS [6, 7].

On the other hand, the latter approach known as an *analog quantum simulator* (AQS) is free from such expensive restrictions since it only tries to resemble the system of interest as closely as possible. It does not need to define individual qubits and/or to perform error correction. In addition, the AQS is constructed for a specific problem not all quantum problems. Accordingly, the physical resources to build an AQS are much less required than those for a DQS. Therefore, scientists in many disciplines have started in establishing their own platforms of AQSs, combining currently available technology and benchmarking some famous toy models [8–12]. Despite technical advantages of the AQS, there are fundamentally serious issues [13]: what specific problems are chosen to be simulated by the AQS?; and, what guarantees the reliability of the obtained solution to unknown problems with the AQS? In other words, it is meaningless to perfectly answer problems irrelevant to any real phenomena even if the AQS solves them accurately. Unfortunately, no methods to single out sources of errors in the AQS, not to mention no error correction protocols, are yet known. Hence, we all should remember the underlying predicament of the AQS when we employ the AQS for significant contributions to our knowledge base.

Keeping this in mind, in this section, we survey hardware and software of currently available AQSs.

5.1.1.1 Hardware of the Analog Quantum Simulators

The term 'hardware' here specifies physical elements of the AQSs akin to classical computer hardware. The AQS research activities were sparked by seminal work in 2002 [14], where I. Bloch and his colleagues observed a superfluid-insulator phase transition in ultracold Bose gases. This observation may be understood by the same physics of the metal-insulator transition in condensed-matter materials [14]. Assessing this successful demonstration, we generalize hardware of the AQS in three aspects: *particles, artificial lattices* and *detection schemes*. For the last decade or so, the AQSs were built upon a variety of particles, diverse methods to engineer lattice geometries, and specialized probe techniques. Figure 5.2 illustrates a few representative AQS platforms.

Current AQSs are based on ultracold bosonic [8] and fermionic [15] atoms, trapped ions [9, 16, 17], electrons [18–21], superconducting qubits [11, 22], linear

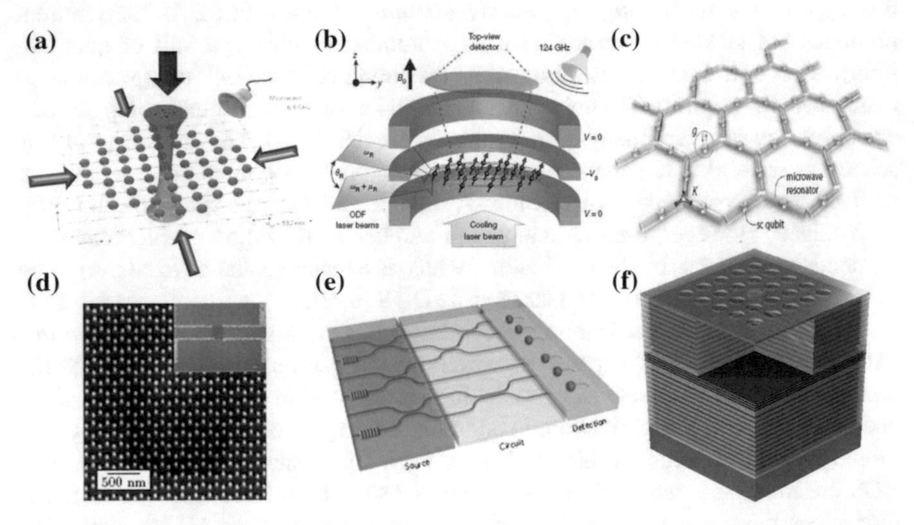

Fig. 5.2 Quantum Simulator Platforms. **a** An illustration of ultracold atom gas in a two-dimensional optical lattice taken from Ref. [8]. **b** A schematic of a two-dimensional triangular lattice formed by $^{9}Be^{+}$ ions in a Penning trap adapted from Ref. [17]. **c** A proposed microwave photon lattice using superconducting resonators and qubits in Ref. [22]. Reprinted figure with permission from Koch et al., Phys. Rev. A 82, 043811 (2010). Copyright(2010) by the American Physical Society. **d**, Scanning electron microscopy image of a honeycomb lattice by nanofabrication in a two-dimensional electron gas system presented in Ref. [20]. **e** A schematic picture of a planar photonic quantum simulator integrating essential components proposed in Ref. [10]. **f** An exciton-polariton quantum simulator in GaAs-based semiconductors. Permission of Figs. (a, b, e) is acquired from Nature Publishing group

photons [10], photons and polaritons in a cavity QED arrays [23–25] and exciton-polaritons [26, 27]. Many new particles or quasi-particles will likely join the list in the near future. Next, how are lattice potentials shaped for these particles? To trap cold quantum gases, pairs of oppositely propagating lasers are used to form spatial standing waves, periodic potentials in one-, two- and three-dimensions, also called optical lattices. This technique is superb in almost defect-free lattices and the controllability of the potential amplitude by adjusting the intensity of the participating lasers. Responses of atoms to potential changes are recorded by time-of-flight absorption imaging, which maps the momentum distribution of particles. For example, the presence or absence of diffraction peaks as interferences distinguish the coherent metallic phase from the incohrent insulating phase [8, 14, 15]. Recently, a high-resolution real-space imaging technique was developed to perform a parity measurement of trapped atoms at a single site, which enabled quantification of correlation functions [8]. For charged ionic particles, a Paul trap was used to establish a string of ions [16], and a Penning trap to arrange hundreds of ions in a triangular lattice [17]. Typically the internal states of ions according to external manipulation are detected by fluorescence signals with photodetectors. Among many techniques for confining electrons in semiconductors, the direct application of DC or AC electric and magnetic fields, strain fields for piezoelectric materials, and the direct etching processes are well established and controlled [21]. DC and AC resistance, magnetoresistance and thermal resistances are quantities to characterize electrical transport properties under specific trapping potentials. Other particles, superconducting qubits, photons and polaritons share a common feature in light, and they are classified as photonic quantum simulators, a central topic of Sect. 5.1.2.

5.1.1.2 Software of the Analog Quantum Simulators

Let us turn to what software AQSs can run as applications. Early on, atomic AQSs studied a Hubbard Hamiltonian [28], one of the simplest but important toy models in condensed matter physics, for example, to exploit the electronic metal-insulator phase transition such as in transition-metal oxides [29]. The Hamiltonian consists of two energy terms to describe interacting particles in a lattice: a kinetic term t for a nearest-neighbor hopping energy and an on-site Coulomb interaction energy term U. The Hamiltonian operator $\hat{H}$ is expressed in terms of particle operators $\hat{c}^\dagger$, $\hat{c}$ with a site index i, j, a spin index σ and a particle density $\hat{n} = \hat{c}^\dagger\hat{c}$,

$$\hat{H} = -t \sum_{<i,j>,\sigma} (\hat{c}^\dagger_{i,\sigma}\hat{c}_{j,\sigma} + h.c.) + U \sum_{i=1}^{N} \hat{n}_{i\uparrow}\hat{n}_{i\downarrow},$$

where $<i, j>$ denotes the nearest-neighbor interaction on the lattice and $h.c.$ is the hermitian conjugate. Even though it is simple, the analytical solutions in higher dimensions than one dimension are not available, and obtaining solutions for even a small number of lattice sites in two dimension is numerically expensive. Both Bose and Fermi-Hubbard Hamiltonians have been studied using bosonic [8, 14] and fermionic [15] gases and electrons [21].

The next one is an Ising Hamiltonian to study quantum magnetism dominantly favored in trapped ions [16, 17]. It is related to some of the challenging problems in condensed matter physics, such as the construction of a phase diagram of quantum matter and in particular quantum criticality as sketched in Fig. 5.3a [30, 31]. Recently, stimulated by the discovery of new quantum matter arising from spin-orbital physics, tremendous efforts have been put into its simulation. Synthesizing the effective magnetic fields [32–34] to selectively manipulate the internal states of atoms in lattices provides a route to investigate the spin-orbital couplings in atom AQSs [34].

In quantum chemistry, AQSs will be useful to understand the static [35] and dynamical [36] properties of molecules. Figure 5.3b depicts a proposal on how to

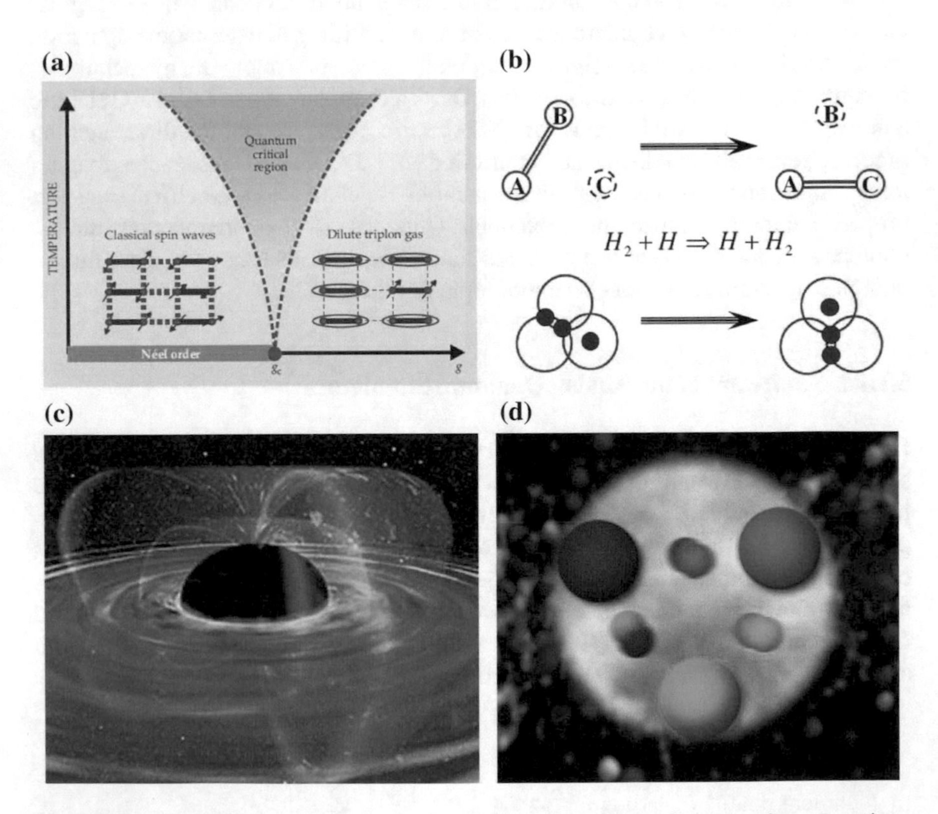

Fig. 5.3 Representative quantum many-body problems as possible applications of quantum simulators. **a** A finite temperature phase diagram for the dimer antiferromagnet near the quantum critical point g_c regimes adapted from Ref. [31]. **b** An exemplary hydrogen chemical reaction (*top*) to be simulated in coupled quantum dot systems (*bottom*) illustrated in Ref. [37]. **c** An artistic view of a Hawking radiation. **d** A nucleon inner structure with quarks in quantum chromodynamics

implement the scattering process between a hydrogen atom and a hydrogen molecule using a coupled quantum dot system. The electron redistribution in three quantum dots under electrical gates would be anticipated to provide insights of the hydrogen dynamics [37].

Phenomena in cosmology and high-energy physics are often explored using gigantic apparatus due to the huge energy scales (>keV) involved. However, the development of AQSs may grant an easy access to investigate such physics in tangible laboratory settings. For instance, a Hawking radiation emitted from black holes (Fig. 5.3c) would be possibly explored with atomic superfluids [38] or exciton-polaritons [39]. Relativistic Dirac physics has been addressed in a trapped ion [40], and artificial Dirac points are created in fermionic atomic gases [41] and copper-oxide molecules [42]. In the future, quark physics inside the nucleus (Fig. 5.3d) would be an interesting quantum problem for AQSs. Since the aforementioned ones are a tiny subsection of quantum many-body problems in condensed matter physics, quantum chemistry, high-energy physics, cosmology, and nuclear physics, the phase space of AQS applications is almost boundless once reliable and functional AQSs are constructed.

5.1.2 Photonic Analog Quantum Simulators

Besides the many promising physical realizations of AQSs discussed earlier, this subsection specializes photonic AQSs, where light is a part of the particles. They include microwave photons coupled to superconducting qubits [11, 22], strongly interacting photons in assorted cavity shapes [10, 23–25] and strongly coupled exciton-polaritons [26, 27]. Most photonic systems would enjoy a planar structure, advantageous en route to large-scale architectures with the help of advanced nanofabrication processing techniques. Photonic AQSs have also been proposed to reveal quantum phase transition of light [23, 25]. In addition, mapping the light polarization to a pseudo-spin '1/2' system allows to explore spin physics and magnetism. The unavoidable loss process through optical cavities makes photonic AQSs well suited to simulate open environment and non-equilibrium physics.

It is true that the exciton-polariton AQS also possesses an inherent photonic nature. However, the non-negligible matter nature from the excitons distinguishes the exciton-polariton AQS from other photonic AQSs. This particular chapter is devoted to this hybrid light-matter exciton-polariton AQS and begins by outlining both hardware and software aspects in Sect. 5.2 and Sect. 5.3, respectively. Section 5.2 explains the fundamentals of exciton-polaritons and their condensation properties followed by technical details of lattice formation and experimental probes. Potential physical problems the exciton-polariton AQS could target to address will be discussed in Sect. 5.3. We give an overview of our experimental progress in constructing an exciton-polariton AQS, and perspective remarks in Sect. 5.4.

5.2 Hardware of Exciton-Polariton Quantum Simulators

We review the physical components of the exciton-polariton quantum simulator hardware: *microcavity exciton-polaritons* as our particle, methods to engineer *periodic potential landscapes* for exciton-polaritons and *microphotoluminescence* imaging and spectroscopy as detection schemes.

5.2.1 Particle:Exciton-Polariton Condensates

5.2.1.1 Microcavity Exciton-Polariton

A 'polariton' is a general term for a quasipatricle as an admixture of a photon and a certain excitation. An exciton-polariton refers in particular to a resulting quasiparticle when a photon is mixed with an exciton in a semiconductor. Here, we consider exciton-polaritons in a two-dimensional (2D) microcavity structure with embedded quantum-wells (QWs) sketched in Fig. 5.4a. A photon is confined in a cavity and an exciton resides in the QWs. In 1992, C. Weisbuch et al. reported the first observation of photon-exciton coupled modes from a λ-cavity, where a single GaAs QW sits in the middle of the cavity [43]. Since its discovery, the microcavity exciton-polariton system has been an attractive solid-state domain to investigate quantum boson statistical effects such as condensation and superfluidity as well as to develop novel devices like polariton lasers or ultrafast spin switches both in theory and experiments [44–46]. In this section, we summarize fundamentals of exciton-polaritons and their condensation properties.

Wannier Exciton

Semiconductors are a class of materials with a gapped electronic excitation. Numerous elemental and compound semiconductors have been the backbone of revolutionary electronic, optical and optoelectronic devices in the 20th century. Atoms are organized in a specific crystal structure, and the electrons of these atoms are brought together to form energy bands. In thermal equilibrium, all electrons are in the valence band, which is often called the system ground state or vacuum state. This stable low energy state is separated from the above excited states by an energy gap, and it is electrically insulating and optically dark. Suppose light shines onto a semiconductor. When the absorbed photon energy is bigger than the gap energy, it can promote electrons from the valence band to the conduction band, leaving positively-charged quasi-particles (holes) behind in the valence band. Similar to the hydrogen atom, where an electron is bound to a nucleus by Coulomb attraction, in perturbed semiconductors, an electron and a hole attract each other, creating an exciton, a Coulombically bound lower energy state [47].

This hydrogen-like entity is parameterized by a Bohr radius a_B, the extent of the electron-hole ($e - h$) pair wavefunction and a binding (or Rydberg) energy E_B, an

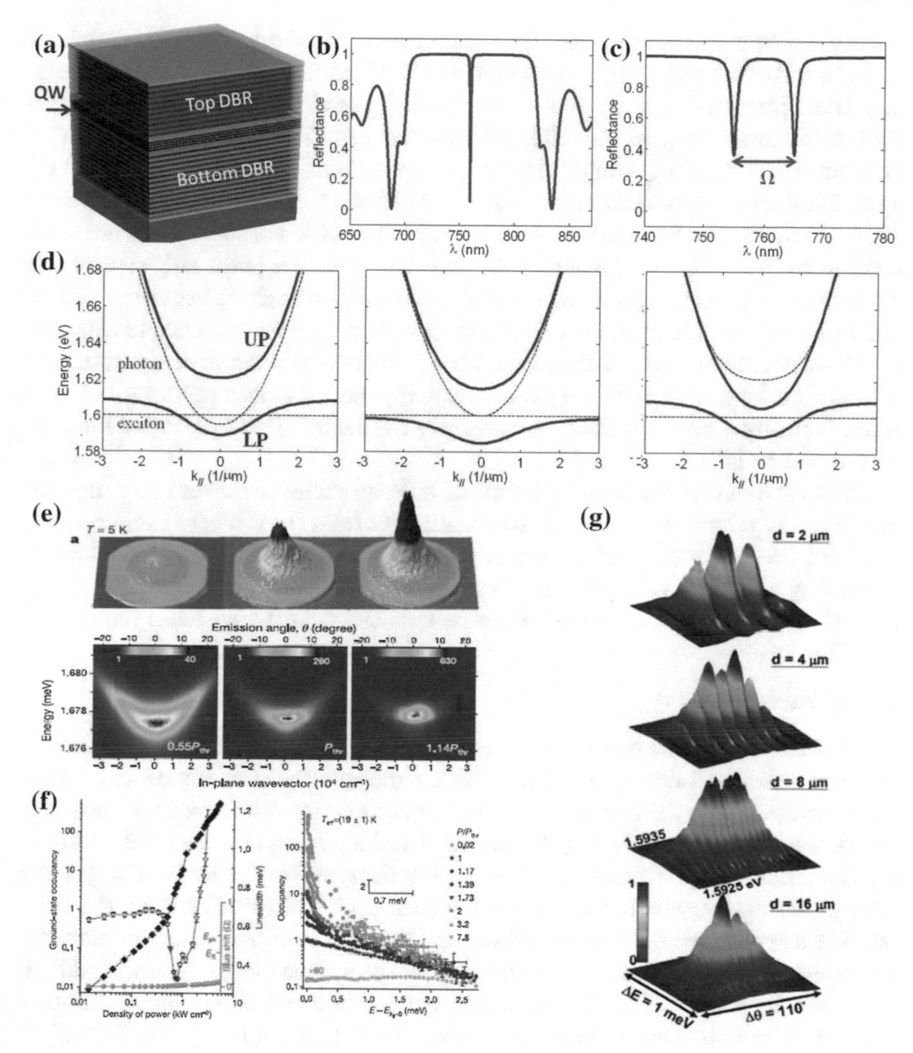

Fig. 5.4 Microcavity exciton-polaritons. **a** A monolithically grown microcavity structure where multi-quantum wells (*red layers*) are embedded. **b** A calculated reflectance spectrum of a λ-sized cavity surrounded by pairs of a distributed Bragg mirror. **c** A reflectance spectrum of two microcavity exciton-polaritons, upper polariton (UP) and lower polariton (LP), resulting from the strong coupling between a cavity photon and a quantum well (QW) exciton (**c**). The energy separation of the double dips is the direct measure of the coupling strength between the photon and exciton denoted as Ω. **d** The energy dispersions of three configurations to mix the photon and the QW exciton modes, more photon-like LP mode (*left, red-detuned*), half-light and half-matter polaritons (*middle*, zero-detuned), and more exciton-like LP (*right, blue-detuned*). **e–g** Experimental signatures of exciton-polariton condensations. **e** Pump-power dependent LP population images (*top*) and spectra (*bottom*) in the momentum space and **f** LP ground-state occupancy as a function of pump power (*left*) and energy (*right*) taken from Ref. [57]. **g** Interference of LP condensates to exhibit spatial coherence from Young's double slit measurements reportd in Ref. [64]. Nature granted permission of Figs. (e, f) from Ref. [57] and Fig. (g) from Ref. [64]

energy cost to dissociate the pair. The complicated many-body system is simplified to a single particle picture by introducing an effective mass and a medium dielectric constant. Effective masses of electrons and holes in semiconductors are smaller than the bare electron mass, and the dielectric constant can be as large as 13 times of the vacuum case. Therefore, compared to he hydrogen atom ($E_B = 13.6$ eV, $a_B = 0.5$ Å), a semiconductor exciton has a larger a_B on the order of 0.1–10 nm and much weaker E_B in the range of a few meV to 1 eV. In large-dielectric-constant inorganic semiconductors like Si, GaAs, CdTe or ZnSe, the electric field is noticeably screened by the surrounding charges, yielding the weaker Coulomb interaction between electrons and holes. As a result, excitons are spread over many lattice sites, and are classified as 'Wannier excitons'. Since the exciton is a primary excitation of semiconductors, it lives for a finite time and the system eventually returns to its equilibrium vacuum state. Typically, GaAs QW excitons have $a_B \sim 6$–10 nm, $E_B \sim 10$ meV, and lifetimes of 100 ps-1 ns [48].

Composed of two fermions, electron and hole, an exciton behaves as a composite boson at low temperatures and in the low-density limit (n_X), where exciton wavefunctions do not easily overlap each other in a given volume. In 2D, the regime is marked by a density $n_X a_B^2 \ll 1$. Excitons in this dilute density regime obey Bose-Einstein statistics, and are expected to be condensed in the system ground state [49, 50].

Cavity Photon

In semiconductors, alternating $\lambda/4$-thick layers with different refractive indices makes a distributed Bragg reflector (DBR), a dielectric mirror. Reflectance of the DBR increases with the number of the layer pairs and the refractive index contrast. At the interface of two layers, photons near wavelength λ will reflect and constructively interfere from all interfaces. The simplest but useful resonator is a Fabry-Perot (FP) planar cavity, where photons are bounced back and forth between the DBR pairs. A λ-sized optical cavity would support only one longitudinal cavity mode for a sub-micron λ value. Such a structure appears as a reflectance dip at the particular photon frequency of the designed cavity mode in broad-band reflectance spectroscopy (Fig. 5.4b). One figure of merit to describe a cavity is the quality-factor (Q), which relates to how long photons reside inside the cavity. The reported Q-factors of an empty FP DBR cavity ranges from a few thousands upto a million, corresponding to cavity photon lifetimes around 1–100 ps.

It is worth mentioning that a cavity photon would have an 'effective mass' m_c in the transverse plane. The growth direction confinement imposes that the longitudinal component $k_\perp$ of the light wave vector is given by $2\pi/\lambda$, which is much bigger than the transverse (or in-plane) wavenumbers $k_\parallel$. Therefore, in the regime of $k_\parallel \ll k_\perp$, the cavity photon energy E_c can be approximated as follows:

$$\begin{aligned} E_c = \hbar v|k| = \hbar v\sqrt{k_\perp^2 + k_\parallel^2} = \hbar v k_\perp \sqrt{1 + \left(\frac{k_\parallel}{k_\perp}\right)^2}, \\ \sim \hbar v k_\perp \left(1 + \frac{k_\parallel^2}{2k_\perp^2}\right) \equiv E_{c0} + \frac{\hbar^2 k_\parallel^2}{2m_c}, \end{aligned}$$

where $\hbar$ is the Planck's constant divided by 2π, v is the velocity in the medium, $E_{c0} = \hbar v k_{\perp}$, and m_c satisfies the following relation $m_c = h/v\lambda$. In the visible to near infrared regions, m_c values around $10^{-5} m_e$, and the energy dispersion ($E_c - k_{\parallel}$ relation) becomes parabolic with a stiff curvature due to extremely light effective mass.

Exciton-Polaritons

Consider a two-level single atom in a high-Q-cavity, which is one of the well-known problems in the cavity quantum electrodynamics. When the atomic transition energy is precisely on resonance with the cavity mode energy, the coherent energy transfer between the atom and the photon occurs reversibly at the "Rabi frequency" provided that the atom and cavity decay rates are much slower than the energy exchange rate. Spectroscopically, the strongly mixed modes appear as an anti-crossing energy gap known as "vacuum Rabi splitting", a manifestation of *strongcoupling* between two modes.

Let us now turn into our case, a QW placed at the maximum of the photon field inside a high-Q DBR cavity. A photon is absorbed to excite an exciton in the QW, and a photon is re-created inside the cavity resulting from the radiative recombination of the $e - h$ pair in the exciton, which can excite an exciton in turn. Such reversible and coherent energy transfer between photon and exciton modes continues within the cavity lifetimes. When this coupling occurs much faster than any other decay process in the system, the QW-cavity system also enters into a *strong coupling* regime. And there emerge new photon-dressed excitons, which we name *exciton-polaritons* (or *cavity polaritons*). As a result, two anti-crossed energy branches, upper polariton (UP) and lower polariton (LP), appear and there exist two dips at UP and LP energy states separated by the interaction strength Ω in reflectance spectroscopy (Fig. 5.4c). In a linear signal regime, the coupled-cavity photon-QW exciton Hamiltonian $\hat{H}$ is written by a cavity photon operator $\hat{a}_k$ with energy $\hbar\omega_{ph}$, a QW exciton operator $\hat{C}_k$ with $\hbar\omega_{exc}$ and their in-between interaction coupling constant Ω_k,

$$\hat{H} = \hbar \sum_k [\omega_{ph} \hat{a}_k^{\dagger} \hat{a}_k + \omega_{exc} \hat{C}_k^{\dagger} \hat{C}_k - i\Omega_k (\hat{a}_k{}^{\dagger} \hat{C}_k - \hat{a}_k \hat{C}_k^{\dagger})],$$

where k is the simplified notation for $k_{\parallel}$. Introducing an exciton-polariton operator at a momentum k as a linear superposition of the cavity photon and the QW exciton modes $\hat{P}_k = u_k \hat{C}_k + v_k \hat{a}_k$, the Hamiltonian is diagonalized, reaching a simplified Hamiltonian $\hat{H}_T$

$$\hat{H}_T = \sum_k \hbar \Omega_k \hat{P}_k^{\dagger} \hat{P}_k,$$

and the exciton-polariton energy dispersion is explicitly given by

$$\hbar\omega_k = \frac{1}{2}\left(\hbar(\omega_{exc} + \omega_{ph}) + i\hbar(\gamma_{ph} + \gamma_{exc})\right)$$
$$\pm \frac{1}{2}\sqrt{(2\hbar\Omega_k)^2 + (\hbar(\omega_{exc} - \omega_{ph}) + i\hbar(\gamma_{exc} - \gamma_{ph}))^2}.$$

γ_{ph} is the photon-mode decay rate through the cavity, and γ_{exc} is the non-radiative decay rate of the exciton.

Figure 5.4d illustrates the energy (E) versus the transverse wavenumber ($k_{\|}$) of independent cavity photon and exciton modes in blue dashed lines and the UP and LP in red straight lines. Effective masses and lifetimes of UP and LP are determined as a weighted average of individual constituents, taking into account of photon and exciton fractions. Near $k_{\|} = 0$, the photon-like LP branch has an extremely light effective mass, but it is exciton-like at large $k_{\|}$ values with a flatter dispersion, corresponding to a heavy mass. Depending on fractions of photons and excitons ($|u_k|^2, |v_k|^2$), three different detuning regimes (red, zero, blue) are possible as shown in Fig. 5.4d. A detuning parameter Δ is defined as $\Delta(k_{\|} = 0) = E_c(k_{\|} = 0) - E_{exc}(k_{\|} = 0)$. For instance, the negative Δ is red detuned, where the photon fraction is bigger, whereas the positive Δ is blue-detuned with exciton dominance.

5.2.1.2 Optical Processes: Radiative Recombination Versus Relaxation Processes

Before delving into the properties of exciton-polariton condensation in the following section, we give a simple picture of optical processes inside a QW-microcavity structure. As seen earlier, when the QW-microcavity is excited by an external pump, $e - h$ pairs are created. These high-energy particles are quickly cooled down to large-momentum excitons or exciton-like polaritons. The dynamics of exciton-polaritons can be primarily understood by the interplay between radiative recombination and energy relaxation processes. The former radiative recombination rate of exciton-polaritons at $k_{\|}$ is simply determined by the lifetime of photon and exciton modes with appropriate fractional values for each mode. Therefore, small $k_{\|}$ exciton-polaritons decay faster due to short-lived photons, whereas those of large $k_{\|}$, more exciton-like ones, have smaller radiative recombination rates, and hence a longer lifetime.

Two dominant exciton-polariton relaxation processes have been extensively studied: one is due to exciton-phonon interaction [48, 51], and the other due to exciton-exciton interaction [52, 53]. Tassone and his colleagues quantified polariton-acoustic phonon scattering rates in a microcavity with a single GaAs QW, which were calculated by the deformation potential interaction using a Fermi golden rule [48, 51]. Setting up semiclassical Boltzmann rate equations, Tassone et al. observed that large-momentum polaritons are efficiently relaxed to lower energy regions through the emission or absorption of acoustic phonons, and that these scattering processes are inelastic and incoherent. However, they also found that phonon scattering is significantly reduced towards low $k_{\|}$ values of the LP branch, where the LP effective mass becomes lighter and the LP density of states is smaller. Hence, exciton-polaritons are rather parked at the deflection of the LP dispersion branch, the *bottleneck* effect.

To overcome the bottleneck effect, people have resorted to another important relaxation process induced by elastic exciton-exciton scattering in the 2D QW [54]. When two excitons encounter, two nearby excitons interact via either direct or exchange interactions among electrons and holes. The direct term comes from dipole-dipole interactions of excitons, which are considered to be negligible in QW. Up to first order, the dominant mechanism is short-ranged Coulomb exchange interactions between electron and electron or between hole and hole [52, 54]. An exchange interaction strength is roughly on the order of $6E_Ba_B^2/S$, where S is the quantization area. Conserving energy and momentum, coherent scatterings are further amplified by the number of LP, which is the bosonic final-state stimulation effect. This process is a crucial mechanism to accumulate LPs near $k_{\|} = 0$, reaching condensation.

5.2.1.3 Exciton-Polariton Condensation

At low temperature and a dilute density, exciton-polaritons are also regarded as composite bosons, governed by Bose-Einstein statistics. Immediately after the discovery of the exciton-polaritons in the microcavity-QW structure [43], people recognized the possibility of exciton-polariton condensation [55]. To validate exciton-polariton condensation at $k_{\|} = 0$, we should clarify system conditions with care: first, the exciton-polariton density is low enough, not exceeding the Mott density, which describes the state where the concept of excitons breaks down due to particle screening. The Mott density is roughly set by $1/a_B^2$ in 2D, and it is critical to maintain the GaAs exciton density per QW at less than 10^{12} cm^{-2}. One way to achieve this is to increase the number of QWs in the cavity, decreasing the average exciton density per QW by the assumption that the created excitons are equally distributed over all QWs. We commonly place 4, 6, 8 or 12 GaAs QWs in a AlAs cavity for this reason. Second, coherence among exciton-polaritons at $k_{\|} = 0$ should occur spontaneously not from the coherence of the pump laser. A safe manner of the excitation scheme is to choose the excitation laser energy to be much higher than the polariton ground state (*incoherent excitation*), which injects either high energy $e - h$ pairs or excitons as we described in the previous Sect. 5.2.1.2. Through incoherent phonon scattering processes, the initial energy and momentum of the injected particles are lost, and LPs are cooled down to the lowest energy and momentum state, establishing coherence through the polariton-polartion coherent scattering processes.

A series of experimental evidence or signatures of exciton-polariton condensation near $k_{\|} = 0$ were reported in GaAs, CdTe and GaN inorganic [56–61] and organic [62, 63] semiconductors. A first piece of evidence is the observation of macroscopic population at the system ground state. Figure 5.4e shows pump-power dependent LP population distribution in momentum space (top panel) and their spectroscopic images (bottom panel) in CdTe-based semiconductors reported by Kasprzak et al. [57]. When the particle density reaches quantum degeneracy threshold, bosons attract to be in the same state, thermodynamically favorable state enhanced by the boson stimulation effect [56, 57, 59]. Next, the essential physical concept of the condensation is off-diagonal long-range order related to spatial coherence in condensation [57, 60, 64]. As a system order parameter, the complex-valued LP wavefunction $\psi(\boldsymbol{r}, \phi)$ is $\psi(\boldsymbol{r}, \phi) = \sqrt{n(\boldsymbol{r})}\exp(-i\phi(\boldsymbol{r}))$ determined by two real values, the LP density $n(\boldsymbol{r})$ and the real-space phase $\phi(\boldsymbol{r})$. By means of an interferometer, the phase information of the order parameter is obtained, and the spatial coherence has been measured [57, 60]. Section 5.2.1 gave an overview of basic mathematical description of exciton-polaritons, the primary optical processes to lead to condensation and its general characterization in a standard QW-microcavity structure. The next subsection will summarize methods of engineering lattices for exciton-polaritons.

5.2.2 Lattice: Photonic and Excitonic Lattices

The second aspect of the exciton-polariton quantum simulator hardware is how to impose artificial crystal lattices on exciton-polaritons. Exploiting the duality of exciton-polaritons emphasized in Sect. 5.2.1, we are able to create the potential landscape for either photon or exciton modes. Within a 2D planar microcavity-QW structure, the engineered lattices lie in zero-, one- or two-dimensions. In this section we describe various designs to create the lateral potential lattices attempted in the exciton-polariton systems.

5.2.2.1 Methods of Creating Polation Lattices

Figure 5.5 collects a variety of methods to produce in-plane potentials for exciton-polaritons. We classify them into two groups, excitonic and photonic traps depending on which component is modified by the method. A mechanical stress induced by a sharp pin (Fig. 5.5a) shifts the QW exciton energy to a lower energy side (red-shifted) acting as an exciton trap [65]. Its induced potential strength can be as large as hundreds of meV, and lateral sizes of the potential are determined primarily by tip sharpness. Because QWs are far below the top surface, QW excitons face a rather broadened potential, whose lateral size is on the order of micrometer. Applying electric fields is widely used to modify the QW exciton energy via the quantum-confined Stark effect [66]. Magnetic fields shift the QW exciton energy as well. Although AC electric and magnetic fields can be applied, to the best of our knowledge all of the above methods have so far been implemented using static sources. Dynamical potentials were successfully launched by surface acoustic waves in piezoelectric GaAs semiconductors, which basically patterns propagating trap potentials [67]. So far, 1D [68] and 2D square [69] lattices have been produced. This method still suffers from limited potential depth on the order of hundreds of μeV and a small number of possible patterns according to the symmetry properties of the host material's piezoelectric tensor.

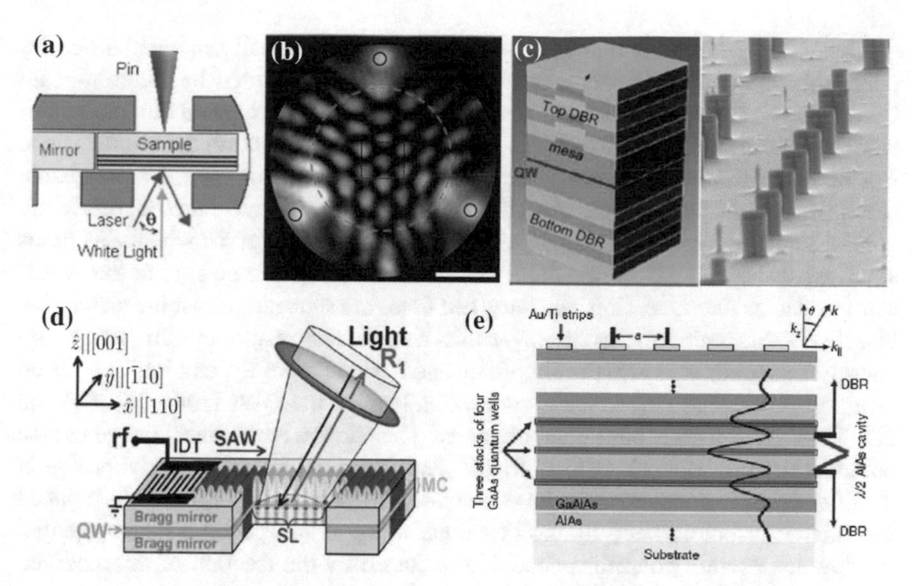

Fig. 5.5 Generations of trapping potentials in microcavity exciton-polaritons. **a** Mechanical strain introduced by a pin near the sample changes the exciton energy (Ref. [59]). **b** A honeycomb lattice of exciton-polaritons induced by a three-spot laser pump profile Ref. [77]. **c** A partial (*left*, Ref. [71]) and complete etching (*right*, Ref. [73]) methods to introduce a photonic trap. Reprinted figures with permission from Nardin et al., Phys. Rev. B 82, 045304 (2010) (*left*), and Galbiati et al., Phys. Rev. Lett. 108, 126403 (2012) (*right*). Copyright(2010, 2012) by the American Physical Society. **d** A dynamic phonon lattice in a piezoelectric GaAs microcavity-QW structure reported in Ref. [67]. Reprinted figure with permission from de Lima et al., Phys. Rev. Lett. 97, 045501 (2006). Copyright(2006) by the American Physical Society. **e** A weakly modulated in-plane one-dimensional photon lattice employing a thin-metal film technique on a grown wafer. A figure is adapted from Ref. [64]

On the other hand, several clever techniques were devised for photon traps. Chemical and dry etchings are common techniques for semiconductor processing. Either partially etching the cavity layer [70, 71] or completely etching all layers, except for the designed area [72] (Fig. 5.5c), is done on microcavity-QW wafers. Partial cavity-layer etching method was developed in B. Deavaud's group in Switzerland, and resulted in tens of meV-strong potential, which quantizes cavity photon modes like zero-dimensional quantum-dot [70, 71]. Recently, the complete etching method for making a single pillar was extended to make 1D wire [74] and 2D pillar-array potentials [75]. Despite recent progress, roughness of the sidewall surface has been the primary concern and caused degradation of the quality of QWs as well as spoiling the strong coupling regime especially for sub-micron sized pillars. Partial cavity-layer etching technique is in this sense less detrimental. Lately, several groups have noticed that the pump laser profile can be designed to trap exciton-polaritons through stimulated scattering gain under the pump profile [76, 77]. This technique can also be implemented to induce vortex lattices of exciton-polaritons due to the particle-particle repulsive interactions shown in Fig. 5.5b [77].

Our group employs a thin metal-film technique to spatially manipulate cavity photon-mode energy illustrated in Fig. 5.5e [64, 78]. Patterned by electron-beam lithography followed by a metal deposition on a grown wafer, a metal film modulates spatially the cavity lengths. The basic principle of such effect on the photon mode can be understood by looking at the boundary conditions of the electromagnetic (EM) fields at the interface between the top semiconductor layer and the metal film. Unlike bare surfaces of the semiconductor layer to vacuum or air where EM fields smoothly decay to an open medium, EM fields are pinned to be zero at the metal-semiconductor interface. Consequently, EM fields are squeezed under the metal film, effectively shortening the cavity-layer thickness, which results in a higher photon energy. This method enjoys the simplicity and design flexibility, and it is completed as a post-processing step to the grown wafer, leaving the QWs intact. 1D [64] and 2D [79–82] geometries have been preprared. Despite the advantages, the generated potential strength is weak (100–400 μeV), and its actual lateral potential profile in the QW planes is broadened for the same reason as mentioned for the strain-induced potential described earlier [78]. In all excitonic and photonic potentials, real potential strength for exciton-polaritons should be adjusted by the fraction of the modified component at different detuning values.

5.2.3 *Probes: Photoluminescence*

Now we turn to the third aspect of the hardware, the measurement schemes to identify optical properties of exciton-polaritons in the QW-microcavity structures.

5.2.3.1 Photoluminscence Setup

Optical properties of semiconductors are probed in various ways: absorption, reflection, transmission and luminescence through responses to optical excitations [47]. Exciton-polariton characteristics are primarily examined via photoluminescence (PL). High energy light shines onto a QW-microcavity structure, creating exciton-polaritons which are eventually leaked out through the cavity in a form of photons. The energy, wavenumber (momentum), polarization of the exciton-polaritons are uniquely transferred to those of the leaked photons. Thus, by observing the leaked photons, we obtain energy, momentum and polarization information of the exciton-polaritons using a spectrometer and the temporal dynamics using a streak camera. Since the QW-exciton binding energy in GaAs and CdTe semiconductors is around 10 meV, experiments are performed in a cryostat held at low temperatures around 4–10 K, where thermal energy cannot dissociate the excitons. In GaN, ZnO or organic semiconductors whose exciton binding energy exceeds the room-temperature thermal energy $\sim$26 meV, their PL setup can be operated at room temperatures without cryostats and cryogenics, which is a huge advantage in cost and space.

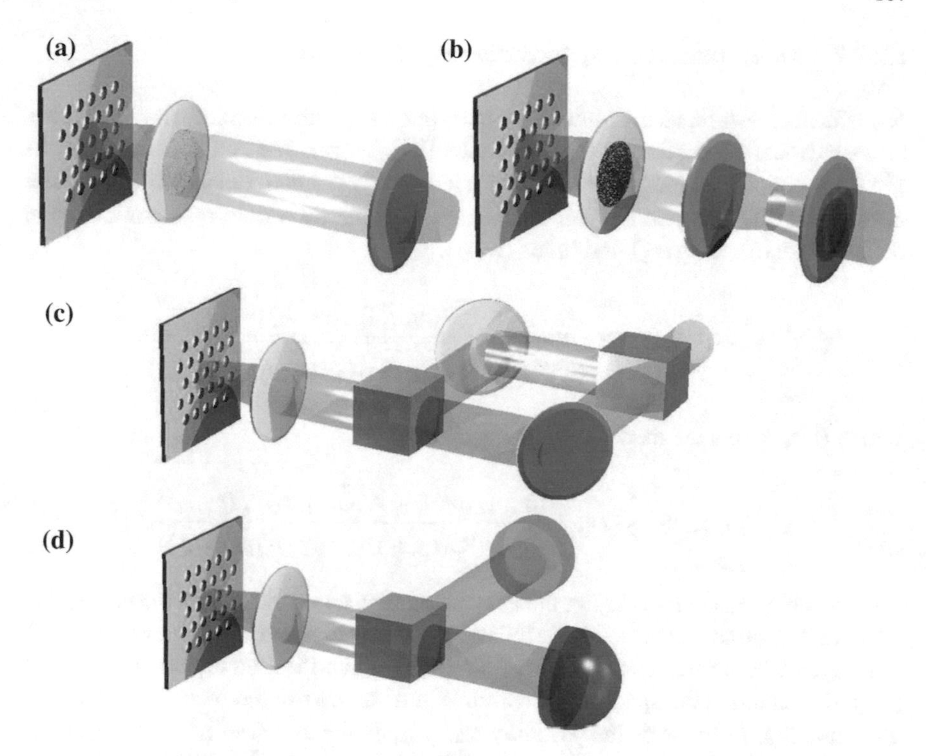

Fig. 5.6 Schematics of Fourier-optic measurement setups for exciton-polariton optical characterization. **a** A real space configuration of the sample plane mapped to an image plane of detectors. **b** A Fourier transformed setup to access the momentum space information by inserting one more lens. **b** A Mach-Zehnder interferometer to access phase in real-space. **d** A Hanbury Brown-Twiss intensity correlator setup

Fourier Optics

A standard micro-photoluminescence setup is constructed by the concept of Fourier optics. It allows us to access exciton-polaritons both in real and reciprocal spaces. In Fourier optics, a lens transforms an image located at one focal plane to its Fourier transformed image at the opposite focal plane [83]. Real space in one side of the lens becomes momentum space on the other side of the lens, and vice versa. We implement a compact and powerful setup in real and momentum spaces with different sets of lenses drawn in Fig. 5.6a and b, respectively, utilizing this transformation property. We put a simple charge-coupled detector to record intensities of each pixel at the arrival of photons, a spectrometer to provide energy-resolved spectra and a streak camera for the temporal responses.

5.2.3.2 Measurement Setup for Coherence Functions

R.J. Glauber established a guiding framework in quantum optics, that characterizes statistical properties of the EM fields [84]. Normalized coherence functions $g^{(n)}(\boldsymbol{r_1}, t_1; \boldsymbol{r_2}, t_2; ...; \boldsymbol{r_n}, t_n)$ for all orders n are clearly defined. Let us pay a particular attention to the first two coherence functions written by a field operator $\hat{\psi}(\boldsymbol{r}, t)$ at different positions $(\boldsymbol{r_1}, \boldsymbol{r_2})$ and times (t_1, t_2):

$$g^{(1)}(\boldsymbol{r_1}, t_1; \boldsymbol{r_2}, t_2) = \frac{\langle \hat{\psi}^\dagger(\boldsymbol{r_1}, t_1))\hat{\psi}(\boldsymbol{r_2}, t_2)\rangle}{\sqrt{\langle \hat{\psi}^\dagger(\boldsymbol{r_1}, t_1)\hat{\psi}(\boldsymbol{r_1}, t_1)\rangle \langle \hat{\psi}^\dagger(\boldsymbol{r_2}, t_2)\hat{\psi}(\boldsymbol{r_2}, t_2)\rangle}},$$

where $\langle\rangle$ indicates the thermal averaging, and

$$g^{(2)}(\boldsymbol{r_1}, t_1; \boldsymbol{r_2}, t_2) = \frac{\langle \hat{\psi}^\dagger(\boldsymbol{r_1}, t_1))\hat{\psi}^\dagger(\boldsymbol{r_2}, t_2)\hat{\psi}(\boldsymbol{r_2}, t_2))\hat{\psi}(\boldsymbol{r_1}, t_1)\rangle}{\langle \hat{\psi}^\dagger(\boldsymbol{r_1}, t_1)\hat{\psi}(\boldsymbol{r_1}, t_1)\rangle \langle \hat{\psi}^\dagger(\boldsymbol{r_2}, t_2)\hat{\psi}(\boldsymbol{r_2}, t_2)\rangle}.$$

It becomes easier to interpret the above formulas when we look at the same position or at the same time. $g^{(1)}(\boldsymbol{r_1}; \boldsymbol{r_2})$ tells us the spatial coherence property of the field, a measure of long-range spatial order. $g^{(1)}(t_1; t_2)$ quantifies the temporal coherence property related to the spectral linewidth. When we rewrite the second-order coherence function in terms of the particle density operator $n(\boldsymbol{r}, t) = \hat{\psi}^\dagger(\boldsymbol{r}, t)\hat{\psi}(\boldsymbol{r}, t)$ and the variance $\Delta n(\boldsymbol{r}, t) = n(\boldsymbol{r}, t) - \langle n(\boldsymbol{r}, t)\rangle$, the meaning of $g^{(2)}(\boldsymbol{r}, t)$ becomes clear,

$$g^{(2)}(\boldsymbol{r}, t; \boldsymbol{r}, t) = \frac{\langle n(\boldsymbol{r}, t)^2\rangle}{\langle n(\boldsymbol{r}, t)\rangle^2} = 1 + \frac{\langle \Delta n(\boldsymbol{r}, t)^2\rangle}{\langle n(\boldsymbol{r}, t)\rangle^2}.$$

That is the variance of the number distribution, giving the information of the intensity-intensity correlation.

The optical properties of exciton-polaritons have also been characterized with these first two coherence functions, and we briefly describe standard setups to measure these quantities: an interferometer for the first-order coherence function and a Hanbury Brown-Twiss setup for the second-order coherence function.

Interferometry

As mentioned earlier, the condensate order parameter is complex-valued with two real variables, particle density $n(\boldsymbol{r})$ and phase $\phi(\boldsymbol{r})$. The imaging intensity is proportional to the exciton-polariton density, but the phase information, which is as important as the density, is lost in imaging. For photons, we construct interferometers as a means to extract this phase information. For exciton-polaritons, Michelson and Mach-Zehnder interferometers (Fig. 5.6c) are used, where the 2D signals interfere with a constant-phase plane wave as reference 2D signal. Analyzing the visibility contrast of this interference image can give a relative phase map of the exciton-polaritons. In combination with the above-mentioned Fourier optics techniques, the interference measurement can be performed both in real and reciprocal

Table 5.1 Exciton-polariton toolbox

	Real space	Momentum space
Density	$n(\boldsymbol{r}; E; t)$	$n(\boldsymbol{k}; E; t)$
Phase	$\phi(\boldsymbol{r}; E; t)$	$\phi(\boldsymbol{k}; E; t)$
First order correlation	$g^{(1)}(\boldsymbol{r}_1, \boldsymbol{r}_2; E; t)$	$g^{(1)}(\boldsymbol{k}_1, \boldsymbol{k}_2; E; t)$
intensity correlation	$g^{(2)}(\boldsymbol{r}_1, \boldsymbol{r}_2; E; t)$	$g^{(2)}(\boldsymbol{k}_1, \boldsymbol{k}_2; E; t)$

spaces. Furthermore, time-resolved and energy-resolved interferograms are also possible together with a spectrometer and a streak camera.

A time-integrated phase map at two different real space coordinates reveals the spatial coherence function $g^{(1)}(\boldsymbol{r}_1; \boldsymbol{r}_2)$, with which we measure the off-diagonal long-range order, a crucial concept of exciton-polariton condensation [57, 60, 85, 86]. The temporal coherence function $g^{(1)}(t_1, t_2)$ can be measured using the streak camera, which tells us the dephasing time of the condensates.

Hanbury Brown-Twiss Setup

A historical Hanbury Brown-Twiss setup [87] consists of two photon detectors (Fig. 5.6d), which record the number of arrival photons. The data of these detectors are analyzed for the second-order coherence functions. Zero-time delay $g^2(0)$ tells us statistical attribute of exciton-polariton condensates. As a well-known fact for a coherent state $g^2(0)$ equals to 1, while it is 2 for a thermal state. Compared to purely coherent or thermal states, $g^2(0)$ of the exciton-polariton condensates is found to be between 1 and 2 [56, 88, 89]; namely exciton-polaritons may not be in a purely coherent nor thermal state. The deviation from $g^2(0) = 1$ in exciton-polaritons would come from polariton-polariton repulsive interactions.

Overall, Table 5.1 summarizes the exciton-polariton toolbox to obtain quantitative information of both static and dynamic variables, energy and time in real and momentum spaces.

5.3 Software of Exciton-Polariton Analog Quantum Simulators

We have discussed that how the exciton-polariton AQS can be physically realized in Sect. 5.2, reviewing the basics of exciton-polaritons and their condensation properties followed by several ways to engineer exciton-polariton crystals. Then, a collection of measurement setups was explained to assess the optical properties of exciton-polaritons. In this section, we describe applications of the exciton-polariton AQS, what physical problems can be studied within single-particle and many-particle physics.

5.3.1 Single-Particle Physics

When we examine a system consisting of many particles in a crystal structure, the first task to know is the band structure of the system, which reflects the topology of the crystal lattices and closely links to the physical properties of the system. It is incredibly difficult to calculate the exact band structures of such a system including all particles as well as their degrees of freedom. Instead of giving up this formidable task, physicists have come up with clever approximations and developed insightful numerical techniques, which illuminate principal nature of the system. A mean-field approximation has been very useful and successful, which reduces the many-body problem into a single-particle physics. It is truly amazing that such a simple single-particle description can faithfully represent the crucial electrical and optical properties of semiconductors assuming a dielectric constant and an effective mass of constituent particles. Learning from this invaluable lesson, we first evaluate the band structures of exciton-polaritons in artificially patterned lattices with four different 2D geometries.

There are salient phenomena in solid-state materials, which root from the orbital nature of electrons, especially high-orbital electrons with *p*-, *d*- and *f*-wave spatial symmetry [29]. Unlike *s*-orbitals, high-orbital wavefunctions share energy degeneracy and exhibit anisotropic distributions in space. Interplaying with spin and charge degrees of freedom, the orbital nature of electrons is responsible for famous phenomena: metal-insulator transition [90], colossal magnetoresistance [29, 91], and the newly discovered iron-pnictide superconductors [92, 93]. Several schemes were proposed to elucidate orbital physics in bosonic atom-lattice AQS [94, 95], and orbital states of coherent boson gases were selectively prepared by population transfer from the ground states [96–99].

We are also captivated by a few fascinating features which fall within single-particle physics. A charged particle in a magnetic field would lead to a quantum Hall effect, resulting from time-reversal symmetry breaking [100]. This physics can be studied in exciton-polariton AQS because the QW exciton is subject to the Zeeman splitting in magnetic fields. Since rotating condensates and the quantum Hall effect are isomorphic in mathematical models [101], exciton-polariton condensates can exhibit quantum Hall physics by rapid rotation. Furthermore, spin-orbital motion of QW excitons can be understood within a single-particle description.

5.3.2 Many-Particle Physics

Despite the success of single-particle approximations, there are a large number of problems in nature beyond single-particle physics. Knowing that materials are composed of many atoms, it is impossible to neglect the many-particle physics completely. The simplest example is a two-body interaction, such as a long-ranged Coulomb interaction between two charged particles, which is prevalent in condensed matter systems. The earlier example, the Hubbard Hamiltonian is widely used to describe properties of transition metal oxides and high-temperature superconductors [2]. At the extremes of the ratio of the two energy terms, the system behaves completely different, either metallic for the stronger kinetic energy case or insulating for the strong interaction case [2].

Motivated by the seminal work in the atom-lattice system [14], Byrnes and colleagues derived the Bose-Hubbard Hamiltonian in exciton-polariton condensates under a 2D periodic potential, identifying a region in phase space to observe superfulid-Mott transitions in exciton-polaritons [26]. Here the on-site interaction term is calculated from exciton-exciton exchange interactions, and the interesting phase transition in 2D is predicted to occur around $U/t \sim 23$, requiring a strong lattice potential of about 6 meV and a small lattice constant on the order of 0.5 μm. A current challenge to demonstrate this proposed phase transition is to generate a sub-micron trap size to increase the interaction term. Unfortunately, aforementioned methods discussed in Sect. 5.2.3 are not able to fulfill this stringent requirement. The thin-metal-film technique can lithographically create a trap size as small as 50 nm; however, its potential is too weak, less than 1 meV. If we are able to realize this phase transition, we open a new world such that the Mott-insulating phase would be a basis for generating single-photon arrays on a macroscopic scale [27], as an initial step towards universal quantum computation.

Terças et al. proposed a scheme of applying a gauge field to a propagating exciton-polariton condensate from the transverse electric and transverse magnetic modes of the cavity and their energy splitting [102]. It is a timely research direction to address spin-orbital interactions as a driving force to search for new quantum matter including topological insulators. Recent progress in this direction of research was reported to observe similar insulating states in photonic systems [103, 104]. The exciton-polariton AQS is an appropriate platform to investigate these phenomena, and we envision that exciton-exciton interactions would go beyond photonic insulating states, which would be much closer to real material dynamics.

One of the challenging problems is dynamics in open environments or a driven-dissipative situation. The non-equilibrium nature of exciton-polariton systems, unavoidable loss and constant replenishment to compensate loss, may be advantageous to investigate difficult open system dynamics. At present, concrete ideas to address this problem in the exciton-polariton AQS are yet to be developed; however, it is worthwhile defining reachable problems which the exciton-polariton AQS can contribute uniquely well.

5.4 Exciton-Polaritons in Two-Dimensional Lattices

We launched the research project to build an exciton-polariton AQS in 2006, impressed by the atomic AQS in 2002. Section 5.4 updates the current status of our exciton-polariton AQS based on GaAs semiconductors, and we present experimental results in exciton-polariton-lattice systems.

5.4.1 Experiment

5.4.1.1 Device

Our wafer under study was grown by molecular beam epitaxy, which has a superb controllability and fidelity of the layer thickness down to atomic length-scales. An AlAs $\lambda/2$-cavity is capped with DBRs by an alternating layer of $Ga_{0.8}Al_{0.2}As$ and AlAs. The 16 and 20 pairs of top and bottom DBRs form a planar Fabry-Perot cavity with a quality factor $Q \sim 3000$, resulting in a cavity photon lifetime of around 2 ps.

At the three antinodes of the EM field in the cavity, a stack of four 7-nm-thick GaAs QWs separated by a 3 nm-thick AlAs barrir are inserted respectively for the purpose of diluting the exciton density per each QW. The whole structure is designed at $\lambda \sim 770$ nm, close to the emission wavelength of the 7-nm-thick GaAs QW exciton. The vacuum Rabi splitting of the 12 GaAs QWs is $\sim$15 meV, which was experimentally confirmed from position-dependent reflectance spectra. The wafer also has spatial detunings, which can vary between -15 and 15 meV. Most of the experiments were performed with devices at near zero or red detuning areas of the wafer (-3 to -5 meV).

5.4.1.2 Lattices

We prepared exciton-polariton 2D lattice devices with the previously grown GaAs wafer. Three basic (square, triangular and honeycomb) and one complex (kagome) lattices were designed and patterned by electron-beam lithography. A 23/3 nm-Ti/Au think film was deposited as a final step of the fabrication. All of these semiconductor processing steps are considered standard and simple. The center-to-center distance of the nearest neighbor sites are fixed between 2 and 20 μm. Figure 5.7a presents a photo of three basic lattices in one device. The photon potential depth ranges between 200–400 μeV, consequently the actual potential strengths for the LPs are around 100–200 μeV at different detuning positions [78]. The kinetic energy at the boundaries of the first Brillouin zone (BZ) is around 0.4–1 meV for lattice constants 2–4 μm. Comparing the potential energy with the kinetic energy, these potentials only weakly perturb the exciton-polaritons.

Owing to weak lattice potentials, justifiably, the band structures of given lattices are calculated using a single-particle plane-wave basis. The complex bands are often displayed along high symmetry points, following the point-group theory. Figure 5.7b, c are BZs of square and honeycomb lattices with high symmetry points denoted as Γ, X, M and $K(K')$. In particular, these points are closely related to the degree of rotational symmetry in the lattice geometries. Γ and M in the square lattice satisfy four-fold rotational symmetry, whereas $X(Y)$ has two-fold rotational degeneracies. A weak periodic potential lifts band degeneracies at high symmetry points, and momentum valleys are protected by the gap energy, comparable to the periodic potential strength on the order of 100 μeV.

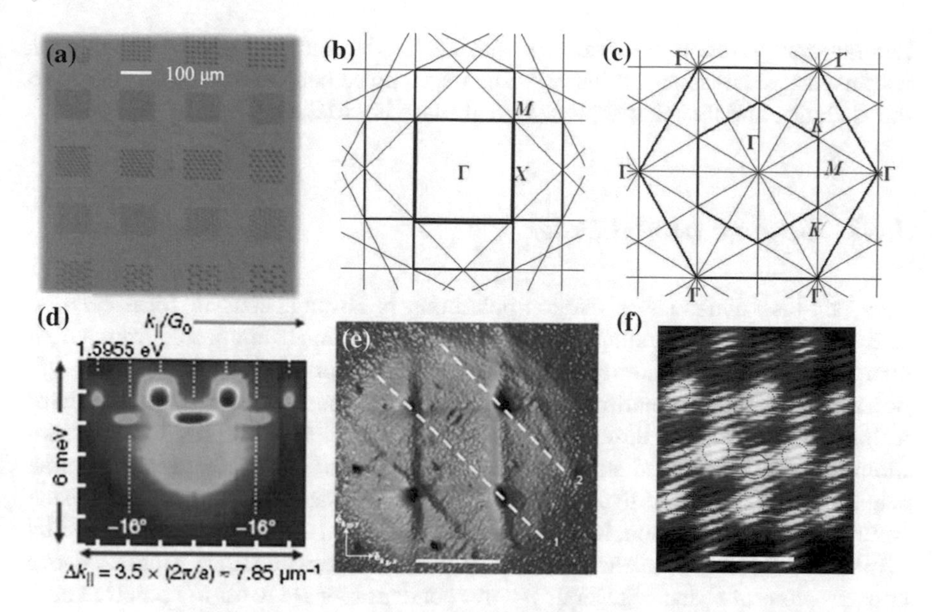

Fig. 5.7 **a** A photograph of the two-dimensional lattices with square-, triangular- and honeycomb-geometries patterned by photolithography and deposited with a thin Ti/Au metal-film. Brillouin zones of the square (**b**) and the honeycomb (**c**) lattices denoted by high symmetry Γ, M, X, K and K' points according to the rotational symmetry of the geometries. **d** In-phased s-orbital (lower energy) and anti-phased coherent p-orbital polaritons in a one-dimensional array [64]. **e** A d_{xy}-orbital polariton condensation occurring at high-symmetry four M points in the first Brillouin zone (BZ) of the square lattice taken from Ref. [79]. The *white bar* is the first BZ size, $2\pi/a$ with a lattice constant $a = 4$ μm. **f** The real-space interferogram of the f-orbital condensation at non-zero Γ points in the honeycomb lattice. The *white bar* is a scale bar of 4 μm

5.4.1.3 Experimental Setup

A low-temperature micro-photoluminescence setup was built as a basic characterization tool to study exciton-polaritons in lattices. Since the GaAs QW exciton binding energy is around 10 meV, we kept the device temperatures below 10 K to avoid thermal dissociation of the excitons. With a replaceable lens, one can select the measurement domain to be either real or reciprocal space. For this sample, we fixed the pump laser energy to 1.61543 eV (~767.5 nm) around 6 meV above the LP ground state energy in the linear regime. 3-ps-long laser pulses enter the sample obliquely at ~60° (~7.4×10^6 m^{-1}) at a 76 MHz repetition rate.

By changing pump power, one can control the density of injected exciton-polaritons in QWs. Using this basic setup, polariton population distributions were mapped in real and momentum spaces. A nonlinear density increase in population marks the quantum degeneracy threshold. Lattice potentials modify a single-particle LP dispersion to band structures of LPs. Complex order parameters of condensates in lattices are constructed by interferometry, which can help us to determine relative phases in real space. A homodyne detection scheme is implemented in a Mach-

Zehnder configuration. A reference signal at a fixed frequency and a wavenumber is combined against signals of interest. The interference contrast is directly related to the visibility, and the relative phase in real space is extracted.

5.4.2 Bosonic Orbital Order

The gain-loss dynamics in exciton-polaritons is advantageous to form condensates beyond zero momentum. Thermal exciton-polaritons are injected from a high energy and large momentum particle reservoir, and undergo relaxation down through polariton-phonon and polariton-polariton scattering processes. Since polaritons have a finite lifetime to leak through the cavity, condensation is stabilized at non-zero momentum states in band structures resulting from the balance between two time scales: the relaxation time to the lower energy states versus the decay time through the cavity. For this very reason, we have observed polariton condensates of high-orbital symmetry: d_{xy}-wave condensation at M points in the square lattice, singlet f-wave condensations at Γ and degenerate p-wave condensations at K (or K') points in both triangle and honeycomb lattices. The degeneracy and anisotropic distribution of high orbital symmetry have been identified with the complex order parameter constructed from interferograms.

The meaning of high-orbital boson order can be found in the following context. We know that high-orbital electrons play a crucial role in special electrical, chemical and mechanical properties of solid state materials. s-orbital fundamental bosons cannot make any complementary picture to such high-orbital fermionic physics. However, the complex-valued high-orbital bosons can be as important as those of fermions. Along this same spirit, we have observed the anti-phased p-orbital condensates in one-dimensional arrays from the diffraction peaks in energy- and angle-resolved spectroscopy shown in Fig. 5.7d [64]. In 2D lattices, meta-stable momentum valleys are available at high symmetry points. At appropriate pump powers, coherent exciton-polaritons are accumulated at non-zero momentum meta-stable bands. d_{xy}-wave symmetry is favorable at M points of the square lattice, as a signature of narrow coherent diffraction peaks captured in the momentum space image (Fig. 5.7e) [79]. In hexagonal geometries, $4f_{y^3-3x^2y}$-like orbital symmetry in space is identified from an interferogram, where alternating phase shifts by π at six lobes in real space is unequivocally detected as shown in Fig. 5.7f.

Degenerate p-orbital condensates are stabilized at the vertices K and K' points in the first hexagonal BZ of triangular-geometry lattices. This observation of degenerate condensates raises a fundamental conceptual question of condensation regarding spontaneous symmetry breaking, condensate fragmentation and the statistical mixture of degenerate condensates. In order to address this important question, we have measured $g^{(2)}(t = 0)$ in momentum space, which is further described in the following subsection.

5.4.3 Degenerate High-Orbital Condensates

Six vertices of the first BZ in the honeycomb lattice are grouped into inequivalent K and K' points, which are connected with reciprocal lattice vectors. K and K' points hold reflection or inversion symmetry, namely, they are mirror imaged. Three degenerate energy states at K and K' points are split into a high energy singlet and lower energy doublets. The upper singlet state with p-orbital symmetry is interesting as a vortex-antivortex state from the linear combination of all three K points, which are rotated by $2\pi/3$ [82]. We call vortex to be a topological defect, whose density is zero at the core and whose phase continuously changes by 2π. An antivortex has the opposite phase-rotation direction. Both vortices and antivortices are located at the trap sites with zero density. The vortex-antivortex order is exactly opposite for K (Fig. 5.8a) and K' (Fig. 5.8a) points revealing the reflection symmetry, confirmed via a modified Mach-Zehnder interferometer.

Next we study the dynamics of the degenerate condensates at K and K' points by measuring intensity correlation functions in momentum space. $g^{(2)}(K, K; \tau = 0)$ and $g^{(2)}(K', K'; \tau = 0)$ are the autocorrelation functions at the same K and K' points with zero time delay $\tau = 0$. We also detect cross-correlation functions $g^{(2)}(K, K'; \tau = 0)$ between K and K' signals. Experimental results of pump-power dependent correlation functions are presented in Fig. 5.8c (left), and theoretical simulation results

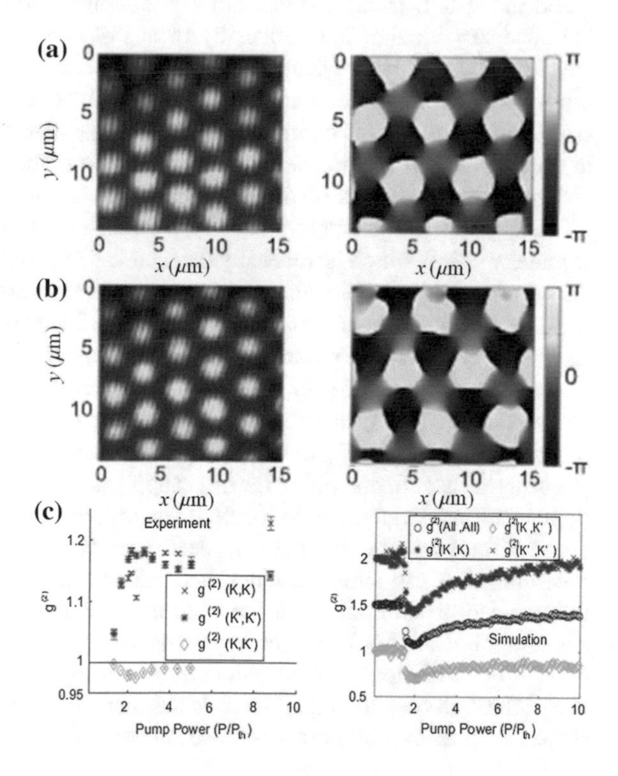

Fig. 5.8 p-orbital degenerate vortex-antivortex condensates in a honeycomb lattice. Raw interferograms (*left*) and extracted phase maps (*right*) to identify a vortex-antivortex single-particle state at K (**b**) and mirror-symmetric K' (**c**) points. **c**, Second-order intensity-intensity auto- and cross-correlation functions of two degenerate p-orbital condensates: experimental data (*left*) and simulation (*right*)

based on complex-number Langevin equations are displayed in Fig. 5.8c (right). A value of less than 1 for $g^{(2)}(K, K'; \tau = 0)$ indicates anti-correlation between the two intensities, which we interpret as the mode competition between K and K' condensates stochastically. At first, one state would be selected randomly between two possibilities, then that chosen path would be more favorable due to subsequent stimulated scatterings. At a given pump power, the total particle number is fixed such that a larger intensity in one state links to a smaller intensity in the other, yielding the anticorrelation. At present, we cannot answer interesting physical questions such as spontaneous symmetry breaking or fragmentation of degenerate condensations, primarily because our measurements are limited to be time averaged over many pulses. A single shot measurement will lead us to reach a conclusive answer, whose setup may be available in the near future.

5.5 Outlook

We have given an overview of the exciton-polariton-lattice system for the application of quantum simulators, specially designed functional quantum machines. Physical elements of the exciton-polariton AQS hardware were concretely identified, and the first generation exciton-polariton AQS has already been experimentally prepared and tested. The initial tasks of the exciton-polariton AQS are to directly map the band structures of 2D lattices by means of angle-resolved photoluminescence spectroscopy. We have identified meta-stable exciton-polariton condensates with high-orbital symmetries at non-zero momentum states. The selectivity of condensate orbital symmetry is controlled by the particle density with pump laser intensity in order to balance lifetime and a relaxation time in the polariton band structures.

The main limitations of the first generation exciton-polariton AQS are a weak potential depth and a micron-sized trap unit, yielding a small gap energy at high symmetry points which is not easily detectable due to a broader spectral linewidth of the generated signals. Hence, how to produce stronger potentials is an imminent goal to be accomplished in order to run interesting software programs. Next, in a micron-sized site, two-particle polariton-polariton interaction is still too weak; estimated value is around tens of μeV for GaAs QW excitons. It is a primary hindrance to study many-body physics. Unless overcoming this challenge, the initial exciton-polariton AQS can solely investigate problems within single-particle physics, such as quantum Hall effects. Either the direct application of magnetic fields, rapid rotation or generation of a gauge field would be incorporated in this direction of researches.

As a final remark, we actively seek a class of unique problems the exciton-polariton AQS can simulate much better than other AQS platforms. Appreciating the non-equilibrium nature of the exciton-polariton system, we anticipate that the exciton-polariton AQS may be a promising testbed to examine dynamical problems in open-dissipative environments. A starting point is to set up a testable toy model through careful and systematic assessment. Witnessing the incredible development of classical computers, we also put our faith in the continuous advancement of quantum simulators in all platforms, which would deepen our knowledge of quantum many-body problems in various areas and would provide crucial insights and novel methods for quantum engineering and technologies.

Acknowledgements We acknowledge Navy/SPAWAR Grant N66001-09-1-2024, the Japan Society for the Promotion of Science (JSPS) through its "Funding Program for World-Leading Innovative R&D on Science and Technology (FIRST Program)". We deeply thank our collaborators: Dr. K. Kusudo and Dr. N. Masumoto for experimental measurement and device fabrication; Prof. A. Forchel, Dr. S. Höfling, Dr. A. Löffler for providing the wafers; Prof. T. Fujisawa, Dr. N. Kumada for supporting the device fabrication; Prof. T. Byrnes, Prof. C. Wu, Dr. Z. Cai for theoretical discussions. N.Y.K thank Dr. C. Langrock for critical reading of the manuscript.

References

1. R. Feynman, Simulating physics with computers. Int. J. Theor. Phys. **21**, 467–488 (1982)
2. G.D. Mahan, *Many-Particle Physics* (Kluwer Academic/Plenum Publishers, New York, 1981)
3. S. Lloyd, Universal quantum simulators. Science **273**, 1073–1078 (1996)
4. I. Buluta, F. Nori, Quantum simulators. Science **326**, 108–111 (2009)
5. J.I. Cirac, P. Zoller, Golas and opportunites in quantum simulation. Nat. Phys. **8**, 264–266 (2012)
6. I.M. Georgescu, S. Ashhab, F. Nori, Quantum simulation. Rev. Mod. Phys. **86**, 153–195 (2014)
7. T.D. Ladd, F. Jelezko, R. Laflamme, Y. Nakamura, C. Monroe, J.L. O'Brien, Quantum computers. Nature **464**, 45–53 (2010)
8. I. Bloch, J. Dalibard, S. Nascimbéne, Quantum simulations with ultracold quantum gases. Nat. Phys. **8**, 267–276 (2012)
9. R. Blatt, C.F. Roos, Quantum simulations with trapped ions. Nat. Phys. **8**, 277–284 (2012)
10. A. Aspuru-Guzik, P. Walther, Photonic quantum simulators. Nat. Phys. **8**, 285–291 (2012)
11. A.A. Houck, H. Türeci, J. Koch, On-chip quantum simulation with superconducting circuits. Nat. Phys. **8**, 292–299 (2012)
12. D. Lu, B. Xu, N. Xu, Z. Li, H. Chen, X. Peng, R. Xu, J. Du, Quantum chemistry simulation on quantum computers: theories and experiments. Phys. Chem. Chem. Phys. **14**, 9411–9420 (2012)
13. P. Hauke, F.M. Cucchietti, L. Tagliacozzo, I. Deutsch, M. Lewenstein, Can one trust quantum simulators? Rep. Prog. Phys. **75**, 082401 (2012)
14. M. Greiner, O. Mandel, T. Esslinger, T. Hänsch, I. Bloch, Quantum phase transition from a superfulid to a Mott insulator in a gas of ultracold atoms. Nature **415**, 39–44 (2002)
15. T. Esslinger, Fermi-Hubbard physics with atoms in an optical lattice. Annu. Rev. Condens. Matter Phys. **1**, 129–152 (2010)
16. K. Kim et al., Quantum simulation of the transverse Ising model. New J. Phys. **13**, 105003 (2011)
17. J.W. Britton, B.C. Sawyer, A.C. Keith, C.-C. Joseph Wang, J.K. Freericks, H. Uys, M.J. Biercuk, J.J. Bollinger, Engineered two-dimensional Ising interactions in a trapped-ion quantum simulator with hundreds of spins. Nature **484**, 489–492 (2012)
18. T. Byrnes, P. Recher, N.Y. Kim, S. Utsunomiya, Y. Yamamoto, Quantum simulator for the Hubbard model with long-range Coulomb interactions using surface acoustic waves. Phys. Rev. Lett. **99**, 016405 (2006)
19. T. Byrnes, N.Y. Kim, K. Kusudo, Y. Yamamoto, Quantum simulation of Fermi-Hubbard models in semiconductor quantum dot arrays. Phys. Rev. B **78**, 075320 (2007)

20. G. De Simoni, A. Singha, M. Gibertini, B. Karmakar, M. Polini, V. Piazza, L.N. Pfeiffer, K.W. West, F. Beltram, V. Pellegrini, Delocalized-localized transition in a semiconductor two-dimensional honeycomb lattice. Appl. Phys. Lett. **97**, 132113 (2010)
21. A. Singha, M. Gibertini, B. Karmakar, S. Yuan, M. Polini, G. Vignale, M.I. Katsnelson, A. Pinczuk, L.N. Pfeiffer, K.W. West, V. Pellegrini, Two-dimensional Mott-Hubbard electrons in an artificial honeycomb lattice. Science **332**, 1176–1179 (2011)
22. J. Koch, A.A. Houck, K. Le Hur, S.M. Girvin, Time-reversal-symmetry breaking in circuit-QED-based photon lattices. Phys. Rev. A **82**, 043811 (2010)
23. D.G. Angelakis, M.F. Santos, S. Bose, Photon-blockade-induced Mott transitions and *XY* spin models in coupled cavity arrays. Phys. Rev. A **76**, 031805(R) (2007)
24. M.J. Hartmann, F.G.S.L. Brandão, M.B. Plenio, Strongly interacting polaritons in coupled arrays of cavities. Nat. Phys. **2**, 849–855 (2006)
25. A.D. Greentree, C. Tahan, J.H. Cole, L.C.L. Hollenberg, Quantum phase transitions of light. Nat. Phys. **2**, 856–861 (2006)
26. T. Byrnes, P. Recher, Y. Yamamoto, Mott transitions of excitons polaritons and indirect excitons in a periodic potential. Phys. Rev. B **81**, 205312 (2010)
27. N. Na, Y. Yamamoto, Massive parallel generation of indistinguishable single photons iva the polaritonic superfulid to Mott-insulator quantum phase transition. New J. Phys. **12**, 123001 (2010)
28. J. Hubbard, Electron correlations in narrow energy bands. Proc. R. Soc. Lond. A **276**, 238–257 (1963)
29. Y. Tokura, N. Nagaosa, Orbital physics in transition-metal oxides. Science **288**, 462–468 (2000)
30. S. Sachdev, Quantum magnetism and criticality. Nat. Phys. **4**, 173–185 (2008)
31. S. Sachdev, B. Keimer, Quantum criticality. Phys. Today **64**(2), 29–35 (2011)
32. D. Jaksch, P. Zoller, Creation of effective magnetic fields in optical lattices: the Hofstadter butterfly for cold neutral atoms. New J. Phys. **5**, 56 (2003)
33. Y.-J. Lin, R.L. Compton, K. Jiménez-García, J.V. Porto, I.B. Spielman, Synthetic magnetic fields for ultracold neutral atoms. Nature **462**, 628–632 (2009)
34. V. Glitski, I.B. Spielman, Spin-orbit coupling in quantum gases. Nature **494**, 49–54 (2013)
35. A. Aspuru-Guzik, A.D. Dutoi, P.J. Love, M. Head-Gordon, Simulated quantum computation of molecular energies. Science **309**, 1704–1707 (2005)
36. I.S. Kassal, S.P. Jordan, P.J. Love, M. Mohseni, A. Aspuru-Guzik, Quantum algorithms for the simulation of chemical dynamics. Proc. Nat. Acad. Sci. **105**, 18681–18686 (2008)
37. AYu. Smirnov, S. Savel'ev, L.G. Mourokh, F. Nori, Modelling chemical reactions using semiconductor quantum dots. Eur. Phys. Lett. **80**, 67008 (2007)
38. S. Giovanazzi, Hawking radiation in sonic black holes. Phys. Rev. Lett. **94**, 061302 (2005)
39. D. Gerace, I. Carusotto, Analog Hawking radiation from an acoustic black hole in a flowing polariton superfluid (2012). arXiv:1206.4276
40. R. Gerritsma, G. Kirchmair, F. Zähringer, E. Solano, R. Blatt, C.F. Roos, Quantum simulation of the Dirac equation. Nature **463**, 68–71 (2009)
41. L. Tarruell, D. Greif, T. Uehlinger, G. Jotzu, T. Esslinger, Creating, moving and merging Dirac points with Fermi gas in a tunable honeycomb lattice. Nature **483**, 302–305 (2012)
42. K.K. Gomes, W. Mar, W. Ko, F. Guinea, H.C. Manoharan, Designer Dirac fermions and topological phases in molecular graphene. Nature **483**, 306–310 (2012)
43. C. Weisbuch, M. Nishioka, A. Ishikawa, Y. Arakawa, Observation of the coupled exciton-photon mode splitting in a semiconductor quantum microcvity. Phys. Rev. Lett. **69**, 3314–3317 (1992)
44. A. Kavokin, J. Baumberg, G. Malpuech, F.P. Laussy, *Microcavities* (Clarendon Press, Oxford, 2006)
45. D. Snoke, P. Littlewood, Polariton condensates. Phys. Today **63**(8), 42–47 (2010)
46. H. Deng, H. Haug, Y. Yamamoto, Exciton-polariton Bose-Einstein condensation. Rev. Mod. Phys. **82**, 1490–1537 (2010)
47. P.Y. Yu, M. Cardona, *Fundamentals of Semciodnuctors*. Springer (1996)

48. F. Tassone, C. Piermarocchi, V. Savona, A. Quattropani, P. Schwendimann, Photoluminescence decay times in strong-coupling semiocnductor microcavities. Phys. Rev. B **53**, R7642–7645 (1996)
49. E. Hanamura, H. Haug, Condensation effects of excitons. Phys. Rep. **33C**, 209–284 (1997)
50. A. Griffin, D.W. Snoke, S. Stringari, *Bose-Einstein Condensation* (Cambridge University Press, Cambridge, 1995)
51. F. Tassone, C. Piermarocchi, V. Savona, A. Quattropani, P. Schwendimann, Bottleneck effects in the relaxation and photoluminescence of microcavity polaritons. Phys. Rev. B **56**, 7554–7563 (1997)
52. F. Tassone, Y. Yamamoto, Exciton-exciton scattering dynamics in a semiconductor microcavity and stimulated scattering into polaritons. Phys. Rev. B **59**, 10830–10842 (1999)
53. D. Porras, C. Ciuti, J.J. Baumberg, C. Tejedor, Polariton dynamics and Bose-Einstein condesnation in semiconductor microcavities. Phys. Rev. B **66**, 085304 (2002)
54. C. Ciuti, V. Savona, C. Piermarocchi, A. Quattropani, P. Schwendimann, Role of the exchange of carriers in elastic exciton-exciton scattering in quantum wells. Phys. Rev. B **58**, 7926–7933 (1998)
55. A. Imamoglu, R.J. Ram, S. Pau, Y. Yamamoto, Nonequilibrium condensates and lasers without inversion: exciton-polariton lasers. Phys. Rev. A **53**, 4250–4253 (1996)
56. H. Deng, G. Weihs, C. Santori, J. Bloch, Y. Yamamoto, Condensation of semiconductor microcavity exciton polaritons. Science **298**, 199–202 (2002)
57. J. Kapsrzak et al., Bose-Einstein condensation of exciton polaritons. Nature **443**, 409–414 (2006)
58. H. Deng, D. Press, S. Götzinger, G.S. Solomon, R. Hey, K.H. Ploog, Y. Yamamoto, Quantum degenerate exciton-polaritons in thermal equilibrium. Phys. Rev. Lett. **97**, 146402 (2006)
59. R.B. Balili, V. Hartwell, D. Snoke, L. Pfeiffer, K. West, Bose-Einstein condensation of microcavity polaritons in a trap. Science **316**, 1007–1010 (2007)
60. H. Deng, G.S. Solomon, R. Hey, K.H. Ploog, Y. Yamamoto, Spatial coherence of a polariton condensate. Phys. Rev. Lett. **99**, 126403 (2007)
61. S. Christopoulos et al., Room-temperature polariton lasing in semiconductor microcavities. Phys. Rev. Lett. **98**, 126405 (2007)
62. S. Kéna-Cohen, S.R. Forrest, Room-temperature polariton lasing in an organic single-crystal microcavity. Nat. Photon. **4**, 371–375 (2010)
63. J.D. Plumhof, T. Stöferle, L. Mai, U. Scherf, R. Mahrt, Room-temperature Bose-Eistein condensation of cavity exciton-polaritons in a polymer. Nat. Mater. **13**, 247–252 (2014)
64. C.W. Lai et al., Coherent zero-state and π-state in an exciton-poalriton condensate array. Nature **450**, 529–533 (2007)
65. R.B. Balili, D.W. Snoke, L. Pfeiffer, K. West, Actively tuned and spatially trapped polaritons. Appl. Phys. Lett. **88**, 031110 (2006)
66. D.A.B. Miller, D.S. Chemla, T.C. Damen, A.C. Gossard, W. Wiegmann, T.H. Wood, C.A. Burrus, Band-edge electroabsorption in quantum well structures: the quantum-confined stark effect. Phys. Rev. Lett. **53**, 2173–2176 (1984)
67. M.M. De Lima, M. van der Poel Jr., P.V. Santos, J.M. Hvam, Phonon-induced polariton superlattices. Phys. Rev. Lett. **97**, 045501 (2006)
68. E.A. Cerda-Méndez et al., Polariton condensation in dynamic acoustic lattices. Phys. Rev. Lett. **105**, 116402 (2010)
69. E.A. Cerda-Méndez, D. Sarkar, D.N. Krizhanovskii, S.S. Gavrilov, K. Biermann, M.S. Skolnick, P.V. Santos, Exciton-polariton gap solitons in two-dimensional lattices. Phys. Rev. Lett. **111**, 146401 (2013)
70. O. El Daïf et al., Polariton quantum boxes in semiconductor microcvities. Appl. Phys. Lett. **88**, 061105 (2006)
71. G. Nardin, Y. Léger, B. Pietka, F. Morier-Genoud, B. Deveaud-Plédran, Phase-resolved imaging of confined exciton-poalriton wave functions in elliptical traps. Phys. Rev. B **82**, 045304 (2010)

72. J. Bloch, F. Boeuf, J.M. Gérard, B. Legrand, J.Y. Marzin, R. Planel, V. Thierry-Mieg, E. Costard, Strong and weak coupling regime in pillar semiconductor microcavities. Phys. E **2**, 915 (1998)
73. M. Galbiati, L. Ferrier, D.D. Solynshkov, D. Tanese, E. Wertz, P. Senellart, I. Sagnes, A. Lemaître, E. Galopin, G. Malpuech, J. Bloch, Polariton condensation in photonic molecules. Phys. Rev. Lett. **108**, 126403 (2012)
74. E. Wertz et al., Spontaneous formation and optical manipulation of extended polariton condensates. Nat. Phys. **6**, 860–864 (2010)
75. T. Jacqmin et al., Direct observation of Dirac cones and a flatband in a honeycomb lattice for polartions. Phys. Rev. Lett. **112**, 116402 (2014)
76. G. Roumpos, W.H. Nitsche, S. Höfling, A. Forchel, Y. Yamamoto, Gain-induced trapping of microcavity exciton polariton condensates. Phys. Rev. Lett. **104**, 126403 (2010)
77. G. Tosi, G. Christmann, N.G. Berloff, P. Tsotsis, T. Gao, Z. Hatzopoulos, P.G. Lagoudakis, J.J. Baumberg, Geometrically locked vortex lattices in semiconductor quantum fluids. Nat. Commun. **3**, 1243 (2012)
78. N.Y. Kim et al., GaAs microcavity exciton-polaritons in a trap. Phys. Rev. Lett. **105**, 116402 (2010)
79. N.Y. Kim et al., Dynamical *d*-wave condensation of exciton-polaritons in a two-dimensional square-lattice potential. Nat. Phys. **7**, 681–686 (2011)
80. N. Masumoto, N.Y. Kim, T. Byrnes, K. Kenichiro, A. Löffler, S. Höfling, A. Forchel, Y. Yamamoto, Exciton-poalriton condensates with flat banda in a two-dimensional kagome lattice. New J. Phys. **14**, 065002 (2012)
81. N.Y. Kim, K. Kenichiro, A. Löffler, S. Höfling, A. Forchel, Y. Yamamoto, Exciton-poalriton condensates near the Dirac point in a triangular lattice. New J. Phys. **15**, 035032 (2013)
82. K. Kusudo, N.Y. Kim, A. Löffler, S. Höfling, A. Forchel, Y. Yamamoto, Stochastic formation of polariton condensates in two degenerate orbital states. Phys. Rev. B **87**, 214503 (2013)
83. E. Hecht, *Optics*, Addison-Wesley (2001)
84. R.J. Glauber, The quantum theory of optical coherence. Phys. Rev. **130**, 2529–2539 (1963)
85. G. Roumpos, M. Lohse, W.H. Nitsche, J. Keeling, M.H. Szymanska, P.B. Littlewood, A. Löffler, S. Höfling, L. Worschech, A. Forchel, Y. Yamamoto, Power-law decay of the spatial correlation function in exciton-polariton condensates. Proc. Nat. Acad. Sci. **109**, 6467–6472 (2012)
86. W.H. Nitsche, N.Y. Kim, G. Roumpos, C. Schneider, M. Kamp, S. Höfling, A. Forchel, Y. Yamamoto, Algebraic order and the Berezinskii-Kosterlitz-Thouless transition in an exciton-polariton gas (2014). arXiv:1401.0756
87. B.R. Hanburry, R.Q. Twiss, The test of a new type of stella interferometer on Sirrus. Nature **177**, 27–29 (1956)
88. T. Horikiri, P. Schwendimann, A. Quattropani, S. Höfling, A. Forchel, Y. Yamamoto, Higher order coherence of exciton-polariton condensates. Phys. Rev. B **81**, 033307 (2010)
89. M. Aßmann et al., From polariton condensates to highly photonic quantum degenerate states of bosonic matter. Proc. Nat. Acad. Sci. **108**, 1804–1809 (2011)
90. M. Imada, A. Fujimori, Y. Tokura, Metal-insulator transition. Rev. Mod. Phys. **70**, 1039–1263 (1998)
91. M.B. Salamon, M. Jaime, The physics of manganites: Structure and transport. Rev. Mod. Phys. **73**, 583–628 (2001)
92. K. Ishida, Y. Nakai, H. Hosono, To what extent iron-pnictide new superconductors have been clarifies: a progress report. J. Phys. Soc. Jpn. **78**, 062001 (2009)
93. I.I. Mazin, J. Schmalian, Pairing symmetry and pairing state in ferropnictides: theoretical overview. Phys. C **469**, 614–627 (2009)
94. A. Isacsson, S.M. Girvin, Multiflavor bosonic Hubbard models in the first excite Bloch band of an optical lattice. Phys. Rev. A **72**, 053604 (2005)
95. L.W. Vincent, C. Wu, Atomic matter of nonzero-momentum Bose-Einstin condensation and orbital current order. Phys. Rev. A **74**, 013607 (2006)

96. T. Müller, S. Fölling, A. Widera, I. Bloch, State separation and dynmics of ultracold atoms in higher lattice orbitals. Phys. Rev. Lett. **99**, 200405 (2007)
97. G. Wirth, M. Ölschläger, A. Hemmercih, Evidence for orbital superfluidity in the *P*-band of a bipartite optical square lattice. Nat. Phys. **7**, 147–153 (2011)
98. M. Ölschläger, G. Wirth, A. Hemmercih, Unconventional superfluid order in the *F* band of a bipartite optical square lattice. Phys. Rev. Lett. **106**, 015302 (2011)
99. P. Soltan-Panahi, D. Lühmann, J. Struck, P. Windpassinger, K. Sengstock, Quantum phase transition to unconventional multi-orbital superfluidity in optical lattices. Nat. Phys. **8**, 71–75 (2012)
100. J.H. Davies, *The Physics of Low-dimensioanl Semiconductors: An Introduction* (Cambridge University Press, Cambridge, 1997)
101. M. Roncaglia, M. Rizzi, J. Dalibard, From rotating atomic rings to quantum Hall states. Sci. Rep. **1**, 43 (2011)
102. H. Terças, H. Flayac, D.D. Solnyshkov, G. Malpuech, Non-abelian gauge fields in photonic cavities. Phys. Rev. Lett. **112**, 066402 (2014)
103. M.C. Rechtsman, J.M. Zeuner, Y. Plotnik, Y. Lumer, D. Podolsky, F. Dreisow, S. Nolte, M. Segev, A. Szameit, Photonic Floquet topological insulators. Nature **496**, 196–200 (2013)
104. A.B. Khanikaev, Mousavi S. Hossein, W.-K. Tse, M. Kargarian, A.H. MacDonald, G. Shvets, Photonic topological insulators. Nat. Mater. **12**, 233–239 (2013)

Chapter 6
Strongly Correlated Photons in Nonlinear Nanophotonic Platforms

D. Gerace, C. Ciuti and I. Carusotto

Abstract Modern nano-fabrication technologies allow to realize photonic propagation and confinement to unprecedented degree of compactness, and very close to lossless conditions. Such figures of merit are inherently driving the possibility to reach a strong enhancement of optical nonlinearities in ordinary semiconductor platforms, which have been mainly used for opto-electronics purposes so far. After reviewing the basic nanophotonic platforms that are used today in integrated quantum photonics, with a focus on photonic crystal cavities and cavity arrays, we will give an overview of recent theoretical descriptions of the strongly correlated photonic concepts in such systems. The focus will be on small-scale systems, compatible with modern nanofabrication capabilities, and on physical quantities of direct experimental access, such as field intensity and second-order correlation function. A few topical cases that will be reviewed include novel quantum photonic devices of increasing system size and complexity, from the quantum optical Josephson interferometer in a three-cavity system, to the out-of-equilibrium phase crossover from delocalized to strongly interacting many-body states in cavity arrays.

6.1 Introduction

Strongly correlated systems have been a benchmark of condensed matter physics for several decades. Traditionally, the definition applies to many-body systems where the mutual interactions between its constituting particles are strong enough to prevent a

D. Gerace (✉)
Dipartimento di Fisica, Università di Pavia, via Bassi 6, 27100 Pavia, Italy
e-mail: dario.gerace@unipv.it

C. Ciuti
Laboratoire Matériaux et Phénomènes Quantiques, Université Paris
Diderot-Paris 7, Bâtiment Condorcet, 10 rue Alice Domon et Léonie Duquet,
75205 Paris Cedex 13, France
e-mail: cristiano.ciuti@univ-paris-diderot.fr

I. Carusotto
INO-CNR BEC Center and Dipartimento di Fisica, Università di Trento,
38123 Povo, Italy
e-mail: carusott@science.unitn.it

D.G. Angelakis (ed.), *Quantum Simulations with Photons and Polaritons*,
Quantum Science and Technology, DOI 10.1007/978-3-319-52025-4_6

separable description of its many-body wavefunction. In a strongly correlated system the magnitude of interactions, such as the Coulomb repulsion between electrons in a solid, is typically sizable as compared to the kinetic energy per particle. Examples belonging to the realm of strongly correlated systems in condensed matter are superconducting materials, where the Coulomb interaction between electrons and ions gives rise to a new many-body ground state responsible for the dissipationless current flow, or diluted magnetic impurities in metals, i.e., the so-called Kondo systems, where a many-body singlet state hybridized with a free electron gas is formed [1]. More recently, nanostructured materials and artificially engineered many particle systems, such as semiconductor quantum dots on one side, or trapped alkali atoms on the other, have allowed the exploration of the rich many-body physics of strongly correlated systems in controllable experimental settings. In mesoscopic transport experiments, a laterally confined region in a two-dimensional electron gas allows to reach the Coulomb blockade regime: when the energy cost to add a single electron to the many-body confined state is larger than $k_B T$ and the quantized energy between single-particle states, single-electron tunneling is blocked [2]. Another example is given by the superfluid to Mott-insulator transition predicted for Bose particles trapped at the minima of a periodic potential, which was experimentally realized with ultra-cold atoms in optical lattices [3, 4], described by the well known Bose-Hubbard model [5]. Depending on the ratio between the hopping energy between neighboring sites, J, and the on-site interaction energy, U, the ground state of the N-particle system at integer filling ranges from a coherently delocalized state describing the superfluid (for $J \gg U$) to a Mott-insulator state (for $J \ll U$), which is a quantum phase transition [6]. Restricting to the one-dimensional geometry, the so-called Tonks-Girardeau gas of impenetrable bosons is a further case of a strongly correlated state, where the many-body wave function is characterized by nodes whenever two particles spatially approach each other. As it was shown by Girardeau [7], a rigorous mapping between the one-dimensional gas of impenetrable bosons and a gas of noninteracting spinless fermions occurs. Again, experimental evidence for such a strongly correlated system was given in a gas of ultra-cold atoms in one-dimensional optical lattices [8, 9] (see also [10] for a recent review).

Following these exciting demonstrations, the analogy between two-body interparticle interactions and photon-photon scattering in strongly nonlinear systems has been consequently triggering an intense theoretical effort to explore the analogies between nonlinear cavity arrays and the physics of strongly correlated systems. The basic building block in this direction is the so-called photon blockade effect, i.e., the photonic analog of the Coulomb blockade. This terminology was first proposed in 1997 [11] to describe the following situation: in the presence of a strong enough optical nonlinearity, photons in a single-mode cavity effectively behave as impenetrable bosonic particles. In the last decade, such an effect has been probed in different cavity quantum electrodynamics settings, from single atoms flying in optical resonators [12], to photonic crystal integrated devices [13, 14]. In fact, the analog of the superfluid-to-Mott insulator quantum phase transition was first proposed in 2006 [15–17], relying on the capability of experimentally realizing multi-cavity arrays where strong photon-photon interactions in each cavity could drive the photonic system from a coherent, superfluid state to a correlated Mott-insulator state of photons.

The observability of this phase transition in realistic devices was first theoretically proposed in a quasi-equilibrium scenario, i.e., in very weakly dissipative systems and neglecting the intrinsic driven-dissipative nature of open photonic systems. The role of losses was later recognized as a crucial one to be taken into account for the effective observability of an analog quantum phase transition in open photonic systems, and simultaneously triggered the motivation to study the interplay between coherent and strongly correlated many-body states in a naturally out-of-equilibrium framework, which is interesting both from theoretical and experimental point of views. In this respect, the next crucial step was to include in the theoretical modeling either photon dissipation or pumping mechanisms, which is an essential element to replenish the photon population and reach a dynamical equilibrium in steady state. The first works addressing and possibly taking advantage of the non-equilibrium nature in the context of strongly correlated photonic systems were in 2009 [18, 19]. These works, some of which were reviewed earlier in this volume and will come across here again from the nanophotonic perspective, essentially pioneered the concept of using photon correlation measurements as a unique probe of the crossover to the strongly correlated regime, even when the interplay between tunneling and interactions occurs in the steady state of a driven-dissipative open system.

In this chapter we explore the potentialities of integrated photonic platforms, namely photonic crystal integrated circuits, to investigate the rich physics of strongly correlated photons. The focus will be on small-scale systems that could be experimentally realized with state-of-the art technology. For this reason, we will first review the main physical properties of nanophotonic platforms based on photonic crystals, where record figures of merit in terms of light propagation and confinement can currently be achieved. In the second part, we will review a few examples of strongly correlated photonic devices, pointing out analogies and differences with respect to the corresponding systems in equilibrium.

6.2 Nanophotonic Platforms: Light Propagation and Confinement

Modern semiconductor technologies allow to manufacture and process different materials with unprecedented degree of control. By applying mature fabrication approaches such as epitaxial growth, high-resolution lithography, and advanced material etching, it is currently possible to realize fully integrated nanophotonic platforms for the control of light propagation and confinement. Within the context of the present chapter, photonic platforms are particularly suited for enhancing nonlinear optical interactions, specifically the third-order nonlinear susceptibility, through confinement of the electromagnetic field in a diffraction-limited volume (i.e., on the order of the wavelength cubed), as it will be discussed in the following. Moreover, in the spirit of integrated quantum photonic technologies [20] it is important to keep

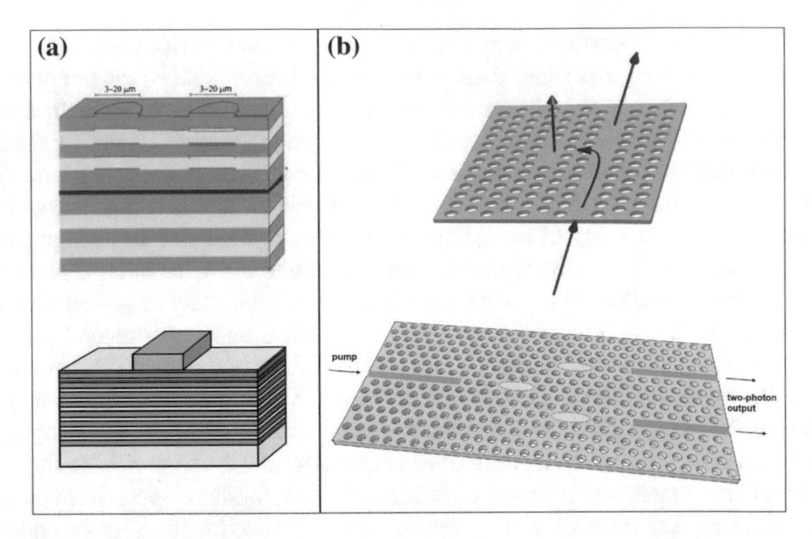

Fig. 6.1 Schematic representation of modern nanophotonic platforms. **a** Semiconductor microcavities, where photons are bound to the microcavity plane by two Bragg mirrors, and laterally confined through local variation of the microcavity thickness (*upper sketch*, from [21]); surface states can be engineered and guided through low-index claddings (*lower sketch*, from [22]). **b** Photonic crystal integrated circuits with single and multi-cavity systems coupled to access waveguides

propagation of light signals within the same photonic systems, very much like in microelectronic circuits where metallic wires allow for the propagation of electrons between different components.

Here we will mainly refer to semiconductor-based nanophotonic platforms, where photons can propagate over distances much larger than their wavelength, and at the same time be trapped in three-dimensional resonators enhancing optical nonlinearities. Nanophotonic resonators can be obtained by laterally etching a planar microcavity, i.e., in a micropillar geometry [23, 24]. These structures are particularly suited for out-of-plane emission. However, in order to keep the photonic field confined in the plane of the high-index cavity layer as in an integrated circuit, shallow photon confinement can be achieved by small diameter mesas, as schematically pictured in Fig. 6.1a. These nanostructures are usually employed in the context of strong light-matter coupling between the planar microcavity photons and elementary electron-hole excitations bound to a two-dimensional quantum well layer placed at the field antinodes. Such geometry is able to enhance the nonlinear properties by generating mixed light-matter excitations, called exciton-polaritons [25], which inherit the intrinsically strong nonlinear interaction derived from Coulomb scattering between electron-hole pairs [26], while simultaneously retaining the light photon mass and most of the desired properties of a nanophotonic platform. Recently, polariton states have been observed at the surface of a planar microcavity without top mirror [27, 28], so called Bloch-surface wave polaritons [22]. Nanophotonic platforms based on propagating polaritons with very low-loss can be envisioned in surface engineered structures, such as the one schematically sketched in the lower panel of Fig. 6.1a, typically working in the near-infrared electromagnetic spectrum. Alternatively, a different concept of compact, efficient and low-loss nanophotonic platforms can be engineered in a planar photonic crystal slab, e.g., a thin layer of a high-index semiconductor periodically textured with holes, where line and point defects can be designed in the periodic lattice to allow for waveguides, cavities, and coupled resonators in the same planar structure, as schematically shown in Fig. 6.1b. The most remarkable features of photonic crystal cavities are the unprecedented figures of merit, such as extremely high values of the Q-factor and the extremely small mode volumes, as it will be detailed in the following. Moreover, large flexibility in materials choice make these platforms a particularly versatile tool to investigate strongly correlated photonic systems on chip.

6.2.1 Photonic Crystal Waveguides, Cavities, and Polaritons

We focus our attention on the specific type of nanophotonic platforms constituted by semiconductor photonic crystals patterned in thin film planar waveguides, either suspended (membraned photonic crystal slabs) or placed on a low-index substrate (lower cladding layer), as already sketched in Fig. 6.1b. Recent progress has allowed these systems to achieve unprecedented figures of merit in terms of light control on-chip [29]. In particular, extremely low loss waveguiding is now possible in systems made of both III–V and group IV semiconductor materials [30–32], which are the most relevant ones for applications in quantum optoelectronics. Nanophotonic resonators in photonic crystals allow to confine photons in volumes that are on the order of their cubic wavelength for more than a nanosecond [33, 34]. Given the relevance that such nanostructures are likely to play in the integrated quantum photonics scenario, we briefly introduce the main physical aspects of light propagation and confinement in photonic crystal slabs.

Photonic crystals are artificially nanostructured materials where the refractive index is periodic along one, two or three spatial directions. The simple analogy with electronic band dispersion in a periodic potential, such as the one in crystalline solids, has allowed to transfer to the photonic realm a few basic concepts such as *photonic bands*, and *photonic band gaps*. In essence, the dielectric modulation acts as an effective periodic potential for photon propagation, which can open forbidden frequency regions in the electromagnetic energy dispersion for wave vectors along the periodicity direction [35]. From a theoretical point of view, the photonic band dispersion in such a structure can be calculated by solving the linear Maxwell equations as a linear eigenvalue problem in steady state, which can be directly deduced in the approximation of harmonic time dependence, and for the magnetic field explicitly reads

$$\hat{\mathbf{O}}(\mathbf{r})\mathbf{H}(\mathbf{r}) = \Omega\mathbf{H}(\mathbf{r}), \tag{6.1}$$

with a differential operator defined as

$$\hat{\mathbf{O}}(\mathbf{r}) = \nabla \times \left(\frac{1}{\varepsilon(\mathbf{r})} \nabla \times \right., \quad (6.2)$$

and the eigenvalues $\Omega = \omega^2/c^2$, where $c = 1/\sqrt{\varepsilon_0 \mu_0}$ in SI units. After the solution of the second-order equation for the magnetic field, the electric field can be obtained from the obvious relation

$$\mathbf{E}(\mathbf{r}) = \frac{i}{\omega \varepsilon_0 \epsilon(\mathbf{r})} \nabla \times \mathbf{H}(\mathbf{r}). \quad (6.3)$$

The operator $\hat{\mathbf{O}}$ is hermitian and positive definite [35], thus it has real and positive eigenvalues with a complete set of orthonormal eigenvectors. The eigenvectors $\mathbf{H}(\mathbf{r})$ are the field patterns of the harmonic modes whose frequencies are obtained by the corresponding eigenvalues as $\omega = c\sqrt{\Omega}$, with the periodicity condition on the spatially-dependent dielectric constant expressed as $\varepsilon(\mathbf{r}) = \varepsilon(\mathbf{r} + \mathbf{R})$. Here, $\mathbf{R}$ is a vector defined by the linear combination of primitive lattice vectors $\mathbf{a}_i$. The photonic crystal structure is invariant for any discrete translation defined by a direct lattice vector $\mathbf{R}$, with dimensionality of the vectorial space defined by the number of dimensions in which $\varepsilon(\mathbf{r})$ is periodic. As we can see, the operator formalism is very similar to the Hamiltonian formulation of quantum mechanics, and the spatial periodicity of the effective potential for this electromagnetic problem allow for a direct application of concepts and theorems known from solid state theories, such as Bloch's theorem. The discrete translation operator $\hat{T}_{\mathbf{R}}$ commutes with the "Hamiltonian" $\hat{\mathbf{O}}$ and the two operators possess a common set of eigenvectors; the wave vector $\mathbf{k}$ is a good "quantum number" for this problem.

Ideally, only three-dimensional photonic crystals allow for a full control over light propagation and confinement. However, it is difficult to manufacture fully three-dimensional photonic crystals at visible or near-infrared wavelengths, although technology is progressing fast in this direction [36, 37]. For this reason, in the past few years much attention has been devoted to developing and optimizing light propagation and confinement in two-dimensional photonic crystal slabs. Hence, here we specify to photonic crystals in planar geometry, also for their direct analogy with planar circuits. In Fig. 6.2a we show a typical lattice of air holes textured in a high-index (e.g., silicon or GaAs) thin film, with a thickness d on the order of half the pitch of the photonic lattice, a. The holes have a radius r/a. The main symmetry points in reciprocal space are indicated in the upper panel of the figure. Equation 6.1 can be solved with usual linear eigenvalue diagonalization, after expanding the solution on a proper basis set of eigenmodes. In the specific case of photonic crystal slabs, a possible and convenient choice is to expand the electromagnetic field in terms of a separable basis of plane waves in the plane, and guided modes of the vertical dielectric stack. Within this formalism, the photonic modes of the photonic crystal slab are radiatively coupled to propagating modes in the upper and lower (semi-infinite) cladding by perturbation theory, which allows to calculate losses as an imaginary part

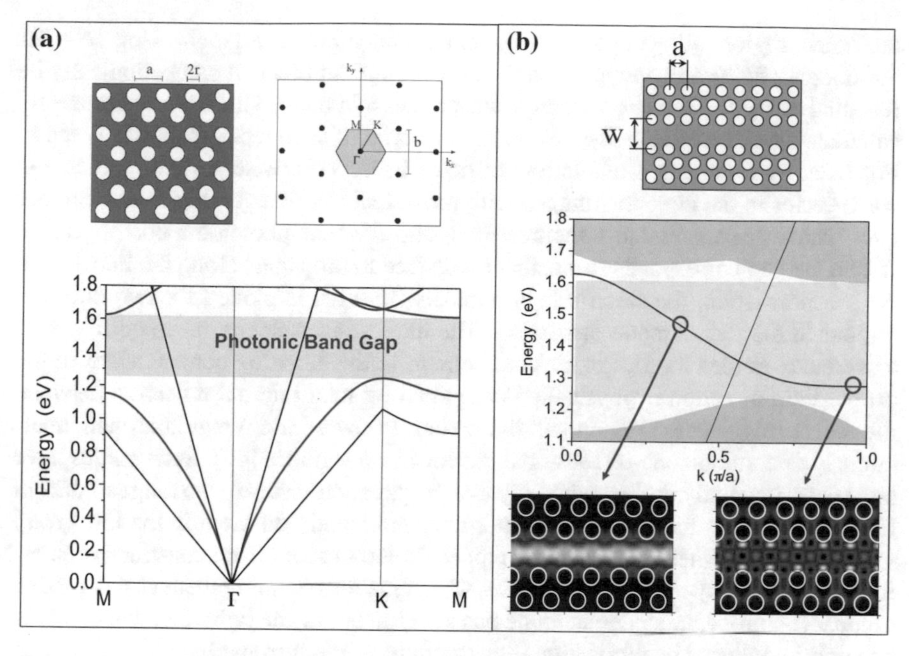

Fig. 6.2 Photonic band structure and line-defects in a triangular lattice: **a** direct and reciprocal triangular lattices in two dimensions, where the high symmetry points are defined, and photonic mode dispersion in a membrane waveguide of thickness $d/a = 0.5$, with a hole radius $r/a = 0.3$, and a lattice constant $a = 260$ nm, which gives a photonic gap centered around 1.4 eV (about 880 nm in wavelength). The *two continuous lines* in the figure represent the light dispersion in the uniform cladding layers (air in this case) and in the high-index core layer (a dielectric constant $\varepsilon = 12$ was used in this case). **b** Line-defect in the triangular lattice along the ΓK symmetry direction (so-called W1 waveguide), with the corresponding guided mode dispersion and electric field intensity profiles in the large (small) group velocity region, respectively

of photonic eigenmodes [38]. This approach has been named "guided-mode expansion" (GME) [39]. An example of photonic mode dispersion is given in Fig. 6.2a for the triangular lattice. Modes whose dispersion falls below the light line of upper and lower claddings (in this case there is a unique light line defined by $\omega = c|\mathbf{k}|/n$) are truly guided within the patterned layer by total internal reflection, i.e., they are ideally lossless, while modes falling above the light line are guided resonances, which acquire a finite linewidth due to coupling to the continuum of radiative modes in the claddings. It can be clearly identified that the spatial modulation of the dielectric profile strongly modifies the dispersion of the otherwise uniform planar waveguide, introducing regions where propagation of photons in the plane of the photonic crystal layer is inhibited (photonic band gap). It is precisely this property that allows to introduce very efficient defect states within the gap, with a mechanism similar to the one producing surface or localized electronic defect states in semiconductors or insulators. Defect states can easily be calculated by GME, by using the technique of supercell expansion: the unit cell is chosen in such a way to fully include the defect region, allowing for a sufficient distance between neighboring cells for the hopping between confined modes to be negligible; this cell can periodically be repeated along the direction where translational symmetry still holds, to correctly calculate the defect mode dispersion. An example of such a calculation is given in Fig. 6.2b: by removing a whole row of holes in the otherwise periodic lattice, the wave vector in the direction aligned with the defect is still a good quantum number over which dispersion can be represented, and the new photonic mode appearing within the gap corresponds to a defect mode free to propagate along the line defect. As a confirmation, the electric field intensity, $|\mathbf{E}(\mathbf{r})|^2$, is plotted for two different regions in the defect mode dispersion. The high group velocity mode corresponds to an index-guided mode, i.e., its confinement in the direction perpendicular to the propagation direction is physically determined by total internal reflection between the defect (high-index) region and the textured (lower index) region around, analogously to common one-dimensional dielectric waveguides [40]. Such modes have been experimentally probed with very low propagation losses in the range of dB/cm [30]. On the other hand, the field pattern is significantly different in the low group velocity region, where the nature of modes directly comes from constructive interference induced by the photonic lattice, which justifies defining them as gap guided modes. The propagating defect mode has a region below the light line, where losses can only be induced by fabrication imperfections of the surrounding photonic lattice [41]. In fact, demonstrations of very low-loss propagation has been reported in the main semiconductor material platforms, such as silicon [32] or GaAs [31].

A key building block for integrated nanophotonics can be realized by exploiting the photonic band gap in a two-dimensional photonic crystal membrane to create a three-dimensional resonator with extremely optimized figures of merit in terms of photon lifetime and spatial confinement. A point-defect in a photonic crystal lattice, which is obtained by removing one or more holes from the underlying lattice, gives rise to fully localized states spectrally situated within the photonic gap, as shown in Fig. 6.3a. Here, the dispersion of localized modes within the gap is represented in a fictitious Brillouin zone, defined by a two-dimensional supercell that allows to calculate defect modes even in the presence of breaking of translational symmetry (here, the supercell lattice period along the ΓK direction is 10 times the underlying triangular lattice pitch, which is largely sufficient for convergence of defect mode energies within the gap). It should be noted that the nature of these confined modes is related to the propagating mode of the W1 waveguide, as it can be recognized by comparing the electric field intensity profile with the dispersion in the low group velocity region in Fig. 6.2b. Since light cannot escape in the plane, being confined by the photonic band gap, the main source of intrinsic losses for these cavity modes is radiation into the claddings induced by coupling to radiation modes. In fact, this is the main reason for these resonances to exhibit a finite linewidth, defined from the ratio between resonance frequency and Q-factor for each cavity mode (focusing on the fundamental cavity mode, $\gamma = \omega_0/Q$). This linewidth is related to the photon lifetime within the nanocavity, $\tau = 1/\gamma$. Interestingly, photonic crystal cavity modes can be engineered to strongly suppress Fourier modal components falling above the light line in air, which are the main responsible for out-of-plane leakage. This can

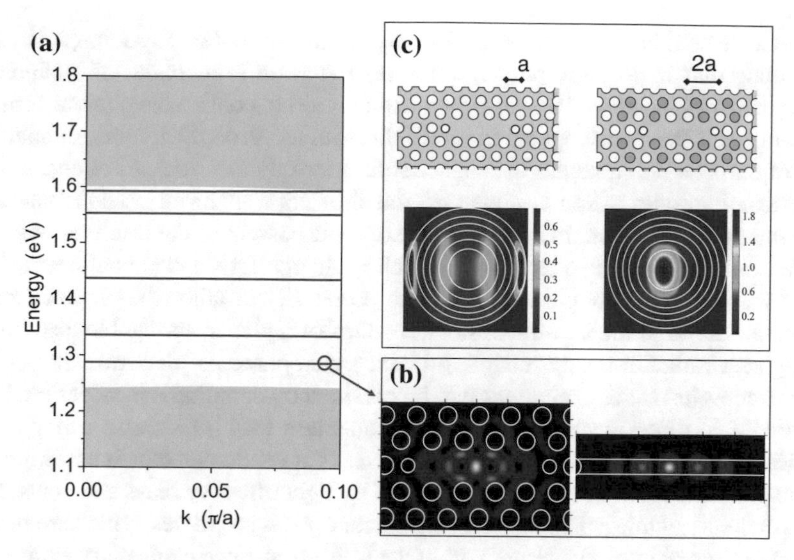

Fig. 6.3 Point defect in a triangular photonic lattice, the so-called L3 cavity (three missing holes along ΓK): **a** defect modes (calculated in a fictitious supercell), **b** fundamental cavity mode field intensity profile in the (x, y) (at $z = 0$) and (x, z) (at $y = 0$) planes, respectively, and **c** far-field optimization of the fundamental cavity mode emission in the upper half-space

be achieved by suitably shifting the holes nearby the cavity region [42, 43], yielding theoretical Q-factors on the order of 10^9 [34]. Measured Q-factors are mainly material and wavelength dependent. Record values have been reported in silicon at resonances in the range $\lambda = 1.5\,\mu$m (typical telecom band), where values of the order of 10^6–10^7 have been shown [44, 45]. In GaAs the measured values are limited to about 7×10^5 [46], which is currently the record value at the same telecom wavelengths. It should be noted that such nanocavities are particularly suited to being excited via evanescent waveguide-cavity coupling, as schematically represented in Fig. 6.1b. On the other hand, such large Q-factors are not compatible with efficient light excitation and/or emission out of plane, since the very principle of Q-factor optimization requires the suppression of Fourier modal components radiating out-of-plane. In fact, the far-field profile calculated for a high-Q cavity is represented in Fig. 6.3c, on the left panel, showing a poor spatial overlap with the typical gaussian profile of a focusing lens. In order to increase light collection in the vertical direction, for specific experiments or applications in which excitation/collection is out-of-plane, clever hole engineering allows to recover a gaussian far-field profile, at the expense of Q-factor [47]: a second-order lattice can be superimposed (i.e., by modifying the holes around the cavity with a periodicity $2a$), thus producing folding of Fourier components around the normal incidence, as shown in the right panel of Fig. 6.3c. In fact, it has been shown that a compromise can be found between having a vertically optimized far-field emission and a Q-factor in the range of 10^5 [48], which is still quite relevant for most applications in integrated nonlinear and quantum photonics.

Since we will be concerned with strongly nonlinear photonic systems, it is worth reminding that in standard photonic devices based on semiconductor technology, the optical nonlinearity of the material medium used to confine the photons remains moderate and the device operation is still far from the strongly nonlinear condition. One of the possible strategies to enhance solid state nonlinearities relies on coupling the photonic modes to some active material showing a sharp material resonance in the same spectral region. In this way, the excitonic content of the resulting polariton modes introduces strong interactions, which could possibly lead to nonlinear behavior at the single or few photon level. The possibility to tailor the photonic mode dispersion through the two-dimensional pattern of a planar waveguide also allows to engineer radiation-matter coupling in such nanophotonic platforms. In particular, mixed light-matter excitations can be obtained by coupling a low-loss photonic mode to a confined electron hole pair in a quantum well (QW), also called a QW exciton. These are the elementary excitations of a semiconductor inheriting most of the photonic modes peculiar properties, i.e., light effective mass and hence high group velocity, while enhancing their nonlinear optical properties, which are mainly due to exciton-exciton scattering within the QW layer. Such elementary excitations are actively exploited in semiconductor microcavities, such as the ones depicted in Fig. 6.1a. Similar theoretical treatments of light-matter coupling in photonic crystal slabs leads to the prediction of new mixed excitations called *photonic crystal polaritons* [49], whose dispersion is shown, e.g., in Fig. 6.4 for a square lattice. The lattice constant can be chosen to match the photonic mode dispersion at a given point in the Brillouin zone with the QW exciton resonance, in this case at $\hbar\omega_{ex} = 1.485$ eV (as typical of InGaAs QW in GaAs barriers). The eigenmode dispersion represented with a full line in the figure describes extended polariton states on the two-dimensional photonic lattice of holes. We notice that around the normal incidence (Γ point in the figure), the photonic crystal polariton dispersion closely mimics the typical microcavity polariton behavior in planar microcavities, where polariton condensation due to the large exciton nonlinearity has been observed [50], and recently stimulated

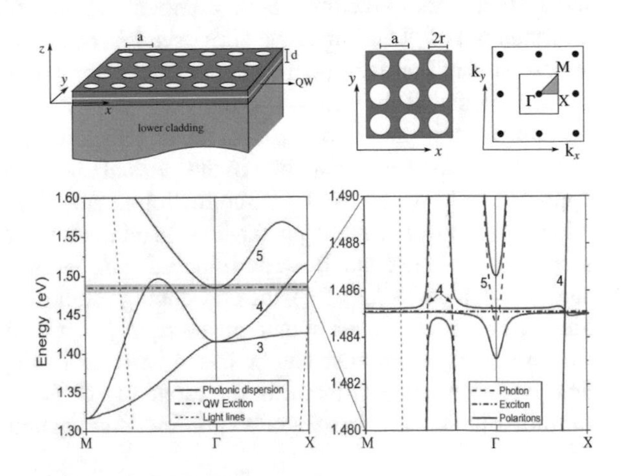

Fig. 6.4 A photonic crystal slab with an embedded quantum well, direct and reciprocal square lattices, and mode dispersion of photonic crystal polaritons (these results were originally published in [49]). The photonic crystal slab has thickness $d/a = 0.3$, hole radius $r/a = 0.34$, and the lattice constant is set to the value $a = 430$ nm. Reprinted with permission from APS

further interest for the study of quantum fluid properties out-of-equilibrium [26]. Tailoring of the exciton-polariton dispersion with a photonic crystal lattice has been shown experimentally [51]. Moreover, photonic crystal polariton confinement in a far-field optimized L3 nanocavity has allowed to demonstrate an ultra-low threshold polariton laser [52].

Nanophotonic platforms with several coupled elements based on photonic crystal slabs have been realized in different materials, from commonly employed III–V semiconductors [46], to CMOS-compatible silicon-on-insulator wafers [29], and diamond [53]. In particular, quantum photonic experiments can be designed on photonic crystal platforms, where efficient coupling of light into the photonic chips and independent control of the different resonator elements can be achieved [54, 55]. Moreover, optical coupling of neighboring photonic crystal resonators via the overlap of the evanescent tails has been demonstrated for two-dimensional arrays of photonic crystal cavities [56, 57], which allows to envision a straightforward realization of a variety of integrated configurations of multi-cavity arrays coupled to input/output photonic crystal waveguiding channels. Also, advanced post-fabrication techniques have been developed for fine tuning of cavity modes and inter-cavity coupling in such arrays [58, 59]. This unprecedented capability of designing, realizing, and controlling a fully integrated platform for quantum photonic experiments is at the basis for motivating the investigation of strongly correlated photonic systems on-chip, which is the main topic of the present chapter.

6.2.2 Effective Photon-Photon Interactions

Strongly correlated photonic systems rely on the ability to enhance the nonlinear interactions in suitably engineered photonic systems. The nonlinear optical response of solids is commonly very weak, such that several photons are typically needed to produce appreciable nonlinear effects. From a theoretical point of view, the nonlinear optical response to the applied electric field of a generic dielectric material is given by [60]

$$\begin{aligned} D_i(\mathbf{r},t) = \varepsilon_0\varepsilon_{ij}(\mathbf{r})E_j(\mathbf{r},t) + \varepsilon_0[\chi^{(2)}_{ijk}(\mathbf{r})E_j(\mathbf{r},t)E_k(\mathbf{r},t) \\ + \chi^{(3)}_{ijkl}(\mathbf{r})E_j(\mathbf{r},t)E_k(\mathbf{r},t)E_l(\mathbf{r},t) + \ldots], \end{aligned} \quad (6.4)$$

where the summation over repeated indices (labeling the three spatial directions) is understood. Equation 6.4 defines the relative dielectric permittivity tensor of the medium, $\varepsilon_{ij}(\mathbf{r}) = \delta_{ij} + \chi^{(1)}_{ij}(\mathbf{r})$. Enhancing optical nonlinearities requires strong electric field intensities, such as the ones that can be obtained in a photonic crystal resonator made of a nonlinear material. In the following, we specify to the case of a single mode of the electromagnetic field confined in a centrosymmetric medium, i.e., we can assume the second-order nonlinear contribution to be negligible, $\chi^{(2)}_{ijk}(\mathbf{r}) = 0$. Nonlinearities of the third-order are usually defined as Kerr-type contributions, due to the $\chi^{(3)}$ tensor elements in Eq. (6.4). Since we will be interested in quantum effects, we can expand the electric field of the cavity mode in a spatially inhomogeneous (i.e., a photonic crystal cavity) nonlinear medium as

$$\hat{\mathbf{E}}(\mathbf{r},t) = i\left(\frac{\hbar\omega_0}{2\varepsilon_0}\right)^{1/2}\left[\hat{a}\frac{\boldsymbol{\alpha}(\mathbf{r})}{\sqrt{\varepsilon(\mathbf{r})}}e^{-i\omega_0 t} - \hat{a}^{\dagger}\frac{\boldsymbol{\alpha}^*(\mathbf{r})}{\sqrt{\varepsilon(\mathbf{r})}}e^{i\omega_0 t}\right], \tag{6.5}$$

with the corresponding magnetic field being given by $\omega_0\mu_0\hat{\mathbf{H}}(\mathbf{r}) = -i\nabla\times\hat{\mathbf{E}}(\mathbf{r})$. In Eq. (6.5) $\hat{a}$ ($\hat{a}^{\dagger}$) defines the destruction (creation) operator of a single photon in the mode, obeying standard commutation relations $[\hat{a},\hat{a}^{\dagger}] = 1$, and $\boldsymbol{\alpha}(\mathbf{r})$ is the normalized three-dimensional cavity field profile, i.e., satisfying $\int|\boldsymbol{\alpha}(\mathbf{r})|^2 d\mathbf{r} = 1$. Canonical quantization is obtained by the time-averaged total energy density in the mode, $\mathcal{H}_{em} = \int d\mathbf{r}\left[\mathbf{E}(\mathbf{r})\cdot\mathbf{D}(\mathbf{r}) + |\mathbf{H}(\mathbf{r})|^2\right]/2$, after which a nonlinear second-quantized hamiltonian can be obtained (neglecting the zero-point energy in the quantization of the electromagnetic field) [61]

$$\hat{H} = \hbar\omega_0\hat{a}^{\dagger}\hat{a} + \hbar U\hat{a}^{\dagger}\hat{a}^{\dagger}\hat{a}\hat{a}. \tag{6.6}$$

As it is evident from the formal expression of Eq. (6.6), the third-order $\chi^{(3)}$ nonlinearity corresponds in the language of Feynman diagrams to four-legged vertices describing, among others, binary collisions between photons pairs, thus fully mimicking the contact-type interaction of charged particles in condensed matter physics, which lies at the heart of the development of strongly correlated photonic analogies. The photon-photon interaction can be approximated as [62]

$$U \simeq \frac{3\hbar\omega_0^2}{4\varepsilon_0}\frac{\overline{\chi}^{(3)}}{\overline{\varepsilon}_r^2}\int|\boldsymbol{\alpha}(\mathbf{r})|^4 d\mathbf{r} = \frac{3\hbar\omega_0^2}{4\varepsilon_0 V_{\text{eff}}}\frac{\overline{\chi}^{(3)}}{\overline{\varepsilon}_r^2}, \tag{6.7}$$

with the effective cavity mode volume defined as $V_{\text{eff}}^{-1} = \int|\boldsymbol{\alpha}(\mathbf{r})|^4 d\mathbf{r}$. In fact, an order of magnitude estimate can be given for each defined combination of materials constituting the nanophotonic resonator, by assuming constant values for the average real part of the nonlinear susceptibility and relative dielectric permittivity, $\overline{\chi}^{(3)}$ and $\overline{\varepsilon}_r$ respectively.

The elementary building block towards strongly correlated photonic systems is the so-called *photon blockade* effect [11, 63], which is schematically described in Fig. 6.5: when two photons inside a resonant system produce a nonlinear shift of its response frequency, U, that is larger than the line-broadening induced by losses, γ, the two-photon state according to Hamiltonian (6.6) is spectrally shifted by an amount $2U \gg \gamma$, and an input radiation field resonant with the linear cavity frequency, ω_0, will not be able to inject into the cavity more than one photon at a time. Hence, in the presence of such a strong single-photon nonlinearity, photons in single-mode cavities effectively behave as impenetrable bosons. The experimental configuration for photon blockade can be modeled by adding a coherent pumping term to obtain the standard Kerr-type hamiltonian that is usually employed in quantum optics [61]

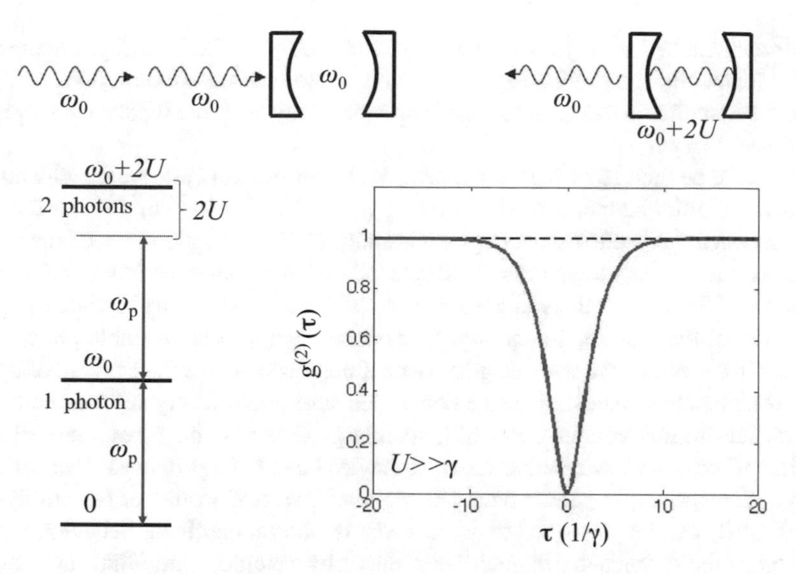

Fig. 6.5 Schematic description of single-photon blockade from a strongly nonlinear system: a train of photons is incident on a single-mode resonator with frequency ω_0 and linewidth γ, a second photon is not able to enter the cavity if the nonlinear coupling $U \gg \gamma$, because of the shift of the two-photon resonance. This blockade can be probed by measuring antibunched light emission at the output of the cavity, as shown in the calculated $g^{(2)}(\tau)$ for $U/\gamma = 100$

$$\hat{H} = \hbar(\omega_0 - i\gamma/2)\hat{a}^\dagger\hat{a} + \hbar U\hat{a}^\dagger\hat{a}^\dagger\hat{a}\hat{a} + \hbar F e^{-i\omega_p t}\hat{a}^\dagger + \hbar F^* e^{i\omega_p t}\hat{a}, \tag{6.8}$$

where F is the coherent pumping rate at the pump frequency ω_p, and losses in the system are quantified through the intrinsic cavity decay rate, γ, which can be added as an imaginary part of the cavity mode resonance frequency and is related to the mode Q-factor as $Q = \omega_0/\gamma$. As a result, the transmitted light across the cavity shows a strongly reduced probability of detecting a single photon soon after a first one has been detected. Such a statistical property, which is strictly related to the quantum nature of the emitted radiation, is called photon antibunching, and it is quantified by the condition $g^{(2)}(0) < g^{(2)}(\tau)$, where the usual definition for the steady state second-order correlation function is [64]

$$g^{(2)}(\tau) = g^{(2)}(t \to \infty, \tau) = \frac{\langle \hat{a}^\dagger(t)\hat{a}^\dagger(t+\tau)\hat{a}(t+\tau)\hat{a}(t)\rangle}{\langle \hat{a}^\dagger(t)\hat{a}(t)\rangle\langle \hat{a}^\dagger(t+\tau)\hat{a}(t+\tau)\rangle}. \tag{6.9}$$

Thus, the figure of merit quantifying the single-photon nonlinear behavior of the cavity mode is the normalized zero-time delay second-order correlation, defined as $g^{(2)}(0) = \langle \hat{a}^{\dagger 2}\hat{a}^2\rangle / \langle \hat{a}^\dagger\hat{a}\rangle^2$, which in the low-excitation limit can be expressed as $g^{(2)}(0) = [1 + 4(\Delta/\gamma)^2]/[1 + 4(\Delta + U)^2/\gamma^2]$, where $\Delta = \omega_0 - \omega_p$. For $g^{(2)}(0) \to 0$ the cavity behaves as an almost ideal single-photon source [64], which occurs when $U/\gamma \gg 1$, with Q^2/V_{eff}^2 being the relevant figure of merit to be optimized. Single photons are released from the cavity at the bare frequency, ω_0. A numerical simulation of the antibunched $g^{(2)}(\tau)$ is shown in Fig. 6.5 for $U/\gamma = 100$. The physical meaning is the following: the probability of detecting more than one photon goes to 1 only after the first photon has been released from the cavity, with the typical cavity photon lifetime $\tau \sim 1/\gamma$.

It should be noted that typical material nonlinearities are quite weak, with nonlinear susceptibility elements on the order of $\chi^{(3)} \sim 10^{-18}$–10^{-17} m^2/V^2 for the main semiconductor crystals used in optoelectronics [60]. A simple order of magnitude estimate for a L3 photonic crystal cavity made of such materials (e.g., GaAs or Si), after Eq. (6.7), gives values in the range of $\hbar U \simeq 10^{-3}$–10^{-2} μeV, which requires Q-factors of the order of 10^8 or larger for observing appreciable single-photon nonlinear effects at telecom wavelengths. Even if unprecedented enhancement of optical nonlinearities has been shown as a consequence of photonic crystal confinement in diffraction-limited volumes [46, 65], such large Q-factors have not been achieved at time of writing. Hence, alternative strategies have been proposed to experimentally realize the single-photon blockade regime. First, cavity quantum electrodynamics (CQED) can be employed to achieve single-photon nonlinear behavior, e.g., by exploiting the strong anharmonicity introduced by a single atomic-like emitter into a high quality factor resonator [63, 66]. After the first experimental demonstration of single-photon blockade employing caesium atoms strongly coupled to a Fabry-perot resonant mode [12], a successful route in this direction employing solid state resonators has been to insert an electronic quantum dot (QD) in a photonic crystal cavity. A QD typically consists of a nanometer-scale semiconductor aggregate surrounded by a larger gap semiconductor in all three spatial directions, such that the electron and hole wave functions are confined and excitonic transitions display a discrete, atomic-like energy spectrum. When one of these transitions is strongly coupled to the high-Q photonic mode, a Jaynes-Cummings (JC) system is recovered [67]. The photon statistics of a JC nonlinear system[1] is very similar to the one represented in Fig. 6.5 [69]. First experimental evidence of photon blockade was reported for a QD embedded in a photonic crystal L3 nanocavity driven by an off-resonant pump [70], followed soon after by an experiment under resonant coherent excitation [13]. Later work showed significantly clearer signatures of strong photon-photon interactions in such kind of systems [14]. From the point of view of realizing strongly correlated states of photons in a many-cavity geometry, QD-based architectures suffer from the significant drawback that neither the spatial position of individual QD nor the exact energy of the electronic transitions can be controlled during the growth stage, as the different dots self-organize at random positions with random sizes. This can become an issue when extending this building block to multi-cavity systems. Hence, a promising route to overcome this difficulty using spatially localized emitters may be related to the use of QW as the active material responsible for the enhanced nonlinearity. In this case, since the thickness of the QW material can be controlled at the mono-layer precision, the experimental problem of correctly positioning the emitter is completely elimi-

[1] We notice that the figure of merit to be optimized for coupled QD-cavity based single-photon nonlinearities scales as $Q/\sqrt{V_{\text{eff}}}$ (see, e.g., Ref. [68]), while antibunching scales as Q^2/V_{eff}^2 for the Kerr-type nonlinearity.

nated, and the electronic contribution to inhomogeneous broadening can be strongly suppressed. A theoretical treatment of the nonlinear dynamics of confined polaritons in a single-mode cavity, such as the one in Fig. 6.3, yields a formal Hamiltonian similar to Eq. (6.6). In this case, the nonlinearity is still of the Kerr-type, with an explicit $U \propto 1/A_{\rm eff}$ scaling (here $A_{\rm eff}$ is the effective cavity mode area, which can be estimated from the cavity mode profile). For details on this approach, we refer to [71]. Further proposals to enhance Kerr-type optical nonlinearities in the solid state involve using the biexciton contribution to the polariton-polariton scattering [72], engineering doubly resonant nanocavities in materials with $\chi^{(2)}$ contribution to the nonlinear response [73], or exploiting quantum interference effects in photonic molecules [74, 75]. Presently, it is still unclear whether the nonlinearity stemming from collisions between delocalized polaritons can eventually lead to the $U \gg \gamma$ regime required for single-photon blockade, but further experimental developments are likely to appear in the near future. In the following, we will not be concerned with the specific nature of the Kerr-type nonlinearity of the single-mode resonator, which will be considered as a building block of a strongly correlated nanophotonic platform assuming that the quantum dynamics of such an elementary system can always be reduced to a Hamiltonian like in Eq. (6.6).

6.3 Strongly Correlated Photons on Chip

We give an overview of a few small-scale systems, compatible with the nanophotonic platforms reviewed in the previous Section, where strong photon correlations may result in the realization of novel quantum devices, or the study of some fundamental physical phenomena in a new scenario, triggered by the driven-dissipative nature of photonic systems. A photonic crystal platform like the one represented in Fig. 6.1b can be described by linear and nonlinear elements: photons (or polaritons) propagating in line-defects are subject to a linear dispersion, the effects of nonlinearities being strongly reduced by the one-dimensional nature of the density of states. This is especially true for index-guided modes with large group velocity (see Fig. 6.2b). On the other hand, photons (or polaritons) confined in photonic crystal cavities in the photon blockade regime may be described by single-photon nonlinear elements, owing to the enhanced nonlinearities, as a result of wavelength-scale mode confinement and extremely reduced losses.

6.3.1 Master Equation Treatment of Open Quantum Systems

Strongly correlated photonic systems can be studied by coupling elementary building elements described by the nonlinear quantum Hamiltonian (6.6), i.e., a generalized Bose-Hubbard model, also called the driven-dissipative Kerr-Hubbard (KH) Hamiltonian

$$\hat{H} = \hbar \sum_i \left(\omega_i \hat{a}_i^\dagger \hat{a}_i + U \hat{a}_i^\dagger \hat{a}_i^\dagger \hat{a}_i \hat{a}_i + F_i e^{-i\omega_p t} \hat{a}_i^\dagger + F_i^* e^{i\omega_p t} \hat{a}_i \right) + \sum_{<i,j>} \hbar J \hat{a}_i^\dagger \hat{a}_j + \hat{H}_{\text{diss}}, \tag{6.10}$$

where the sum is over the number of coupled cavities, and each operator $\hat{a}_i$ represents photonic (or polaritonic) quanta of a single-mode resonator, which for simplicity can be all assumed with the same bare resonant frequency (i.e., $\omega_i = \omega_0$, for every i), while neighboring resonators are evanescently coupled with the same tunnel-rate, J, such as in a photonic crystal cavity array, and the sum is extended to next neighbor sites only. We notice that each resonator is independently pumped at a rate F_i, which can be achieved either through in-plane waveguide coupling, as in the schematic of Fig. 6.1, or through out-of-plane coupling in case of far-field engineered photonic crystal cavities (see Fig. 6.3). In Eq. 6.10, $\hat{H}_{\text{diss}}$ exactly describes coupling of the confined degrees of freedom to the radiation environment, as opposed to describing losses by an imaginary part of mode frequencies as in Eq. (6.8). Since the main source of losses results from cavity photon leakage into radiative modes out of the photonic crystal slab plane (reservoir), they can be correctly taken into account within a quantum Master equation approach in the Born-Markov approximation, i.e., a Liouville-von Neumann equation [76]

$$\frac{\partial}{\partial t}\rho = \frac{i}{\hbar}[\rho, \tilde{H}_{\text{KH}}] + \mathcal{L}(\rho), \tag{6.11}$$

where $\tilde{H}_{\text{KH}}$ is the KH Hamiltonian written in the rotating frame with respect to the pump frequency

$$\begin{aligned} \tilde{H}_{\text{KH}} &= \hat{\text{R}}(t)\hat{H}_{\text{KH}}\hat{\text{R}}^\dagger(t) - i\hbar\hat{R}\frac{\text{d}\hat{R}^\dagger}{\text{d}t} \\ &= \hbar\sum_i [(\omega_0 - \omega_p)\hat{a}_i^\dagger\hat{a}_i + U\hat{a}_i^\dagger\hat{a}_i^\dagger\hat{a}_i\hat{a}_i + F\hat{a}_i^\dagger + F^*\hat{a}_i] + \hbar J \sum_{<i,j>} \hat{a}_i^\dagger\hat{a}_j, \end{aligned} \tag{6.12}$$

with $\hat{\text{R}}(t) = \exp\{i\omega_p t \sum_i (\hat{a}_i^\dagger \hat{a}_i)\}$. After tracing out the environment degrees of freedom responsible for the coupling of the confined modes to the external bath, the Liouvillian can be expressed in the usual Lindblad form [76]

$$\mathcal{L}(\rho) = \sum_i \frac{\gamma_i}{2}[2\hat{a}_i\rho\hat{a}_i^\dagger - \hat{a}_i^\dagger\hat{a}_i\rho - \rho\hat{a}_i^\dagger\hat{a}_i], \tag{6.13}$$

where we are neglecting other decoherence mechanisms (such as pure dephasing), and γ_i are the photon (polariton) dissipation rates in each cavity, for which we will assume $\gamma_i = \gamma$ for every i in the following. In photonic crystal integrated circuits, which we are considering as model platforms in the present chapter, the coefficients γ_i are physically related to the out-of-plane losses of photonic nanocavities (see previous section), and possibly to leakage into access waveguides within the same photonic crystal chip, which will determine a total resonance linewidth to each resonator.

These coefficients can be estimated in perturbation theory, as outlined in Sect. 6.2.1, for each specific design.

We notice that the KH model of Eq. (6.12) is just an example of a wide class of models describing systems of strongly interacting photons, depending on the nature of the single-photon nonlinearity in the elementary building block. For example, an array of coupled photonic crystal cavities each one containing a strongly coupled QD would be described by a Jaynes-Cummings-Hubbard (JCH) model, whose predictions in the low-excitation regime can be shown to be in qualitative agreement with the driven-dissipative KH model, with proper parameter matching between the two models [77]. Differences arise when considering the formation of many-body states [78]. However, due to difficulties in practically realizing a JCH with uniform QD-cavity coupling throughout a multi-cavity array, we will rather restrict our theoretical analysis to different versions (depending on the number of sites or the lattice geometry) of the generic KH model, in which the nonlinearity either comes from the bulk material $\chi^{(3)}$, or rather from a polariton-polariton contact-type coupling due to QW exciton scattering, as outlined before.

6.3.2 *Examples of Strongly Correlated Photonic Systems*

As it was mentioned in Sect. 6.2.2, the second-order correlation function can be used as an operational probe of the single-photon nonlinear behavior in a photon blockade experiment. We argue that the same quantity can also be assessed as a quantitative probe of the interplay between tunneling and on-site interactions in a driven-dissipative KH model, and by extension it can potentially be considered as an analog “order parameter” able to discriminate between globally coherent and strongly correlated quantum phases out-of-equilibrium. The latter point represents one of the main motivating scopes in the context of theoretical research activities connected with strongly correlated photonic systems, requiring the study of ideal configurations with infinite number of lattice sites. In such a limit, an analog of the superfluid-to-Mott insulator quantum phase transition has been explored with advanced mean-field approaches, extended to the out-of-equilibrium nature of strongly correlated photonic systems [79]. Recently, a non-equilibrium version of the celebrated Bose-Hubbard phase diagram has been calculated in steady state under continuous coherent pumping conditions [80], opening the way to deeper investigations over the rich many-body physics of driven-dissipative cavity arrays. In fact, the open nature of photonic systems is responsible for a wealth of unexplored features due to the interplay of nonequilibrium statistical mechanics, quantum optics, and many-body physics, which is likely to bring new exciting theoretical developments in the future.

In this last part of the chapter, we have chosen to review a few topical examples of strongly correlated photonic systems that can be realized on integrated photonic crystal platforms with state-of-art technology. The aim is to reveal new features due to the genuine non-equilibrium interplay of coherent tunneling and on-site interactions, in systems that can be readily accessible and measurable by optical means. To this aim, and to give a quantitative assessment of the above statement on photon correlation measurements, we will review a few proposals involving small coupled-cavity arrays of driven-dissipative elements, and study the interplay between U and J within the master equation framework. In the following, we will mainly be concerned with the zero-time delay correlation function in steady state, which can be calculated by numerically solving the master equation (6.11) with the condition $\mathrm{d}\rho/\mathrm{d}t = 0$, as a function of the relevant parameters ratio U/J in Hamiltonian (6.12).

6.3.2.1 Quantum Photonic Devices: A Quantum Optical Josephson Interferometer

The physics of strongly correlated electrons has been thoroughly explored in transport experiments with mesoscopic devices in the nineties. One of the strongly correlated electronic devices is a phase-biased Cooper pair transistor, which is a quantum version of the Josephson junction. In a common Josephson junction, two macroscopic superconducting states are separated by an insulating layer. The tunnel current flowing at zero bias through the junction has an oscillating dependence on the phase difference between the two macroscopic wave functions at the two sides of the insulating barrier, $\psi(\phi_1) = \sqrt{\rho}\exp(i\phi_1)$ and $\psi(\phi_3) = \sqrt{\rho}\exp(i\phi_3)$, i.e., $I = I_J \sin\phi$ with $\phi = \phi_1 - \phi_3$ (DC Josephson effect). The mesoscopic version of this device is schematically shown in Fig. 6.6a, where the insulating layer is substituted with a gate controlled Cooper pair box. Strong electron correlations can thus inhibit the Josephson tunnel current when the charging energy of the mesoscopic insulating island (a superconducting QD) is much larger than the tunneling rate to and from the superconducting leads, as a direct consequence of strong charge correlations between Cooper pairs confined on the mesoscopic island, in direct analogy with the Coulomb blockade [2]. This Josephson interferometer was first proposed in [81], and later realized in [82]. Besides these striking demonstrations, the interplay between coherent tunnel coupling and on-site interactions typical of such strongly correlated systems has been exploited in many ground-breaking experiments, such as the observation of a quantum phase transition in Josephson junction arrays [83] and the experimental test of the number-phase uncertainty principle [84].

The realization of a quantum optical version of a Josephson interferometer in integrated photonic crystal platforms is shown Fig. 6.6b: the mesoscopic insulating island is replaced by a nonlinear resonator, while the macroscopic superconducting phases have their natural quantum optical analog in coherent states with well-defined phases [64]. The latter can be simulated either with two external linear cavities (e.g., with a larger mode volume than the central cavity, such that the single photon nonlinearity in Eq. 6.7 be much smaller than γ), or tunnel coupled photonic crystal waveguides, as first proposed in [18]. The master equation is solved for the corresponding three sites model adapted from the general Hamiltonian (6.12), where the external sites are labelled with indices 1 and 3, respectively, and the central nonlinear site is 2. While the operation of both electronic and photonic versions is based on the quantum mechani-

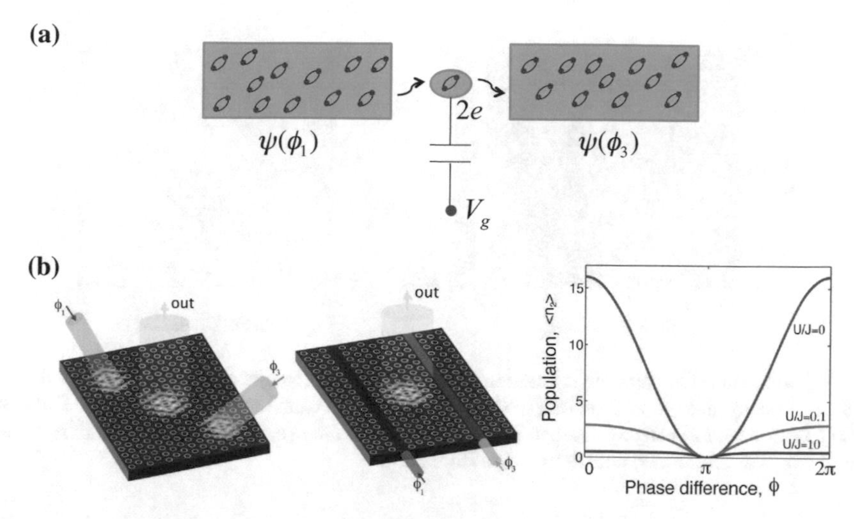

Fig. 6.6 **a** Scheme of a Josephson interferometer based on a Cooper pair island between two macroscopic superconducting phases, where the island charge state can be modulated through an external gate voltage (V_g), corresponding to the mesoscopic superconducting device realized in [82]. **b** The analog quantum optical device in a nanophotonic platform, where the macroscopic superconducting states are replaced by coherent states with well defined phase, represented either by two photonic crystal waveguides or two linear cavities; the charging island is replaced by a nanocavity with a single-photon nonlinearity. The suppression of Josephson-like oscillations as a function of the phase difference is shown for light emitted from the central cavity, on increasing ratio *U/J* (from Ref. [18]). Panel **b** reprinted with permission from NPG

cal conjugation between phase and number variables, the quantum optical Josephson interferometer is an intrinsically open system. Signatures of strong photon correlations can be found either in the light emitted from the central cavity, or in its autocorrelation function. The first quantity is proportional to the average photon population in the central cavity, which can be calculated as $\langle n_2\rangle = \langle \hat{a}_2^\dagger \hat{a}_2\rangle = Tr\{\hat{a}_2^\dagger \hat{a}_2 \rho_{ss}\}$, where ρ_{ss} is the calculated steady state density matrix at weak continuous wave pumping $F/\gamma \ll 1$ (we assume $\gamma_2 = \gamma$ here) and at resonant frequency, $\omega_p = \omega_2$. The result is shown in Fig. 6.6b as a function of the phase difference, $\phi = \phi_1 - \phi_3$, and for increasing value of the ratio *U/J* in the model. For $U = 0$, regular interference between the two coherent states occurs in the emitted light intensity from the central cavity, with an analytic expression $\langle n_2\rangle = n_{\max} \cos^2(\phi/2)$ [18], which is eventually the optical analog of Josephson current oscillations. In analogy with the mesoscopic Josephson interferometer, also here the increasing photon-photon interaction *U/J* yields a suppression of Josephson-like behavior, as clear from the figure. In the strongly correlated limit $U/J \to \infty$, the average cavity population at $\phi = 0$ ($\phi = 2\pi$) goes to the steady state value of $\langle n_2\rangle = 0.5$, as a consequence of the photon number being locked either on the $|0\rangle$ or on the $|1\rangle$ Fock states, respectively. Such behavior is witness of the crossover from the delocalized to the correlated regime: at large hopping ($U \ll J$) the state of the system is approximately a coherent state, and phases are

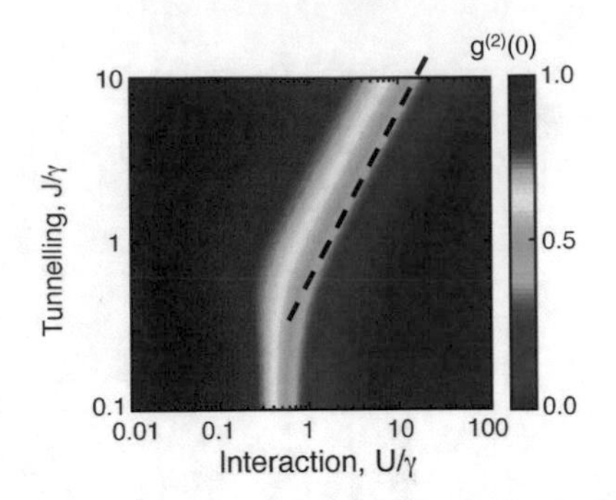

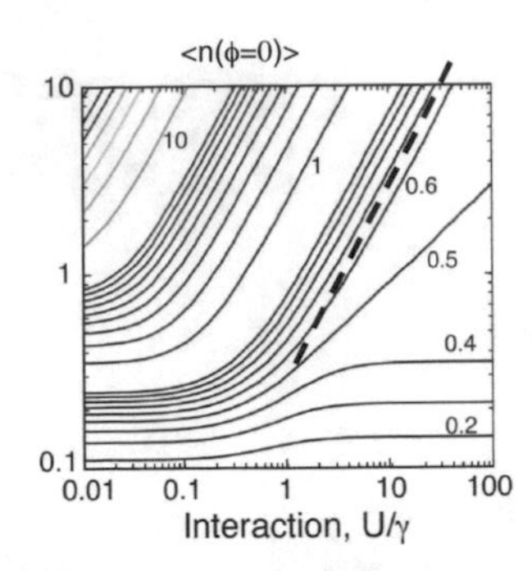

Fig. 6.7 Second-order correlation function at zero-time delay (results originally published in Ref. [18]) and steady state population (at $\phi = 0$) for light emitted from the central cavity, as a function of U/γ and J/γ, respectively. Dashed lines are a guide to the eye, indicating the boundary of the phase crossover. Reprinted with permission from NPG

well defined, in the sense that there are strong fluctuations in the number operator and so phase fluctuations are suppressed. In the opposite situation ($U \gg J$), the number state is better defined (oscillating between 0 and 1 owing to the photon blockade), and large phase fluctuations in the central cavity suppress the global coherence of the system and hence the visibility of Josephson-like oscillations.

It is then interesting to analyze how this crossover reflects into the photon statistics of light emitted from the central cavity. A full picture of this quantitative behavior is shown in Fig. 6.7. Here, $g^{(2)}(0) = Tr\{\hat{a}_2^\dagger \hat{a}_2^\dagger \hat{a}_2 \hat{a}_2 \rho_{ss}\}/\langle n_2 \rangle^2$ and $\langle n_2(\phi = 0) \rangle$ are plotted versus U/γ and J/γ, respectively. The second-order correlation clearly displays a neat transition from Poissonian to sub-Poissonian light (antibunched) statistics as the interaction strength U is increased. The threshold for antibunched light emission is a function of J. For $J \ll \gamma$, the threshold nonlinearity is $U_{th} \sim \gamma$, while for $J \gg \gamma$, U_{th} is linear in J. These two regimes are connected by a smooth crossover region, which can be defined by the condition $g^{(2)}(0) = 0.5$. The crossover from Poissonian to antibunched behavior reflects in a clear way the crossover from delocalized to localized states. As J is increased, the relevant eigenstates of the coupled system are superpositions of center and outer cavity states while in the opposite case the good eigenstates are the Fock states due to the onset of photon blockade. A similar behavior can be recognized in the photon population, for which the strongly correlated region is identified by $\langle n_2(\phi = 0) \rangle = 0.5$ in the phase space defined by U/γ and J/γ, as evident also from the previous Figure.[2]

From a fundamental physical perspective, it is worth emphasizing that the main contribution of the present analogy was to establish the use of photon correlation measurements as effective probes of the interplay between macroscopic coherence and

[2]We notice that a slightly more complicated behavior, which does not hinder the main conclusions outlined above, occurs for the system of three coupled cavities, sketched in Fig. 6.6b, when the external resonators have a comparable dissipation rate as the central cavity $\gamma_{1,3} \sim \gamma$, since the coupled modes affect the physical response of the system at large J/γ [18].

interactions in small-scale strongly correlated photonic systems, which may eventually contain the key to interpret possible phases of driven-dissipative quantum-nonlinear cavity-arrays which operate under non-equilibrium conditions. On the other hand, given the enormous impact of the Josephson devices in electronics, from metrology to quantum information processing, a quantum optical Josephson interferometer might also constitute the building block for a new class of quantum optical devices. As an example, we notice that such a device can also be employed as a single-photon transistor in an integrated quantum photonic circuit. In fact, the same system shown in the schematic pictures of Fig. 6.6b can be seen as a three-point device where a single control photon absorbed by the central cavity from the vertical direction is able to suppress the phase coherence between the two coherent photonic states propagating in each waveguide channel, where information can be encoded. Photonic crystal platforms appear particularly suited to this purpose, since the collection efficiency of the middle cavity can be increased by far-field optimization techniques, as already shown in Fig. 6.3.

6.3.2.2 Out-of-Equilibrium Crossover Between Quantum States: The Driven-Dissipative Photonic Dimer

To generalize the system described in the previous Section towards multi-cavity arrays mimicking the physics of strongly correlated systems, but still keeping the system size at a level that is accessible to exact numerical simulation and/or analytic solution, we consider a two-site KH model [85], or a photonic dimer, represented schematically in Fig. 6.8 with the corresponding version in a photonic crystal platform. When the two resonators, both assumed nonlinear with the same coefficient U as in Eq. 6.7, are symmetrically pumped at the same rate $F/\gamma \ll 1$, the statistical analysis of the light emitted from the system shows a crossover from Poissonian to antibunched radiation similar to the behavior in Fig. 6.7. In particular, we notice that there is a rather sharp boundary dividing the limiting values $g^{(2)}(0) = 1$ from $g^{(2)}(0) = 0$ as a function of U/γ and J/γ: when the inter-cavity tunneling dominates ($J \gg U$) the light emitted is coherent, reflecting the statistics of the coherent laser beams. In fact, in this regime the system state is delocalized over the two cavities (photons can occupy either the symmetric or the antisymmetric combinations of the two isolated cavity modes). In the limit of dominant nonlinearity ($J \ll U$), the emitted light is antibunched as in the single-photon blockade case, since for $U \gg J, \gamma$ the number of photons in each cavity can only fluctuate between 0 and 1. However, it is worth stressing that this case is different from the single cavity photon blockade (which is the limit for $J = 0$ in the present case), since at increasing J/γ the crossover from $g^{(2)}(0) = 1$ to $g^{(2)}(0) = 0$ takes place at larger and larger values of U/γ (see dashed line in the figure). Hence, not only more than a single photon per pump beam will be prevented from entering the corresponding resonator ($U \gg \gamma$), but photons will be prevented from tunneling from one cavity to the other ($U \gg J$), being locked on each cavity.

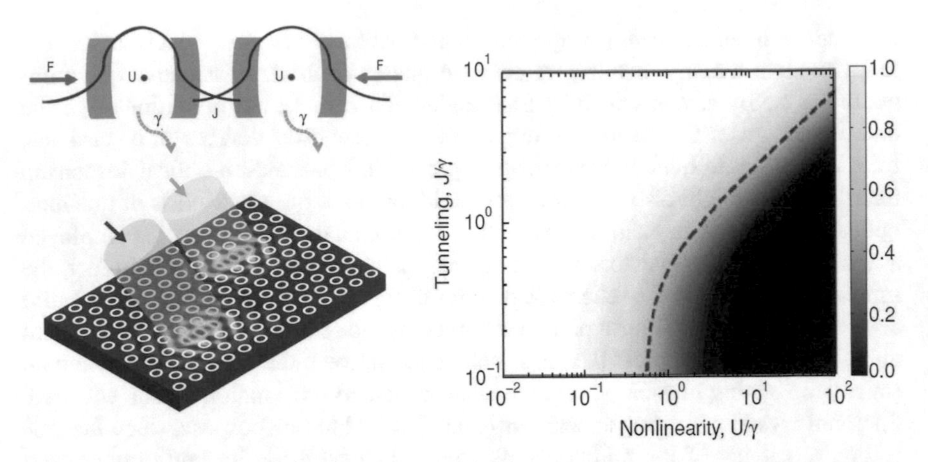

Fig. 6.8 Scheme of the driven-dissipative photonic dimer (*upper panel*), which can be practically realized with a photonic crystal molecule in an integrated photonic platform (*lower panel*). The corresponding second-order correlation function at zero-time delay is plotted as a function of U/γ and J/γ, respectively (results originally published in Ref. [85]). The dashed line is a plot of the function $J(U)$ corresponding to the condition $g^{(2)}(0) = 0.5$, marking the crossover boundary. Reprinted with permission from APS

We can interpret this state as the precursor of a Mott insulating state of photons in an infinite array of cavities, which requires more sophisticated generalization of the mean-field approach to effectively calculate the signatures of an out-of-equilibrium quantum phase transition [79]. However, the simplicity of this system also allows for an analytic solution of the master equation in the steady state condition and for $F/\gamma \ll 1$, where truncation of the Hilbert space to maximum double occupancy in each cavity is justified. In this case, such an analytic solution perfectly matches the full numerical one [85]. Interestingly, this simple two-site system already shows a clear interplay between tunneling and on-site interactions in a steady state condition, in which the second-order auto-correlation function can be again used as an effective probe of the quantum phase of the global system and the relative crossover on varying the system parameters. It has been later shown that, under proper conditions, such quantity can indeed be used to discriminate between different many-body quantum states out of equilibrium in driven-dissipative (infinitely extended) cavity arrays described by the KH model, in close analogy to the mean-field order parameter in the phase diagram of the equilibrium Bose-Hubbard model [80].

6.3.2.3 Photon Fermionization: Out-of-Equilibrium Tonks-Girardeau Gas

To further assess the use of photon correlation measurements to detect out-of-equilibrium strongly correlated photonic states, we describe a few sites analog of a one-dimensional chain of sites occupied by strongly interacting bosons. In the

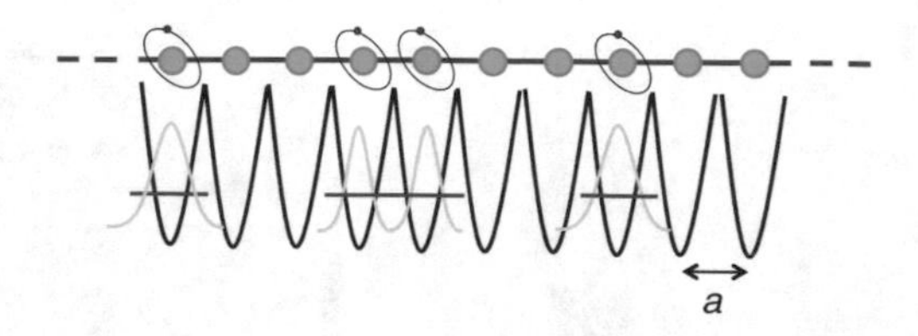

Fig. 6.9 Schematic representation of a one-dimensional lattice of tunnel-coupled wells, and corresponding single-particle wave functions. In the strongly interacting regime, the many-body wave function has nodes corresponding to multiple occupation of the same potential well

limit of impenetrable bosons in an infinite lattice with periodicity a, as in the scheme of Fig. 6.9, such system is known as the Tonks-Girardeau gas. When $U \to \infty$ in the one-dimensional Bose-Hubbard model at thermodynamic equilibrium, the many-body wave function of the Tonks-Girardeau gas is characterized by nodes whenever two particles approach each other in the same lattice site, as a direct consequence of the one-dimensional nature of the lattice. From a mathematical point of view, it was elegantly shown by Girardeau [7] that a rigorous mapping exists between the one-dimensional gas of impenetrable bosons and a gas of non-interacting and spinless fermions. In fact, the energy spectra of the two systems are identical, while a signature of the strong correlations existing between the bosons is evidenced in the momentum distributions, which remain instead quite different from the purely fermionic ones: the *pseudo*-wave vectors $q_i \in [-\pi/a, \pi/a]$ of the fermions occupying the lattice sites (where $i = 1 \ldots N_p$ runs over the total number of particles) do not have a direct meaning in terms of physical observables of the bosonic system. In particular, they do not correspond to their physical momentum k [7]. A truly remarkable fact of the Bose-Fermi mapping is that the periodicity condition on the N-body wavefunction does not directly transfer to the single-particle orbitals of the fermionized wavefunction: depending on the number N_p of particles in the system, the fermionic orbitals have to fulfill either periodic (if N_p is odd) or anti-periodic (if N_p is even) boundary conditions [86, 87]. This reflects on the quantization of the pseudo-wavevector $q = 2\pi n/M$ (if N_p is odd) or $q = 2\pi(n + 1/2)/M$ (if N_p is even), where n is an integer number and M is the number of sites. While it is practically impossible to detect this difference in a standard realization of the one-dimensional strongly interacting boson gas with cold atomic clouds [8, 9], a small number of sites would enable to evidence such a striking feature.

A possible realization of a strongly correlated gas of fermionized photons in an integrated photonic platform is shown in Fig. 6.10a for the smallest possible one-dimensional chain of photonic crystal nanocavities designed in a triangular lattice (e.g., three L3 cavities as in Fig. 6.3). This system was first proposed in [19]. In order to realize a KH model with uniform next neighbor tunneling rates, J, the cavities can be placed at 60° angle from each other (imposed by the triangular symmetry of the underlying photonic crystal lattice), in a necklace configuration such that periodic boundary conditions can be realized (i.e., the last cavity of the array corresponds to the first in the Hamiltonian). As sketched in this figure, such quantum photonic

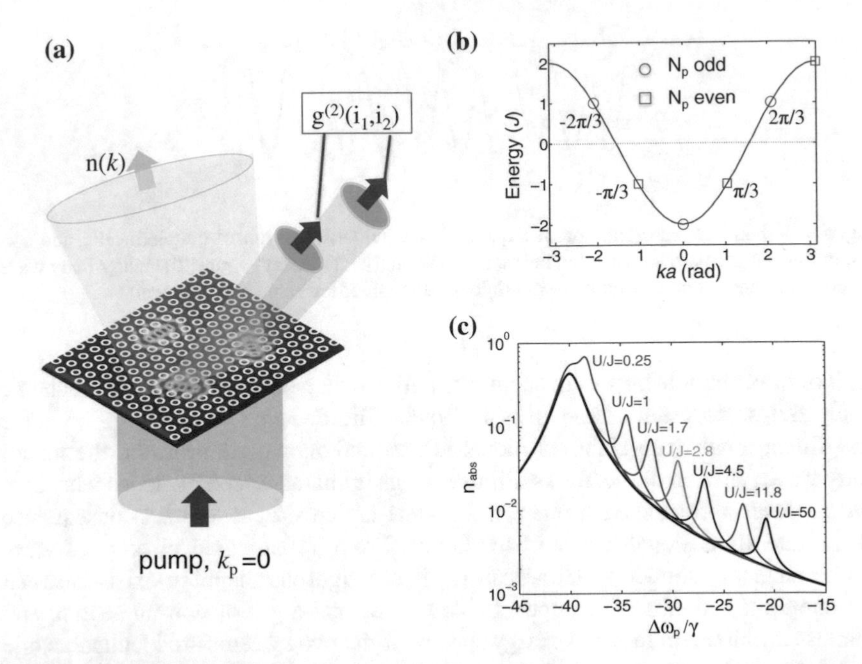

Fig. 6.10 **a** Realization of an out-of-equilibrium Tonks-Girardeau gas of interacting photons in a photonic crystal platform, made of a necklace of three tunnel-coupled nanocavities at 60° in a triangular lattice photonic crystal. **b** The quantized momenta for even and odd number of particles, respectively, is superimposed to the ideal dispersion of an infinitely extended one-dimensional chain of lattices; the one-dimensional band dispersion is plotted in units of J, giving a total bandwidth $4J$ as in usual tight-binding schemes. **c** Absorption spectra of the photonic array for fixed $J/\gamma = 20$ and increasing ratio *U/J*, showing convergence of the high-energy absorption peak towards the $N_p = 2$ state, i.e., the asymptotic value (for $U \to \infty$) $\Delta\omega_p = -J$ (results originally published in Ref. [19]). Reprinted with permission from APS

system provides experimental access to a wider range of observables than in ultra-cold atoms, from spectroscopic measures of absorption and transmission, to correlation measurements of the out-of-place emitted radiation. The peculiar pseudo-wave vector quantization rule outlined above is shown, as an example, in Fig. 6.10b, which exactly describes the case with $M = 3$. The quantized momenta are superimposed to the one-dimensional cosine-like band dispersion of the ideally infinite lattice, with a bandwidth $4J$ around the reference energy level (i.e., the bare cavity frequency ω_0 in this case). After the numerical solution of the master equation in steady state, Fig. 6.10c shows the calculated absorption for different values of the ratio *U/J* (at fixed $J/\gamma = 20$), as a function of the pump frequency detuning from the bare cavities frequency, $\Delta\omega_p = \omega_p - \omega_0$, for a pump rate $F/\gamma = 0.5$ and pump wave vector $k_p = 0$. Evidently, the main peak at $\Delta\omega_p = -2J$ corresponds to absorption from the resonance of the one particle state with $q = 0$. The additional feature that emerges in the strongly interacting regime ($U/J \gg 1$) is clearly related to the two-particle state, which is well captured by the lowest fermionized state with the pairs of photons

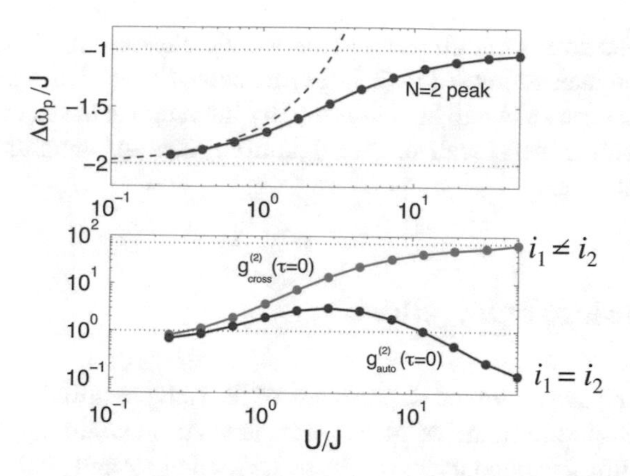

Fig. 6.11 The $N_p = 2$ absorption peak at different $\Delta\omega_p$ is tracked as a function of *U/J*: at large nonlinearity, the absorption peak occurs closer and closer to the limiting pump detuning $\Delta\omega_p = -J$. The *dashed lines* represent the asymptotic values in the dilute ($U/J \ll 1$) and strongly interacting ($U/J \gg 1$) limits, respectively. In the *lower panel*, the corresponding cross- and auto-correlation functions at zero time-delay are shown as a function of *U/J*

($N_p = 2$) occupying pseudo-momenta orbitals with $q = \pm\pi/3$. We notice that the half-integer value of the pseudo-momentum is a clear signature of the fermionization, and reflects in the asymptotic position of the peak at $\Delta\omega_p/J = -2\cos(\pi/3) = -1$.

As already shown in the previous examples, further information on the microscopic nature of the many-body physics of the system can be obtained by the intensity correlations of the emitted light. In the present case, we can generally define the correlation functions on the necklace as

$$g^{(2)}_{i_1,i_2}(\tau = 0) = \frac{\langle \hat{a}^\dagger_{i_1} \hat{a}^\dagger_{i_2} \hat{a}_{i_2} \hat{a}_{i_1} \rangle}{\langle \hat{a}^\dagger_{i_1} \hat{a}_{i_1} \rangle \langle \hat{a}^\dagger_{i_2} \hat{a}_{i_2} \rangle}, \tag{6.14}$$

where $i_{1,2}$ labels the site, and the auto (cross) correlation is for $i_1 = i_2$ ($i_1 \neq i_2$). These two functions are plotted in the lower panel of Fig. 6.11 as a function of *U/J*, for a pump frequency tuned exactly on resonance with the (*U/J*-dependent) two-photon transition, whose peak is tracked in the upper panel of the same Figure. When the system is dilute, or weakly interacting ($U/J \ll 1$), the resonance peaks corresponding to different values of N_p overlap, hence the emitted light inherits the Poissonian statistics of the coherent pump, $g^{(2)}_{i_1,i_2}(\tau = 0) \to 1$. In the opposite, strongly interacting regime ($U/J \gg 1$) the two-particle state has the fermionized form similar to the Tonks-Girardeau gas. As a consequence, for the $N_p = 2$ state there is a large probability that the pair of photons occupies different sites (strong bunching in the cross-correlation function), while little probability that they be on the same site (strong antibunching of autocorrelation). This is a striking operational demonstration of a genuine, out-of-equilibrium fermionization of a strongly interacting gas of photons on-chip. It is worth mentioning that different aspects of the steady state of an array of nonlinear and lossy optical resonators driven by coherent lasers have been later addressed in [88]. In particular, for weak driving intensity the steady state has been shown to be dominated by interactions in such a way that photon crystallize into dimers localized on neighboring sites and anticorrelations appear between distant sites.

6.4 Conclusions and Outlook

In summary, we have presented an overview of the rich possibilities offered by integrated photonic platforms made of two-dimensional photonic crystal lattices patterned in thin film semiconductors for the realization of strongly correlated photonic systems. Such systems allow to design and realize low-loss waveguides and cavities for photon propagation and confinement on-chip. We have described their essential physics, and the enhanced nonlinearities that can be envisioned. Coupled multi-cavity arrays and waveguides in photonic crystal platforms could allow to fully address the interplay between tunneling and photon correlations on chip. The driven-dissipative nature of the photon gas is responsible for a wealth of unexplored features due to the interplay of non-equilibrium statistical mechanics, quantum optics, and many-body physics. Regarding strongly correlated photonic systems potentially realized in such platforms, we have described small scale systems, fully compatible with state-of-the art fabrication capabilities, where the interplay between inter-site interactions and on-site correlations can be meaningfully probed. In particular, we have concentrated on quantities of direct experimental access in photonics, such as field intensities, transmission or reflection, photon correlations. With respect to the latter, we have shown in systems of increasing complexity how the second-order correlation function of emitted photons can represent a key probe allowing to operationally discriminate between delocalized and strongly correlated regimes, even under typical driven-dissipative conditions of open quantum photonic systems.

The field of strongly correlated photonic systems is a relatively new one, and as such a large number of questions are still calling for a theoretical assessment and/or new experimental ways of being addressed. In this respect, nanophotonic platforms are likely to play an important role in the emerging activity aimed at probing the first theoretical predictions made in the past few years, as well as preparing the ground for the new questions that will certainly arise. In perspective, we can identify a few research direction that will definitely benefit from developing strongly correlated photonic systems on-chip, besides unravelling the open questions regarding out-of-equilibrium statistical mechanics. On one side, the interplay of artificial gauge fields and strongly correlated photonic systems might lead to the generation of topologically nontrivial states of a photon gas, such as out-of-equilibrium Laughlin [89] or Majorana [90] states. On another side, driven-dissipative cavity arrays might become the preferential platform to engineer continuous-variable bipartite and quadripartite entanglement, and eventually cluster entangled states [91], with potential impact on a photonics-based quantum information architecture in the near future. Finally, the increasing interest in quantum simulators is a further arena where nanophotonic circuits of strongly correlated photons could play a key role [92]. In fact, even if the idea of simulating quantum matter with other well-controlled quantum systems has found its most straightforward realization with ultra-cold atoms in optical lattices, it is nevertheless true that open systems-based quantum simulators provide a novel complement by naturally accessing non-equilibrium physics.

Acknowledgements This chapter was meant to provide a short review of some of our research works on strongly correlated photonic systems in integrated photonic platforms. For all these contributions, and for fruitful inspiration and several useful discussions, we are indebted to L.C. Andreani, S. De Liberato, R. Fazio, S. Ferretti, M. Galli, V. Giovannetti, A. Imamoğlu, V. Savona, and H.E. Türeci.

References

1. A.C. Hewson, *The Kondo Problem to Heavy Fermions* (Cambridge University Press, Cambridge, 1993)
2. U. Meirav, E.B. Foxman, Semicond. Sci. Technol. **11**, 255 (1996)
3. D. Jaksch, C. Bruder, J.I. Cirac, C.W. Gardiner, P. Zoller, Phys. Rev. Lett. **81**, 3108 (1998)
4. M. Greiner, O. Mandel, T. Esslinger, T.W. Hänsch, I. Bloch, Nature **415**, 39 (2002)
5. M.P.A. Fisher, P.B. Weichman, G. Grinstein, D.S. Fisher, Phys. Rev. B **40**, 546 (1989)
6. S. Sachdev, *Quantum Phase Transitions* (Cambridge University Press, New York, 2011)
7. M.D. Girardeau, J. Math. Phys. **1**, 516 (1960)
8. B. Paredes et al., Nature **429**, 277 (2004)
9. T. Kinoshita, T. Wenger, D.S. Weiss, Science **305**, 1125 (2004)
10. M.A. Cazalilla, R. Citro, T. Giamarchi, E. Orignac, M. Rigol, Rev. Mod. Phys. **83**, 1405 (2011)
11. A. Imamoğlu, H. Schmidt, G. Woods, M. Deutsch, Phys. Rev. Lett. **79**, 1467 (1997)
12. K.M. Birnbaum, A. Boca, R. Miller, A.D. Boozer, T.E. Northup, H.J. Kimble, Nature **436**, 87 (2005)
13. A. Faraon, I. Fushman, D. Englund, N. Stoltz, P. Petroff, J. Vučković, Nat. Phys. **4**, 859 (2008)
14. A. Reinhard, T. Volz, M. Winger, A. Badolato, K.J. Hennessy, E.L. Hu, A. Imamoğlu, Nat. Photon. **6**, 93 (2012)
15. M.J. Hartmann, F.G.S.L. Brandão, M.B. Plenio, Nat. Phys. **2**, 849 (2006)
16. A.D. Greentree, C. Tahan, J.H. Cole, L.C.L. Hollenberg, Nat. Phys. **2**, 856 (2006)
17. D.G. Angelakis, M.F. Santos, S. Bose, Phys. Rev. A **76**, 031805(R) (2007)
18. D. Gerace, H.E. Türeci, A. Imamoğlu, V. Giovannetti, R. Fazio, Nat. Phys. **5**, 281 (2009)
19. I. Carusotto, D. Gerace, H.E. Türeci, S. De Liberato, C. Ciuti, A. Imamoğlu, Phys. Rev. Lett. **103**, 033601 (2009)
20. J.L. O'Brien, A. Furusawa, J. Vučković, Nat. Photon. **3**, 687 (2009)
21. O. El Daïf, A. Baas, T. Guillet, J.-P. Brantut, R. Idrissi, Kaitouni, J.L. Staehli, F. Morier-Genoud, B. Deveaud, Appl. Phys. Lett. **88**, 061105 (2006)
22. M. Liscidini, D. Gerace, D. Sanvitto, D. Bajoni, Appl. Phys. Lett. **98**, 121118 (2011)
23. J.M. Gérard, D. Barrier, J.Y. Marzin, R. Kuszelewicz, L. Manin, E. Costard, V. Thierry-Mieg, T. Rivera, Appl. Phys. Lett. **69**, 449 (1996)
24. G. Panzarini, L.C. Andreani, Phys. Rev. B **60**, 16799 (1999)
25. L.C. Andreani, in *Electron and Photon Confinement in Semiconductor Nanostructures*, ed. by B. Deveaud, A. Quattropani, P. Schwendimann (IOS Press, Amsterdam, 2003), p. 105
26. I. Carusotto, C. Ciuti, Rev. Mod. Phys. **85**, 299 (2013)
27. S. Pirotta, M. Patrini, M. Liscidini, M. Galli, G. Dacarro, G. Canazza, G. Guizzetti, D. Comoretto, D. Bajoni, Appl. Phys. Lett. **104**, 051111 (2014)

28. G. Lerario, A. Cannavale, D. Ballarini, L. Dominici, M. De Giorgi, M. Liscidini, D. Gerace, D. Sanvitto, G. Gigli, Opt. Lett. **39**, 2068 (2014)
29. M. Notomi, A. Shinya, A. Mitsugi, S. Kuramochi, H.Y. Ryu, Opt. Express **12**, 551 (2004)
30. S.J. McNab, N. Moll, Y.A. Vlasov, Opt. Express **11**, 2927 (2003)
31. Y. Sugimoto, Y. Tanaka, N. Ikeda, Y. Nakamura, K. Asakawa, K. Inoue, Opt. Express **12**, 1090 (2004)
32. L. O'Faolain, X. Yuan, D. McIntyre, S. Thoms, H. Chong, R.M. De La Rue, T.F. Krauss, Electron. Lett. **42**, 1454 (2006)
33. Y. Takahashi, H. Hagino, Y. Tanaka, B.S. Song, T. Asano, S. Noda, Opt. Express **15**, 17206 (2007)
34. M. Notomi, Rep. Prog. Phys. **73**, 096501 (2010)
35. J.D. Joannopoulos, S.G. Johnson, J.N. Winn, R.D. Meade, *Photonic Crystals: Molding the Flow of Light* (Princeton University Press, Princeton, 2008)
36. K. Aoki, D. Guimard, M. Nishioka, M. Nomura, S. Iwamoto, Y. Arakawa, Nat. Photon. **2**, 688 (2008)
37. J.M. van den Broek, L.A. Woldering, R.W. Tjerkstra, F.B. Segerink, I.D. Setija, W.L. Vos, Adv. Mater. **22**, 25 (2012)
38. T. Ochiai, K. Sakoda, Phys. Rev. B **64**, 045108 (2001)
39. L.C. Andreani, D. Gerace, Phys. Rev. B **73**, 235114 (2006)
40. A. Yariv, P. Yeh, *Photonics: Optical electronics in modern communications* (Oxford University Press, New York, 2007)
41. D. Gerace, L.C. Andreani, Opt. Lett. **29**, 1897 (2004)
42. Y. Akahane, T. Asano, B.S. Song, S. Noda, Nature **425**, 944 (2003)
43. L.C. Andreani, D. Gerace, M. Agio, Photon. Nanostruct. Fundam. Appl. **2**, 103 (2004)
44. H. Sekoguchi, Y. Takahashi, T. Asano, S. Noda, Opt. Express **22**, 916 (2014)
45. Y. Lai, S. Pirotta, G. Urbinati, D. Gerace, M. Minkov, V. Savona, A. Badolato, M. Galli, Appl. Phys. Lett. (to be published, May 2014)
46. S. Combrié, A. De Rossi, Q.V. Tran, H. Benisty, Opt. Letters **33**, 1908 (2008)
47. N.-V.-Q. Tran, S. Combrié, A. De Rossi, Phys. Rev. B **79**, 041101(R) (2009)
48. S.L. Portalupi, M. Galli, C. Reardon, T.F. Krauss, L. O'Faolain, L.C. Andreani, D. Gerace, Opt. Express **18**, 16064 (2010)
49. D. Gerace, L.C. Andreani, Phys. Rev. B **75**, 235325 (2007)
50. J. Kasprzak, M. Richard, S. Kundermann, A. Baas, P. Jeambrun, J.M.J. Keeling, F.M. Marchetti, M.H. Szymańska, R. André, J.L. Staehli, V. Savona, P.B. Littlewood, B. Deveaud, L.S. Dang, Nature **443**, 409 (2006)
51. D. Bajoni, D. Gerace, M. Galli, J. Bloch, R. Braive, I. Sagnes, A. Miard, A. Lemaître, M. Patrini, L.C. Andreani, Phys. Rev. B **80**, 201308(R) (2009)
52. S. Azzini, D. Gerace, M. Galli, I. Sagnes, R. Braive, A. Lemaître, J. Bloch, D. Bajoni, Appl. Phys. Lett. **99**, 111106 (2011)
53. A. Faraon, C. Santori, Z. Huang, V.M. Acosta, R.G. Beausoleil, Phys. Rev. Lett. **109**, 033604 (2012)
54. D. Englund, I. Fushman, A. Faraon, J. Vučković, Photon. Nanostruct. Fundam. Appl. **7**, 56 (2009)
55. R. Bose, D. Sridharan, G. Solomon, E. Waks, Opt. Express **19**, 5398 (2011)
56. N. Caselli, F. Intonti, F. Riboli, A. Vinattieri, D. Gerace, L. Balet, L.H. Li, M. Francardi, A. Gerardino, A. Fiore, M. Gurioli, Phys. Rev. B **86**, 035133 (2012)
57. A. Majumdar, A. Rundquist, M. Bajcsy, V.D. Dasika, S.R. Bank, J. Vučković, Phys. Rev. B **86**, 195312 (2012)
58. N. Caselli, F. Intonti, C. Bianchi, F. Riboli, S. Vignolini, L. Balet, L.H. Li, M. Francardi, A. Gerardino, A. Fiore, M. Gurioli, Appl. Phys. Lett. **101**, 211108 (2013)
59. T. Cai, R. Bose, G.S. Solomon, E. Waks, Appl. Phys. Lett. **102**, v141118 (2013)
60. R. Boyd, *Nonlinear Optics* (Academic Press, California, 1992)
61. P.D. Drummond, D.F. Walls, J. Phys. A **13**, 725 (1980)
62. S. Ferretti, D. Gerace, Phys. Rev. B **85**, 033303 (2012)

63. M.J. Werner, A. Imamoğlu, Phys. Rev. A **61**, 011801(R) (1999)
64. R. Loudon, *The Quantum Theory of Light* (Oxford University Press, Oxford, 2003)
65. M. Galli, D. Gerace, K. Welna, T.F. Krauss, L. O'Faolain, G. Guizzetti, L.C. Andreani, Opt. Express **18**, 26613 (2010)
66. L. Tian, H.J. Carmichael, Phys. Rev. A **46**, 6801(R) (1992)
67. E.T. Jaynes, F.W. Cummings, Proc. IEEE **51**, 89 (1963)
68. L.C. Andreani, G. Panzarini, J.-M. Gérard, Phys. Rev. B **60**, 13276 (1999)
69. S. Rebić, A.S. Parkins, S.M. Tan, Phys. Rev. A **69**, 035804 (2004)
70. K. Hennessy, A. Badolato, M. Winger, D. Gerace, M. Atatüre, S. Gülde, S. Fält, E. Hu, A. Imamoğlu, Nature **445**, 896 (2007)
71. A. Verger, C. Ciuti, I. Carusotto, Phys. Rev. B **73**, 193306 (2006)
72. I. Carusotto, T. Volz, A. Imamoğlu, Europhys. Lett. **90**, 37001 (2010)
73. A. Majumdar, D. Gerace, Phys. Rev. B **87**, 235319 (2013)
74. T.C.H. Liew, V. Savona, Phys. Rev. Lett. **104**, 183601 (2010)
75. M. Bamba, A. Imamoğlu, I. Carusotto, C. Ciuti, Phys. Rev. A **83**, 021802(R) (2011)
76. H.J. Carmichael, *An Open Systems Approach to Quantum Optics* (Springer, Berlin, 1993)
77. T. Grujic, S.R. Clark, D. Jaksch, D.G. Angelakis, New J. Phys. **14**, 103025 (2012)
78. F. Nissen, S. Schmidt, M. Biondi, G. Blatter, H.E. Türeci, J. Keeling, Phys. Rev. Lett. **108**, 233603 (2012)
79. A. Tomadin, V. Giovannetti, R. Fazio, D. Gerace, I. Carusotto, H.E. Türeci, A. Imamoğlu, Phys. Rev. A **81**, 061801(R) (2010)
80. A. La Boité, G. Orso, C. Ciuti, Phys. Rev. Lett. **110**, 233601 (2013)
81. K.A. Matveev, M. Gisselfält, L.I. Glazman, M. Jonson, R.I. Shekhter, Phys. Rev. Lett. **70**, 2940 (1993)
82. P. Joyez, P. Lafarge, A. Filipe, D. Esteve, M.H. Devoret, Phys. Rev. Lett. **72**, 2458 (1994)
83. L.J. Geerligs, L.E.M. de Groot, A. Verbruggen, J.E. Mooji, Phys. Rev. Lett. **63**, 326 (1989)
84. W.J. Elion, M. Matters, U. Geigenmüller, J.E. Mooji, Nature **371**, 594 (1994)
85. S. Ferretti, L.C. Andreani, H.E. Türeci, D. Gerace, Phys. Rev. A **82**, 013841 (2010)
86. E.H. Lieb, W. Liniger, Phys. Rev. **130**, 1605 (1963)
87. I. Carusotto, Y. Castin, New J. Phys. **5**, 91 (2003)
88. M.J. Hartmann, Phys. Rev. Lett. **104**, 113601 (2010)
89. R.O. Umucalilar, I. Carusotto, Phys. Rev. Lett. **108**, 206809 (2012)
90. C.E. Bardyn, A. Imamoğlu, Phys. Rev. Lett. **109**, 253606 (2012)
91. T.C.H. Liew, V. Savona, New J. Phys. **15**, 025015 (2013)
92. A.A. Houck, H.E. Türeci, J. Koch, Nat. Phys. **8**, 292 (2012)

Chapter 7
Quantum Simulations with Circuit Quantum Electrodynamics

Guillermo Romero, Enrique Solano and Lucas Lamata

Abstract Superconducting circuits have become a leading quantum platform for the implementation of quantum information tasks. Here, we revise the basic concepts of circuit network theory and circuit quantum electrodynamics for the sake of analog and digital quantum simulations with microwave photons in superconducting circuit lattices. We prove that superconducting circuits are a promising quantum technology for building scalable quantum simulators, enjoying unique and distinctive properties when compared to more advanced platforms as trapped ions and optical lattices.

7.1 Introduction

Nowadays, the field of quantum simulations [1–7] is one of the most active in quantum information science. Following the original idea by Feynman [8], and its subsequent development by Lloyd [9], this field has experienced a significant growth in the last decade. The motivation is the fact that a large quantum system cannot be efficiently simulated with a classical computer due to the exponential growth of the Hilbert space dimension with the number of quantum subsystems. On the other hand, it should be feasible to reproduce the dynamics of quantum systems making use of other, controllable, quantum platforms, which constitute a quantum simulator.

G. Romero (✉)
Departamento de Física, Universidad de Santiago de Chile (USACH),
Avenida Ecuador 3493, 9170124 Santiago, Chile
e-mail: guillermo.romero@usach.cl

E. Solano · L. Lamata
Department of Physical Chemistry, University of the Basque Country UPV/EHU,
Apartado 644, 48080 Bilbao, Spain
e-mail: enr.solano@gmail.com

L. Lamata
e-mail: lucas.lamata@gmail.com

E. Solano
IKERBASQUE, Basque Foundation for Science, Maria Diaz de Haro 3,
48013 Bilbao, Spain

D.G. Angelakis (ed.), *Quantum Simulations with Photons and Polaritons*,
Quantum Science and Technology, DOI 10.1007/978-3-319-52025-4_7

Superconducting circuits [10–12] and circuit quantum electrodynamics (QED) [13–15] represent prime candidates to implement a quantum simulator with interacting microwave photons because of their scalability and controllability. There have been already some proposals for quantum simulations in superconducting qubits, as is the case of the quantum simulation of Anderson and Kondo lattices [16], sudden phase switching in a superconducting qubit array [17], molecular collisions [18], quantum phases [19], Holstein polarons [20], and quantum magnetism [21, 22]. Moreover, two pioneering experiments on digital quantum simulators of fermions [23] and spins [24] have been performed.

In this chapter, we introduce the main concepts of superconducting circuits and circuit QED, and their performance as a quantum simulator with interacting microwave photons. In particular, in Sect. 7.2, we will describe the circuit network theory and the Hamiltonian description of a quantum circuit. In Sect. 7.3, we provide an introduction to circuit QED and cavity-cavity coupling mechanism, pointing out the coupling regimes of light-matter interaction as building blocks for circuit QED lattices. In Sect. 7.4, we discuss the analog quantum simulation of many-body states of light and relativistic phenomena. In addition, in Sect. 7.5, we present our recent proposals of digital quantum simulations in circuit QED, such as spin chains and quantum field theories. Finally, in Sect. 7.6, we present our concluding remarks.

7.2 Circuit Network Theory

Nowadays, integrated quantum circuits [10–12] have become a leading technology for quantum information processing and quantum simulations. These devices present noticeable features such as scalability, controllability, and tunable physical parameters which in turn allow us to engineer complex Hamiltonians. In this sense, it is important to understand how an integrated circuit shows its quantum nature, and how to design *two-level systems* or *qubits*. To achieve it, there are three features that one should point out: (i) ultra-low dissipation provided by *superconductivity*, (ii) ultra-low noise reached by *low temperatures*, and (iii) nonlinear, non dissipative elements implemented by Josephson junctions.

(i) *Ultra-low dissipation*. In an integrated circuit, all the metallic parts have to be made out of superconducting materials with negligible resistance at the qubit operating temperature and at the qubit frequency transition. In particular, current experiments make use of low temperature superconductors [25] such as aluminum or niobium which in turn allow quantum signals to propagate without experiencing dissipation, thus any encoded quantum information may preserve its coherence.
(ii) *Ultra-low noise*. In a physical realization one can access qubit energies $\hbar\omega_q$ that belong to the range 1–10 GHz. In order to avoid thermal fluctuations that may spoil the quantum coherence of qubits, integrated circuits must be cooled down to temperatures of about $T \approx 20\,\text{mK}$. In general, the energy scales that

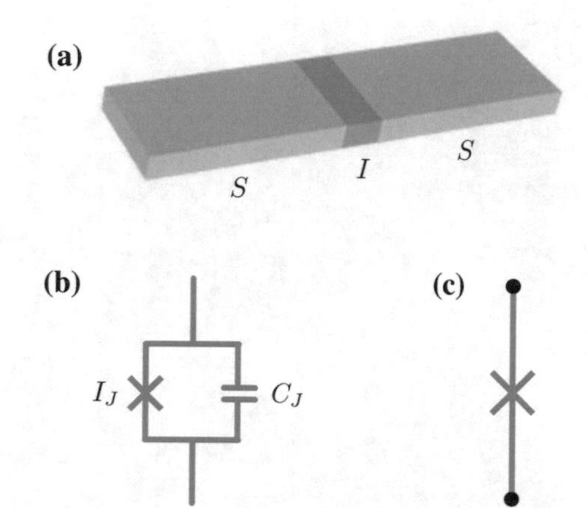

Fig. 7.1 **a** Schematic representation of a Josephson junction. It consists of two bulk superconductors linked by a thin insulator layer (≈2 nm). **b** In the zero-voltage state, a JJ can be characterized by a critical current I_J and a capacitance C_J. **c** In circuit network theory, a JJ can be described by a single cross that links two nodes

appear in the system should satisfy the conditions $kT \ll \hbar\omega_q$ and $\hbar\omega_q \ll \Delta$, where Δ is the energy gap of the superconducting material.

(iii) *Nonlinear, non dissipative elements*. In order to engineer and manipulate two-level systems, it is necessary the access to a device that allows unequal spaced energy levels at the zero-voltage state, where no dissipative current flows through it. These conditions are matched by tunnel junctions [26] as depicted in Fig. 7.1. The Josephson junction (JJ) is a device that consists of two bulk superconductors linked by a thin insulator layer, typically of 1–2 nm.

7.2.1 Hamiltonian Description of a Circuit Network

In general, an electrical circuit network is formed by an array of branches and nodes as depicted in Fig. 7.2a. Each branch may contain linear or non linear devices such as inductors, capacitors, or Josephson junctions, and it can be characterized by a branch flux (ϕ_b) and a branch charge (q_b) which are defined in terms of the branch voltages and branch currents (see Fig. 7.2b) by

$$\phi_b(t) = \int_{t_0}^{t} dt'\, V_b(t'), \tag{7.1}$$

$$q_b(t) = \int_{t_0}^{t} dt'\, I_b(t'). \tag{7.2}$$

However, these variables do not constitute the degrees of freedom of the circuit because they are linked by the circuit topology through Kirchhoff's laws. Indeed, the sum of all voltages around a closed path Γ has to be zero $\sum_{\Gamma[b]} V_b = 0$. In addition, the sum of all currents of branches tied to a node has to be zero $\sum_{\nu[b]} I_b = 0$, where ν determines a specific node of the network.

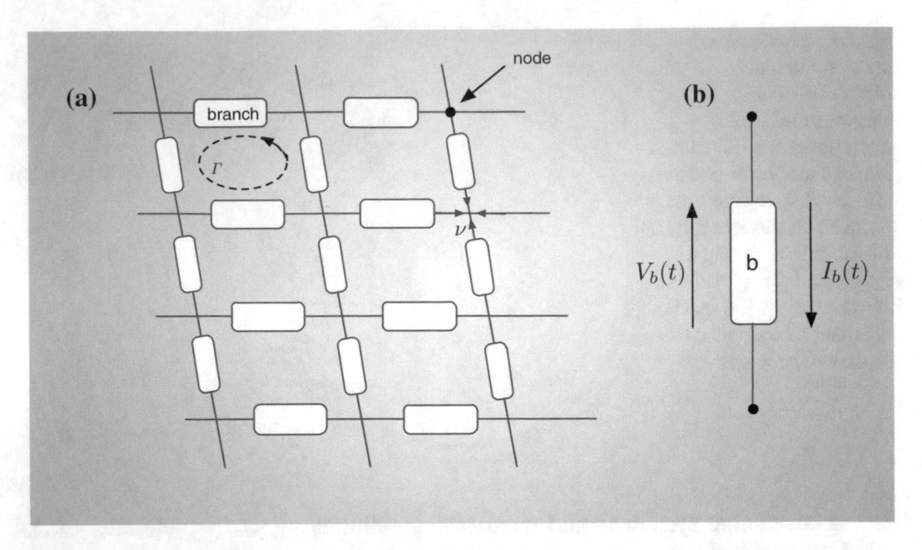

Fig. 7.2 **a** Electrical circuit network formed by an array of branches and nodes. Each branch may contain an inductor, a capacitor, or a Josephson junction. **b** A single branch is characterized by a voltage drop $V_b(t)$ and a current $I_b(t)$. As in a classical electrical circuit, one has to choose the sign convention for currents and voltages, and the Kirchhoff's laws allow us to compute the classical motion equations

In order to describe the dynamics of an electrical circuit, one should identify the independent coordinates ϕ_n, their associated velocities $\dot{\phi}_n$, and to formulate the corresponding Lagrangian $L(\phi_n, \dot{\phi}_n) = T - V$, where T and V stand for the kinetic and potential energies, respectively. A detailed analysis of this procedure can be found in Refs. [27–29], and here we summarize the main concepts.

In circuit theory, one can define *node fluxes* (ϕ_n) which are variables located at the nodes of the network. These variables depend on a particular description of the topology of the circuit, and such a description is based on the *spanning tree* concept presented in Ref. [27]. Specifically, one of the nodes is chosen to be the reference by letting them to act as the ground, say $\phi_N = 0$. From the ground node one chooses a unique path that connects to the *active nodes* without closing loops. Figure 7.3 shows two possible choices of *node fluxes* and the *spanning tree* (T). The latter defines two kind of branches in the network: the set of branches that belong to the spanning tree (blue branches), and the set of closure (C) branches each associated with an irreducible loop (grey branches). In terms of node fluxes, each branch belonging to the spanning tree can be defined as $\phi_{b(T)} = \phi_{n+1} - \phi_n$. In addition, the closure branches must be treated in a special way because of the constraint imposed by the flux quantization [25]. The above condition establishes that for a closed loop threaded by an external magnetic flux $\Phi_{\rm ext}$, the sum of all flux branches that belong to that loop satisfies the rule $\sum_{\Gamma[b]} \phi_b - \Phi_{\rm ext} = m\Phi_0$, where $\Phi_0 = h/2e$ is the flux quantum, and m is an integer. In this sense, the treatment for a closure branch follows the relation $\phi_{b(C)} = (\phi_{n+1} - \phi_n) - \Phi_{\rm ext}$.

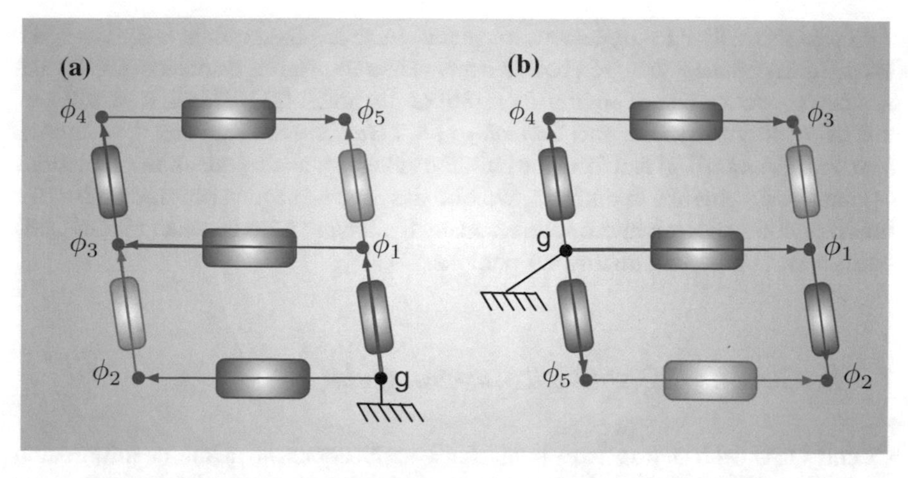

Fig. 7.3 Two possible choices of the spanning tree in an electrical circuit. From the ground node $\phi_N = 0$, one chooses single paths that pass through a branch (*blue branches*) to connect actives nodes. The remaining branches (*grey branches*) that close a loop have a special treatment due to the flux quantization condition

The building blocks of a general electrical circuit are capacitors, inductors, and Josephson junctions. From these elements it is possible to formulate the corresponding Lagrangian by taking into account their corresponding energies as follows

$$T = \frac{C_j}{2}(\dot{\phi}_{n+1} - \dot{\phi}_n)^2, \tag{7.3}$$

$$V = \frac{1}{2L_j}(\phi_{n+1} - \phi_n)^2, \tag{7.4}$$

$$V_{JJ} = \frac{C_{j,J}}{2}(\dot{\phi}_{n+1} - \dot{\phi}_n)^2 - E_{j,J}\cos\left(\frac{\phi_{n+1} - \phi_n}{\varphi_0}\right), \tag{7.5}$$

where C_j, L_j, $C_{j,J}$, and $E_{j,J}$ stand for the jth capacitance, inductance, Josephson capacitance, and Josephson energy, respectively. This procedure allows us to know the matrix capacitance of the system which, in turn, allows us to formulate the Hamiltonian through a Legendre transformation.

7.3 Circuit Quantum Electrodynamics

Nowadays, quantum technologies offer a testbed for fundamentals and novel applications of quantum mechanics. In particular, circuit QED [13–15] has become a leading platform due to its controllability and scalability, with different realizations involving the interaction between on-chip microwave resonators and superconducting circuits which in turn allow to implement transmon, flux, or phase qubits [12]. Here, we briefly describe the interaction between a coplanar waveguide resonator (CWR) and a superconducting circuit be transmon [30] or flux qubit [31]. This in turn will permit us to introduce the *Jaynes-Cummings (JC) Hamiltonian* [32] and the *quantum Rabi Hamiltonian* [33] that form the building blocks for analog quantum simulations of many-body physics. In addition, we also describe some physical mechanism for the interaction between microwave resonators, allowing us to consider circuit QED lattices with interacting microwave photons.

7.3.1 Circuit QED with a Transmon Qubit

Circuit QED with a *transmon* qubit finds applications in quantum information processing (QIP), where it is possible to implement single- and two-qubit quantum gates [34–40], and three-qubit entanglement generation [41–43]. These proposals find a common basis in the light-matter interaction described by the Jaynes-Cummings model [32].

Figure 7.4a represents the electrostatic coupling between a quarter-wave cavity and a transmon qubit, as implemented in current experiments [44]. For the lowest cavity eigenfrequency, the spatial distribution of the voltage $V(x)$ follows the profile shown in Fig. 7.4c. Here, the transmon has to be located at the cavity end to assure a maximum coupling strength. Figure 7.4b shows the effective circuit for the above situation. Two Josephson junctions with capacitance C_J and Josephson energy E_J are shunted by an additional large capacitance C_I. This system is coupled to the cavity, with capacitance and inductance C_r and L_r, by a comparably large gate capacitance C_c. In addition, the transmon is coupled to an external source of dc voltage V_g through the capacitance C_g. It is noteworthy to mention that the capacitances C_I, C_c, and C_g represent effective quantities seen by the transmon, see Ref. [30].

The effective Hamiltonian of the joint cavity-transmon system reads

$$H = 4E_C(n - n_g)^2 - E_J \cos\varphi + \hbar\omega_r a^\dagger a + 2i\beta e V^0_{\text{rms}} n(a - a^\dagger), \tag{7.6}$$

where n and φ stand for the number of Cooper pairs transferred between the islands and the gauge-invariant phase difference between them, respectively. In addition, $a(a^\dagger)$ annihilates(creates) a single photon of frequency ω_r, $E_C = e^2/2C_\Sigma$ is the charging energy, with $C_\Sigma = C_I + C_J + C_c + C_g$ is the total capacitance associated with the transmon, $n_g = C_g V_g/2e$ is the effective offset charge, and $V^0_{\text{rms}} = \sqrt{\hbar\omega_r/2C_r}$.

The transmon qubit is less sensitive to charge noise due to an added shunting capacitance C_I between the superconducting islands [30]. This lowers E_C, resulting in an energy ratio of $E_J/E_C \approx 50$. In this case, the two lowest energy levels have an energy splitting that can be approximated by

$$\hbar\omega_q \approx \sqrt{8E_C E_J^{\text{max}} |\cos(\pi\Phi_{\text{ext}}/\Phi_0)|} - E_C. \tag{7.7}$$

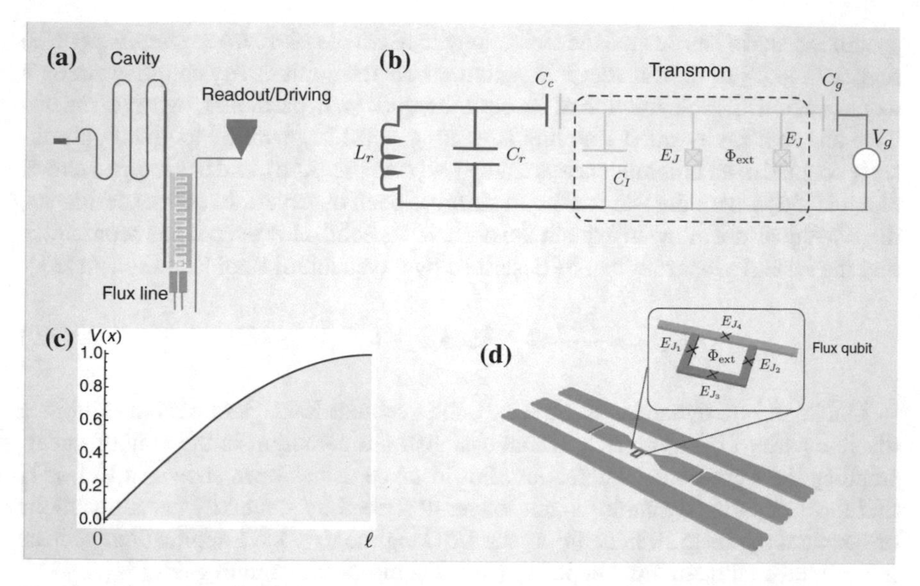

Fig. 7.4 Schematic of circuit QED with a transmon qubit. **a** A $\lambda/4$ cavity is capacitively coupled to a transmon device. It is also shown the flux line that provides an external magnetic field threading a SQUID loop, and an additional cavity for measuring and driving the qubit. **b** Effective lumped circuit element of the cavity-transmon system. **c** Voltage distribution for a $\lambda/4$ cavity, where ℓ stands for the cavity length. **d** A flux qubit formed by three Josephson junctions is galvanically coupled to an inhomogeneous cavity by means of a fourth junction with Josephson energy E_{J4}. This kind of coupling allows us to reach the ultrastrong coupling regime

Notice that the qubit energy can be tuned by an external magnetic flux Φ_{ext} applied to the superconducting quantum interference device (SQUID), see Fig. 7.4b. In this two-level approximation, the dynamics of the cavity-transmon system can be described by the Jaynes-Cummings Hamiltonian [32]

$$H_{\text{JC}} = \frac{\hbar\omega_q}{2}\sigma_z + \hbar\omega_r a^\dagger a + \hbar g(\sigma^+ a + \sigma^- a^\dagger), \tag{7.8}$$

which exhibits a continuous $U(1)$ symmetry. Here, the two-level system is described by the Pauli matrices σ_j $(j = x, y, z)$, and g is the cavity-qubit coupling strength. Notice that working at or near resonance ($\omega_q \sim \omega_r$), the above Hamiltonian holds true in the rotating-wave approximation (RWA) where the system parameters satisfy the condition $\{g, |\omega_q - \omega_r|\} \ll \omega_q + \omega_r$.

7.3.2 *Circuit QED with a Flux Qubit*

Circuit QED can also be implemented by means of the inductive interaction between the persistent-current qubit [31], or flux qubit, and a microwave cavity [15]. This specific scenario has pushed the technology to reach the *ultrastrong coupling* (USC) regime [45–50] of light-matter interaction, where the qubit-cavity coupling strength reaches a considerable fraction of the cavity frequency. In particular, two experiments have shown a cavity-qubit coupling strength $g = 0.12\omega_r$ with a flux qubit galvanically coupled to an inhomogeneous cavity [49] (see Fig. 7.4d), and to a lumped circuit element [50] where the Bloch-Siegert shift has been observed. In both experiments, the RWA does not allow to explain the observed spectra. However, it has been shown that the system properties can be described by the quantum Rabi Hamiltonian [33]

$$H_{\text{Rabi}} = \frac{\hbar\omega_q}{2}\sigma_z + \hbar\omega_r a^\dagger a + \hbar g \sigma_x (a + a^\dagger). \tag{7.9}$$

Unlike the JC dynamics in Eq. (7.8), the quantum Rabi Hamiltonian exhibits a discrete parity (Z_2) symmetry which establishes a paradigm in the way of understanding the light-matter interaction. For instance, it has been shown in Ref. [51] that the dissipative dynamics is not longer described by standard master equations of quantum optics [52]. In addition, the USC regime may have applications such as parity-protected quantum computing [53], and ultrafast quantum gates [54].

The quantum Rabi Hamiltonian allows also to describe the *deep strong coupling* (DSC) regime [55], where the coupling strength is similar or larger than the cavity frequency. This coupling regime has interesting consequences in the breakdown of the Purcell effect [56], and it has also been simulated in a waveguide array [57].

7.3.3 Cavity-Cavity Interaction Mechanisms

The versatility of superconducting circuits allow us to design complex arrays involving the interaction among several microwave cavities. There are two possible physical mechanisms to couple them, that is, by means of the capacitive coupling of two half wave cavities (Fig. 7.5a), or two quarter wave cavities via current-current coupling mediated by a SQUID (Fig. 7.5b). In both cases, it is possible to show that the cavity-cavity interaction can be described by a nearest-neighbor Hamiltonian

$$H = \hbar \sum_n \omega_n a_n^\dagger a_n + \hbar \sum_{\langle n,n' \rangle} J_{nn'} a_n^\dagger a_{n'}, \tag{7.10}$$

where $a_n(a_n^\dagger)$ stands for the annihilation(creation) operator associated with the nth mode of frequency ω_n. Here, we assume the RWA in the hopping term provided by the condition $\{J_{nn'}, |\omega_n - \omega_{n'}|\} \ll \omega_n + \omega_{n'}$, and the single-mode approximation in each cavity. These are valid assumptions for realistic parameters that determine the hopping amplitude. For instance, if we consider the coupling capacitance $C_c \ll C_r$, where C_r is the total cavity capacitance, it can be shown that [58]

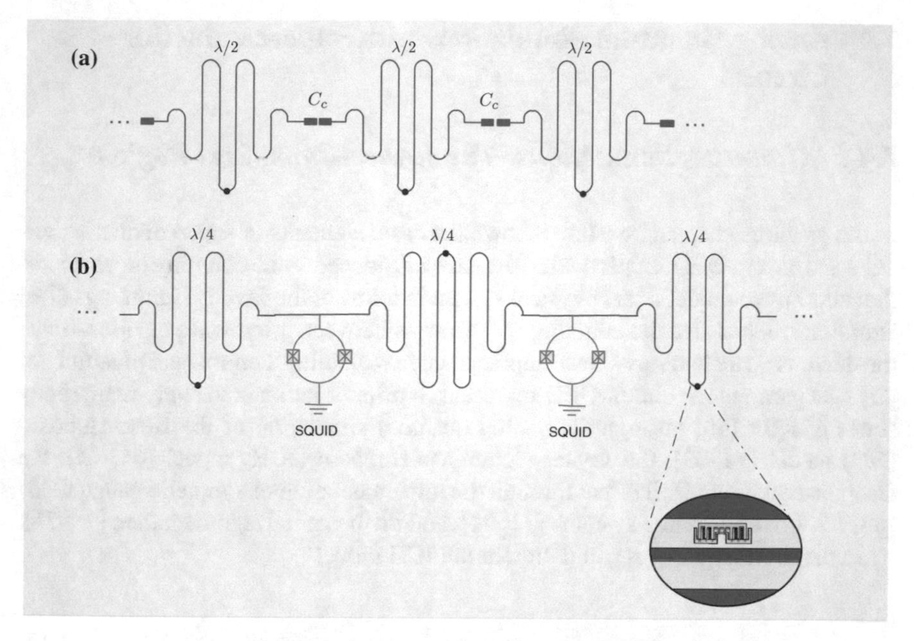

Fig. 7.5 Schematic of cavity-cavity interaction. **a** Capacitive coupling of half wave cavities. **b** Current-current coupling of quarter wave cavities. The latter is mediated by a SQUID device. *Inset red dots* represent a transmon qubit capacitively coupled to a microwave cavity

$$J_{nn'} = \frac{1}{2}\sqrt{\omega_n\omega_{n'}}C_c u_n(x)u_{n'}(x')\Big|_{\text{ends}}, \tag{7.11}$$

where $u_n(x)$ stands for the spatial dependence of the charge distribution along the cavity.

In the case of Fig. 7.5b, where the coupling is mediated by the SQUID, it has been shown that the hopping amplitude reads [59]

$$J_{nn'} \propto \frac{1}{2}\frac{L_J(\Phi_{\text{ext}})}{\sqrt{\omega_n\omega_{n'}}}\partial_x u_n(x)\partial_x u_{n'}(x')\Big|_{\text{ends}}, \tag{7.12}$$

where $L_J(\Phi_{\text{ext}})$ is the flux-dependent Josephson inductance associated with the SQUID. In analogy to the capacitive coupling case, it is assumed the weak coupling regime $L_J^0 \ll L_r$ where $L_J^0 = \varphi_0^2/E_J$ is the bare Josephson inductance of the SQUID, L_r is the total inductance of the cavity, and $\varphi_0 = \Phi_0/2\pi$ is the reduced flux quantum. It is noteworthy to mention that this kind of coupling mechanism also generates single-mode squeezing in each cavity that may spoil the implementation of Eq. (7.10). This can be avoided by considering an array of cavities with different lengths as depicted in Fig. 7.5b, and by tuning the system parameters to fulfill the RWA [59].

7.4 Analog Quantum Simulations with Superconducting Circuits

7.4.1 Quantum Simulations: The Jaynes-Cummings Regime

In the previous chapters, we have shown the basic elements of superconducting circuits and circuit QED. In particular, we have introduced some coupling mechanisms that allow us to model a real physical system in terms of the Jaynes-Cummings (7.8) Hamiltonian, but also the coupling mechanisms between microwave cavities. Since the microwave technology shows unprecedented scalability, control, and tunability of physical parameters, circuit QED represents a prime candidate to study many-body states of light [60] through the analog quantum simulation of the Bose-Hubbard (BH) model [61–63], the Jaynes-Cummings-Hubbard (JCH) model [64, 65], the fractional quantum Hall effect through the implementation of synthetic gauge fields [58, 66–70], spin lattice systems [71, 72], and the bosonic Kagome lattice [73–75].

In the analog quantum simulation of the JCH model

$$H_{\mathrm{JCH}} = \sum_{j=1}^{N} \omega_0 \sigma_j^+ \sigma_j^- + \sum_{j=1}^{N} \omega a_j^\dagger a_j + J \sum_{\langle ij \rangle}^{N} (a_i a_j^\dagger + \mathrm{H.c.}), \tag{7.13}$$

the whole system is composed of elementary cells that may consist of a transmon qubit capacitively coupled to a microwave resonator [76], where the dynamics is described by the JC Hamiltonian. In addition, the connection between neighboring cells is achieved by the capacitive coupling between half-wave cavities in a linear array, see Fig. 7.5a. Notice that the two-site JCH model has been already implemented in the lab [77]. Remark that more complex geometries [58] may be achieved, thus establishing an additional advantage over quantum optics platforms [78].

Unlike standard setups of cavity QED in the optical or microwave regimes, circuit QED allows us to engineer nonlinear interactions between microwave photons provided by nonlinear elements such as Josephson junctions. In particular, the cavity-cavity coupling mediated by SQUID devices, as depicted in Fig. 7.5b, may represent a prime candidate on the road of simulating the Bose-Hubbard model with attractive interactions [79, 80], but also the full Bose-Hubbard and extended models may also be simulated [81]. It is noteworthy that the Bose-Hubbard-dimer model has been already implemented in a circuit QED setup [82]. The latter proposal and the experiment presented in Ref. [77] encourage the theoretical work for the sake of simulating many-body states of light and matter. In addition, the driven-dissipative dynamics of many-body states of light [83–89], and applications of polariton physics in quantum information processing [90, 91] may also be simulated with *state-of-the-art* circuit QED technologies. The coupling mechanisms appearing in circuit QED allow us to study interesting variants of standard models of condensed matter physics. For instance, photon solid phases have been analyzed in the out of equilibrium dynamics of nonlinear cavity arrays described by the Hamiltonian

$$H = \sum_i [-\delta a_i^\dagger a_i + \Omega(a_i + a_i^\dagger)] - J\sum_{\langle i,j\rangle}(a_i a_j^\dagger + \text{H.c.}) + U\sum_i n_i(n_i - 1) + V\sum_{\langle i,j\rangle} n_i n_j, \tag{7.14}$$

which exhibits Bose-Hubbard interaction as well as nearest-neighbor Kerr nonlinearities [92]. The latter is a direct consequence of the nonlinearity provided by the Josephson energy in the SQUID loop.

Circuit QED technologies allow also to study two-dimensional arrays of coupled cavities, as stated in Refs. [73, 74], with the implementation of the bosonic Kagome lattice. This provides room to the application of powerful numerical techniques such as the *projected entangled-pair states* (PEPS) [75, 93, 94] with the aim of studying the interplay between light and matter interactions, as well as predicting new many-body states of light. It is noteworthy to mention that the study of a quantum simulator for the Kagome lattice may predict new physics that otherwise would not be accessible with classical simulations.

7.4.2 Quantum Simulations: The USC Regime of Light-Matter Interactions

In the previous section, we have shown the ability of superconducting circuits to simulate many-body states of light, where the building block or unit cell corresponds to a cavity interacting with a two-level system in the strong coupling (SC) regime. In this case, the cavity-qubit coupling strength exceeds any decay rate of the system such as photon losses, spontaneous decay, and dephasing of the qubit [95]. Circuit QED has also reached unprecedented light-matter coupling strength with the implementation of the USC regime [45–50] and potentially the DSC regime [55, 56]. In this sense, it would be interesting to exploit these coupling regimes aiming at building new many-body states of light, where the building block consists of a cavity-qubit system described by the quantum Rabi model in Eq. (7.9).

Recently, it has been pointed out the importance of the counter-rotating terms in order to describe many-body effects in the *Rabi-Hubbard* model [96–98]

$$H_{\text{RH}} = \sum_i H_{\text{Rabi}}^{(i)} - J\sum_{\langle i,j\rangle}(a_i a_j^\dagger + \text{H.c.}). \tag{7.15}$$

This model exhibits a Z_2 parity symmetry-breaking quantum criticality, long order-range superfluid order, as well as the break of the conservation of local polariton number at each site. This leads to the absence of Mott lobes in the phase diagram as compared with the Bose-Hubbard model. The extension of the Rabi-Hubbard model to the two-dimensional case may represent an additional example where a quantum simulator could outperform classical simulations.

The above results make it necessary to introduce a quantum simulator that provides the quantum Rabi model (QRM) in a controllable way. As stated in Ref. [99], this task can be done by making use of a two-tone driving on a two-level system of frequency

ω_q, that interacts with a single mode of a microwave cavity of frequency ω. In the rotating-wave approximation, the Hamiltonian describing the above situation reads

$$\begin{aligned} H = {} & \frac{\hbar\omega_q}{2}\sigma_z + \hbar\omega a^\dagger a - \hbar g(\sigma^\dagger a + \sigma a^\dagger) \\ & - \hbar\Omega_1(e^{i\omega_1 t}\sigma + e^{-i\omega_1 t}\sigma^\dagger) - \hbar\Omega_2(e^{i\omega_2 t}\sigma + e^{-i\omega_2 t}\sigma^\dagger). \end{aligned} \tag{7.16}$$

Here, Ω_j and ω_j represent the Rabi amplitude and frequency of the jth microwave signal, respectively. The simulation of the QRM can be accomplished in a specific rotating frame as follows. First, we write the Hamiltonian (7.16) in the reference frame that rotates the frequency ω_1. This leads to

$$\begin{aligned} H^{R_1} = {} & \hbar\frac{(\omega_q - \omega_1)}{2}\sigma_z + \hbar(\omega - \omega_1)a^\dagger a - \hbar g\left(\sigma^\dagger a + \sigma a^\dagger\right) \\ & -\hbar\Omega_1\left(\sigma + \sigma^\dagger\right) - \hbar\Omega_2\left(e^{i(\omega_2-\omega_1)t}\sigma + e^{-i(\omega_2-\omega_1)t}\sigma^\dagger\right). \end{aligned} \tag{7.17}$$

Second, we go into the interaction picture with respect to $H_0^{R_1} = -\hbar\Omega_1\left(\sigma + \sigma^\dagger\right)$ such that $H^I(t) = e^{iH_0^{L_1}t/\hbar}\left(H^{R_1} - H_0^{R_1}\right)e^{-iH_0^{R_1}t/\hbar}$. The above transformation can be implemented by means of a Ramsey-like pulse as described in Ref. [99]. In the dressed-spin basis, $|\pm\rangle = (|g\rangle \pm |e\rangle)/\sqrt{2}$, the interaction Hamiltonian we can be written as

$$\begin{aligned} H^I(t) = {} & -\hbar\frac{(\omega_q - \omega_1)}{2}\left(e^{-i2\Omega_1 t}|+\rangle\langle-| + \text{H.c.}\right) + \hbar(\omega - \omega_1)a^\dagger a \\ & - \frac{\hbar g}{2}\left(\{|+\rangle\langle+| - |-\rangle\langle-| + e^{-i2\Omega_1 t}|+\rangle\langle-|\right. \\ & \left. - e^{i2\Omega_1 t}|-\rangle\langle+|\}a + \text{H.c.}\right) \\ & - \frac{\hbar\Omega_2}{2}\left(\{|+\rangle\langle+| - |-\rangle\langle-| - e^{-i2\Omega_1 t}|+\rangle\langle-|\right. \\ & \left. + e^{i2\Omega_1 t}|-\rangle\langle+|\}e^{i(\omega_2-\omega_1)t} + \text{H.c.}\right). \end{aligned} \tag{7.18}$$

Third, if we tune the external driving frequencies as $\omega_1 - \omega_2 = 2\Omega_1$, and for a strong first driving, Ω_1, the Hamiltonian (7.18) can be represented as

$$H_{\text{eff}} = \hbar(\omega - \omega_1)a^\dagger a + \frac{\hbar\Omega_2}{2}\sigma_z - \frac{\hbar g}{2}\sigma_x\left(a + a^\dagger\right), \tag{7.19}$$

which corresponds to the quantum Rabi model. For values $\Omega_2 \sim (\omega - \omega_1) \sim g/2$, the original cavity-qubit system is capable of simulating the dynamics associated with the USC/DSC regime. The simulated interaction strength corresponds to the ratio $g_{\text{eff}}/\omega_{\text{eff}}$, where $g_{\text{eff}} \equiv g/2$ and $\omega_{\text{eff}} \equiv \omega - \omega_1$. Figure 7.6 shows the two-level system dynamics for an effective ratio $g_{\text{eff}}/\omega_{\text{eff}} = 1$ and two different values of the effective qubit frequency Ω_2. The initial condition is the ground state of the qubit

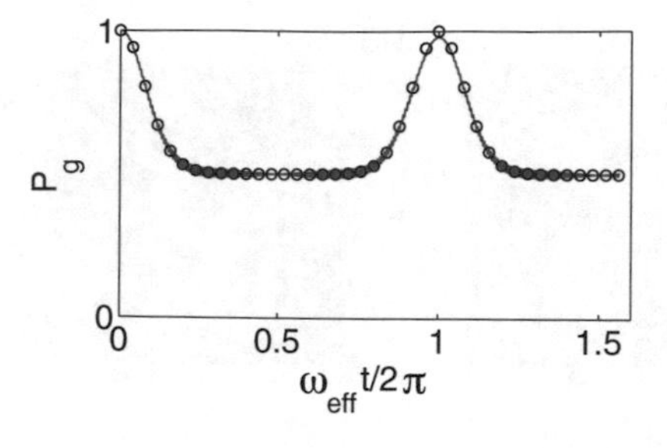

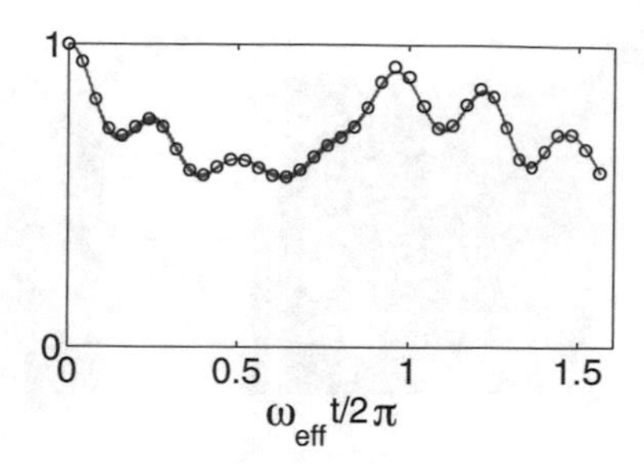

Fig. 7.6 The population of the ground state for the two-level system, $P_g(t)$, has been calculated by integrating the exact (*solid line*) dynamics in Eq. (7.16) and the effective (*circles*) Hamiltonian dynamics in Eq. (7.19). We have considered two different cases: (*left panel*) $\Omega_2 = 0$; (*right panel*) $\Omega_2 = 2\pi \times 10\,\text{MHz}$. The simulated interaction strength corresponds to $g_{\text{eff}}/\omega_{\text{eff}} = 1$. Figure from Ref. [99], used under the terms of the Creative Commons Attribution 3.0 licence

and the vacuum state for the field, $|\psi(0)\rangle = |g\rangle \otimes |0\rangle$. This dynamics corresponds to the one predicted in Ref. [55] for the deep strong coupling regime.

This quantum simulation may pave the way for building a complete toolbox of complex cavity arrays where the unit cell can be tuned, at will, from the strong coupling regime, described by the Jaynes-Cummings model, to the USD/DSC regime described by the quantum Rabi model.

7.4.3 *Quantum Simulations of Quantum Relativistic Mechanics*

The quantum simulation of the quantum Rabi model [99] allows us to access a wide range of physical phenomena such as cat-state generation and simulating relativistic quantum mechanics on a chip [100]. The latter can be achieved in a similar way as the QRM by applying three classical microwaves, that is, a two-tone driving on a two-level system interacting with a single cavity mode in the SC regime, and a driving on the cavity mode. This process is modeled by the Hamiltonian

$$\begin{aligned} H = {} & \frac{\hbar\omega_q}{2}\sigma_z + \hbar\omega a^\dagger a - \hbar g(\sigma^+ a + \sigma^- a^\dagger) - \hbar\Omega(\sigma^+ e^{-i(\omega t+\varphi)} + \sigma^- e^{i(\omega t+\varphi)}) \\ & - \lambda(\sigma^+ e^{-i(\nu t+\varphi)} + \sigma^- e^{i(\nu t+\varphi)}) + \hbar\xi(a e^{i\omega t} + a^\dagger e^{-i\omega t}), \end{aligned} \tag{7.20}$$

where Ω, λ, and ξ stand for the driving amplitudes, ω is the resonator frequency, and ν is the driving frequency, respectively. As stated in Ref. [100], if we consider a strong microwave driving $\Omega \gg \{g, \lambda\}$ and the condition $\omega - \nu = 2\Omega$, the effective Hamiltonian, in the rotating frame, reads

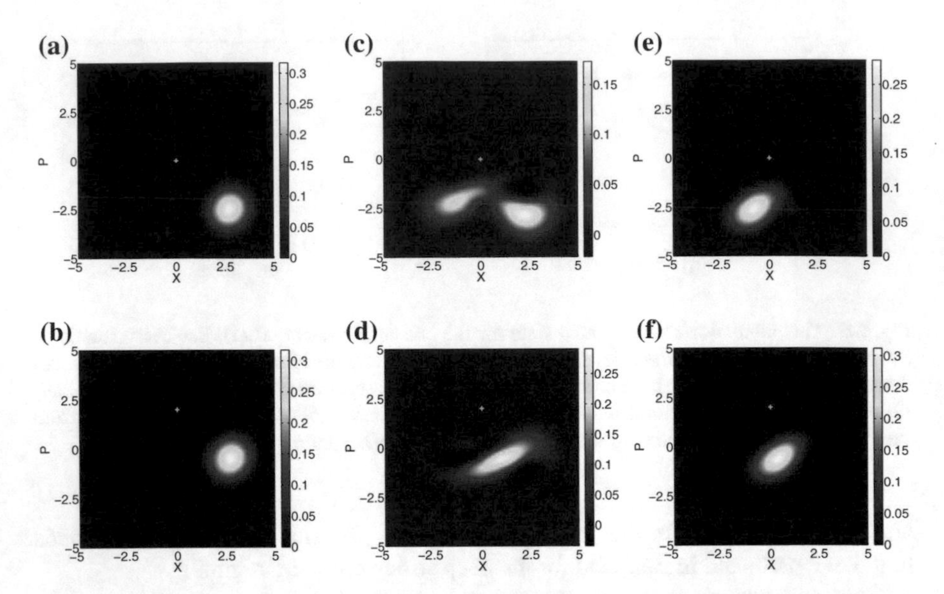

Fig. 7.7 Wigner function $W(x, p)$ representation of the field state in the microwave cavity. We have computed the time evolution by means of the Hamiltonian (7.20) for a time of 60 ns. We used realistic parameters $g = 2\pi \times 10$ MHz, $\Omega = 2\pi \times 200$ MHz, $\xi = g/2$. **a** $\lambda = 0$ with the initial state $|+, 0\rangle$; **b** $\lambda = 0$ with the initial state $|+, \sqrt{2}i\rangle$; **c** $\lambda = \sqrt{2}g$ with the initial state $|+, 0\rangle$; **d** $\lambda = \sqrt{2}g$ with the initial state $|+, \sqrt{2}i\rangle$; **e** $\lambda = 4\sqrt{2}g$ with the initial state $|e, 0\rangle$; and **f** $\lambda = 4\sqrt{2}g$ with the initial state $|e, \sqrt{2}i\rangle$. Figure from Ref. [100], used under the terms of the Creative Commons Attribution 3.0 licence

$$H_{\rm eff} = \frac{\hbar\lambda}{2}\sigma_z + \frac{\hbar g}{\sqrt{2}}\sigma_y \hat{p} + \hbar\xi\sqrt{2}\hat{x}, \tag{7.21}$$

where $\hat{x} = (a + a^\dagger)/\sqrt{2}$ and $\hat{p} = i(a^\dagger - a)/\sqrt{2}$ are the field quadratures satisfying the commutation relation $[\hat{x}, \hat{p}] = i$. The above Hamiltonian describes a $1 + 1$ Dirac particle in a linear external potential $U = \hbar\xi\sqrt{2}\hat{x}$, where the terms $\hbar g/\sqrt{2}$ and $\hbar\lambda/2$ represent the speed of light and the mass of the particle, respectively.

Adding an external potential $U(x)$ allows us to simulate the scattering of a single relativistic particle. In particular, we can start by considering the case of a massless Dirac particle whose Hamiltonian is given by $H_{\rm K} = \hbar g/\sqrt{2}\sigma_y \hat{p} + \hbar\xi\sqrt{2}\,\hat{x}$. Figure 7.7a and b show the evolution of the initial states $|+, 0\rangle$ and $|+, \sqrt{2}i\rangle$, respectively, where $|+\rangle$ stands for the positive eigenstate of σ_y and $|0\rangle$ the vacuum state for the field. Here the field state remains coherent while experiencing two independent displacements along the $\hat{x}$-quadrature proportional to $g/\sqrt{2}$, and along the $\hat{p}$-quadrature. It is remarkable that the external potential does not modify the rectilinear movement in position representation. This phenomenon corresponds to the *Klein paradox*, which states that a massless Dirac particle may propagate through the potential barrier with probability different from zero.

As the quantum simulation allows us to tune the physical parameters at will, we can also study the scattering of a massive nonrelativistic Schrödinger particle. In this case, the dynamics is governed by the Hamiltonian $H_{\mathrm{NRel}} = \hbar\sigma_z \hat{p}^2/\lambda + \hbar\xi\sqrt{2}\,\hat{x}$. Note that this Hamiltonian assures that any initial Gaussian state remains Gaussian as time elapses. This can be seen in Fig. 7.7e and f where the initial states are $|e, 0\rangle$ and $|e, \sqrt{2}i\rangle$, respectively. The mass of the particle has been chosen such that $\hbar\lambda/2 = 4 \times \hbar g/\sqrt{2}$. Figure 7.7e shows how the particle is scattered backwards by the potential. For the case of Fig. 7.7f, the particle has an initial positive kinetic energy that allowed it to enter the external potential, though after 60 ns it moves backwards.

These two limiting cases have shown a total transmission or reflection. It is natural that a particle with an intermediate mass features only partial transmission/reflection. Figure 7.7c shows the scattering of a massive particle with $\hbar\lambda/2 = \hbar g/\sqrt{2}$ prepared in the initial state $|+, 0\rangle$. We see how the wave packet splits into spinor components of different signs which move away from the center. Furthermore, If some initial kinetic energy is provided to the wavepacket, as shown in Fig. 7.7d with initial state $|+, \sqrt{2}i\rangle$, the particle enters the barrier to stop and break up sooner or later.

The simulation of the Dirac Hamiltonian, together with all available technology in circuit QED, may have interesting consequences in the study of many-body states of light. For instance, one may have access to Dirac lattices and their possible extension to Dirac materials [101].

7.5 Digital Quantum Simulations with Superconducting Circuits

In many situations, the quantum simulator does not evolve according to the dynamics of the system to be simulated. Therefore, it is appropriate to employ digital techniques to emulate a wider variety of quantum systems [9]. Digital quantum simulators are akin to universal quantum computers, with the advantage that in principle with a small number of qubits one will already be able to outperform classical computers. Thus, one does not need to reach thousands of qubits to perform interesting quantum simulations of mesoscopic quantum systems.

The digital quantum simulators are based on the fact that most model Hamiltonians are composed of a finite number of local terms, $H = \sum_{k=1}^{N} H_k$, where each of them acts upon a reduced Hilbert space, or at least is efficiently implementable with a polynomial number of gates. In these cases, the system dynamics can be obtained via digital decomposition into stroboscopic steps, via Trotter techniques,

$$e^{-iHt} = (e^{-iH_1 t/n} \ldots e^{-iH_N t/n})^n + O(t^2/n). \tag{7.22}$$

By making n large, the error can be made in principle as small as desired. Naturally there will be a limit to the size of n that will be given by the finite fidelity of the local gates.

There have been already a number of experiments on digital quantum simulators, either in quantum photonics [102], or in trapped ions [103]. Regarding superconducting circuits, some theoretical proposals for digital quantum simulations have been put forward [104, 105], and we will review these in the next sections.

7.5.1 Digital Quantum Simulations of Spin Systems with Superconducting Circuits

In this section, we analyze the realization of digital quantum simulations of spin Hamiltonians in a superconducting circuit setup consisting of several transmon qubits coupled to a microwave resonator [104]. Although our protocol is appropriate for every superconducting qubit with sufficiently long coherence time, we consider specifically a transmon qubit device. This kind of qubits are typically used because of its insensitivity to charge fluctuations [30]. Nevertheless, depending on the specific phenomena to simulate, one can consider other superconducting qubits for quantum simulations. First, we show how one can simulate the Heisenberg model in a circuit QED setup with state-of-the-art technology. Then, we consider typical simulation times and their associated fidelities with current superconducting qubit technology, showing the potential of superconducting qubits in terms of digital quantum simulators. Finally, we study the necessary resources with realistic parameters for a versatile quantum simulator of spin models able to emulate a general many-body spin dynamics.

Digital methods can be employed to emulate the Heisenberg model with current circuit QED technology. Even though the latter does not feature the Heisenberg interaction from first principles, one can nevertheless analyze a digital quantum simulation of this model. We show that a set of N transmon qubits coupled through a resonator is able to simulate Heisenberg interactions of N spins, which in the symmetric-coupling case is given by

$$H = \sum_{i=1}^{N-1} J \left(\sigma_i^x \sigma_{i+1}^x + \sigma_i^y \sigma_{i+1}^y + \sigma_i^z \sigma_{i+1}^z \right). \tag{7.23}$$

Here σ_i^j, $j \in \{x, y, z\}$ are Pauli matrices that refer to the first two levels of the ith transmon qubit.

We start by the simplest case, with only two spins. The XY exchange interaction can be implemented by dispersive coupling of two transmon qubits with a common resonator [13, 34, 106], $H_{12}^{xy} = J\left(\sigma_1^+ \sigma_2^- + \sigma_1^- \sigma_2^+\right) = J/2\left(\sigma_1^x \sigma_2^x + \sigma_1^y \sigma_2^y\right)$. The XY interaction can be mapped through local rotations of the qubits onto the effective Hamiltonians

$$H_{12}^{xz} = R_{12}^x(\pi/4) H_{12}^{xy} R_{12}^{x\dagger}(\pi/4) = J/2\left(\sigma_1^x \sigma_2^x + \sigma_1^z \sigma_2^z\right), \tag{7.24}$$

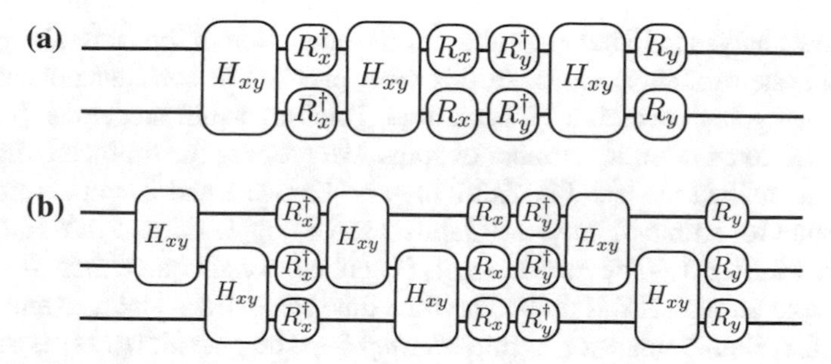

Fig. 7.8 Schemes for the proposed digital quantum simulations with superconducting transmon qubits. **a** Heisenberg model for two qubits. **b** Heisenberg model for three qubits. Here, $R_{x(y)} \equiv R^{x(y)}(\pi/4)$ and $\overline{R}_x \equiv R^x(\pi/2)$. We point out that exchanging each of the R matrices with its adjoint does not affect the protocols. Reprinted with permission from [104], Copyright (2014) American Physical Society

$$H_{12}^{yz} = R_{12}^{y}(\pi/4) H_{12}^{xy} R_{12}^{y\dagger}(\pi/4) = J/2 \left(\sigma_1^y \sigma_2^y + \sigma_1^z \sigma_2^z\right). \tag{7.25}$$

Here, $R_{12}^{x(y)}(\pi/4) = \exp[-i\pi/4(\sigma_1^{x(y)} + \sigma_2^{x(y)})]$ is a local rotation of the first and second qubits with respect to the $x\,(y)$ axis. The XYZ Heisenberg Hamiltonian H_{12}^{xyz} can thus be performed according to the following protocol (see Fig. 7.8a). *Step 1*—The two qubits interact with the XY Hamiltonian H_{12}^{xy} for a time t. *Step 2*—Single qubit rotations $R_{12}^{x}(\pi/4)$ are applied to both qubits. *Step 3*—The two qubits interact with H_{12}^{xy} Hamiltonian for a time t. *Step 4*—Single qubit rotations $R_{12}^{x\dagger}(\pi/4)$ are applied to both qubits. *Step 5*—Single qubit rotations $R_{12}^{y}(\pi/4)$ are applied to both qubits. *Step 6*—The two qubits interact according to the H_{12}^{xy} Hamiltonian for a time t. *Step 7*—Single qubit rotations $R_{12}^{y\dagger}(\pi/4)$ are applied to both qubits. Accordingly, the final unitary operator reads

$$U_{12}(t) = e^{-iH_{12}^{xy}t} e^{-iH_{12}^{xz}t} e^{-iH_{12}^{yz}t} = e^{-iH_{12}t}. \tag{7.26}$$

This evolution emulates the dynamics of Eq. (7.23) for two transmon qubits. Furthermore, arbitrarily inhomogeneous couplings can be engineered by performing different evolution times or couplings for the different digital gates. Here we point out that a single Trotter step is needed to obtain a simulation with no digital errors, because of the fact that H_{12}^{xy}, H_{12}^{xz}, and H_{12}^{yz} operators commute. Accordingly, in this case the only error source will be due to the accumulated gate errors. We consider two-qubit gates with a process fidelity error of about 5% and eight $\pi/4$ single-qubit gates with process fidelity errors of about 1%. Therefore, we will have a total process fidelity of this protocol of about 77%. Furthermore, the total protocol time for a $\pi/4$ Heisenberg phase is around $0.10\,\mu$s. Throughout this section, we estimate the protocol times by adding the respective times of all the gates, for which we take into account standard superconducting qubit values.

We now analyze a digital algorithm for the emulation of the Heisenberg dynamics for a system of three spins. In this case, one has to consider noncommuting Hamiltonian gates, involving Trotter errors. This three-spin model can be directly extrapolated to an arbitrary number of spins. We propose the following digital protocol for its realization (see Fig. 7.8b). *Step 1*—Qubits 1 and 2 couple through XY Hamiltonian for a time *t/l*. *Step 2*—Qubits 2 and 3 couple through XY Hamiltonian for a time *t/l*. *Step 3*—The gate $R_i^x(\pi/4)$ is applied to each qubit. *Step 4*—Qubits 1 and 2 couple through XY Hamiltonian for a time *t/l*. *Step 5*—Qubits 2 and 3 couple through XY Hamiltonian for a time *t/l*. *Step 6*—The gate $R_i^{x\dagger}(\pi/4)$ is applied to each qubit. *Step 7*—The gate $R_i^y(\pi/4)$ is applied each qubit. *Step 8*—Qubits 1 and 2 couple through XY Hamiltonian for a time *t/l*. *Step 9*—Qubits 2 and 3 couple through XY Hamiltonian for a time *t/l*. *Step 10*—The gate $R_i^{y\dagger}(\pi/4)$ is applied to each qubit. Finally, the global unitary evolution operator per Trotter step is given by

$$U_{123}(t/l) = e^{-iH_{12}^{xy}t/l}e^{-iH_{23}^{xy}t/l}e^{-iH_{12}^{xz}t/l}e^{-iH_{23}^{xz}t/l}e^{-iH_{12}^{yz}t/l}e^{-iH_{23}^{yz}t/l}.$$

Here, the sequence has to be repeated l times following Eq. (7.22), to implement an approximate dynamics of Eq. (7.23) for the three qubits. Each of these Trotter steps consists of four single-qubit gates at different times (performed collectively upon different sets of qubits) and six XY gates, with a total step time around 0.16 μs, well below typical decoherence times in transmon qubits [107]. In Fig. 7.9a and b, we depict the fidelity loss associated with the digital error of the emulated XYZ dynamics for three transmon qubits, together with straight horizontal lines showing the error of the imperfect gates multiplied by the number of Trotter steps, i.e., the total accumulated gate imperfection. One can appreciate time intervals dominated by the digital Trotter error and time intervals where the largest error source in the digital quantum simulation is produced by experimental gate imperfections. One can take into account Hamiltonians with open and periodic boundary conditions, adding an extra coupling between the first and the last spin. Extending our protocol to N

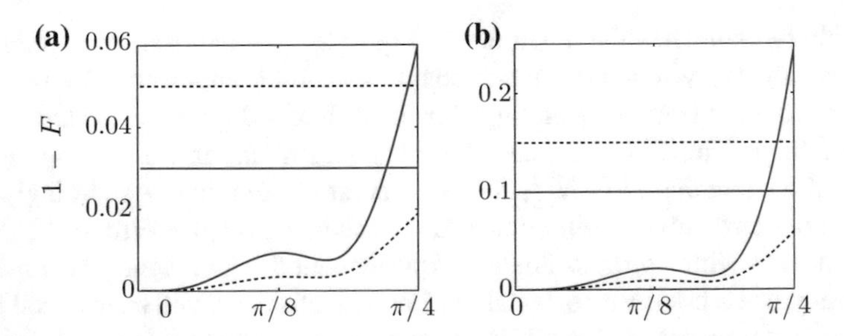

Fig. 7.9 Loss of fidelity for the emulated Heisenberg model for three qubits, in the range $\theta = [0, \pi/4]$, $\theta \equiv Jt$. The *wavy lines* represent digital errors, while the *straight lines* represent the accumulated gate error due to a step error of ε. *Red solid* (*black dotted*) *lines* are associated with lower (higher) digital approximations l. **a** $\varepsilon = 10^{-2}$, $l = 3, 5$, and **b** $\varepsilon = 5 \times 10^{-2}$, $l = 2, 3$. Reprinted with permission from [104], Copyright (2014) American Physical Society

transmon qubits with open or periodic boundary conditions, we estimate an upper bound on the second-order digital Trotter error, given by $E_{\rm open} = 24(N-2)(Jt)^2/l$ and $E_{\rm periodic} = 24N(Jt)^2/l$.

In order to assess the proposals in a realistic circuit QED setup, we made numerical simulations for the Heisenberg dynamics between two qubits in the transmon regime coupled to a stripline waveguide resonator. We estimate the influence on the proposal of a state-of-the-art XY dynamics, given as an effective dispersive Hamiltonian, obtained at second order from the first order one,

$$H_{\rm t} = \sum_{i=0}^{2}\sum_{j=1}^{2} \omega_i^j |i,j\rangle\langle i,j| + \omega_r a^\dagger a + \sum_{i=0}^{2}\sum_{j=1}^{2} g_{i,i+1}(|i,j\rangle\langle i+1,j| + \text{H.c.})(a+a^\dagger). \tag{7.27}$$

Here, ω_r is the resonance frequency of the resonator, and ω_i^j is the transition frequency of the ith level, with respect to the ground state, of the jth transmon qubit. We take into account the first three levels for each qubit, and an anharmonicity factor given by $\alpha_r = (\omega_2^j - 2\omega_1^j)/\omega_1^j = -0.1$, standard for transmon qubits [30]. We consider equal transmons with frequencies $\omega_1^{1,2} \equiv \omega_1 = 2\pi \times 5$ GHz. The frequency of the resonator is fixed to $\omega_r = 2\pi \times 7.5$ GHz. We take into account the coupling strength between the different levels of a single transmon qubit [30] $g_{i,i+1} = \sqrt{i+1}g_0$, being $g_0 = 2\beta e V_{\rm rms} = 2\pi \times 200$ MHz. The experimental parameters we consider are standard for circuit QED platforms and they can be optimized for each specific platform. The transmon-resonator Hamiltonian, in interaction picture with respect to the free energy $\sum_{i,j} \omega_i^j |i,j\rangle\langle i,j| + \omega_r a^\dagger a$, produces an effective interaction between the first two levels of the two qubits $H_{\rm eff} = [g_{01}^2\omega_1/(\omega_1^2 - \omega_r^2)](\sigma_1^x\sigma_2^x + \sigma_1^y\sigma_2^y)$, where we have neglected the cavity population $\langle a^\dagger a\rangle \approx 0$ and renormalized the qubit energies to include Lamb shifts. Here, we have considered the set of Pauli matrices for the subspace spanned by the first two levels of each transmon qubit, e.g. $\sigma_{1(2)}^x \equiv |0,1(2)\rangle\langle 1,1(2)| + \text{H.c.}$

In order to analyze the influence of decoherence in a state-of-the-art circuit QED setup, we compute the master equation evolution,

$$\dot{\rho} = -i[H_{\rm t},\rho] + \kappa L(a)\rho + \sum_{i=1}^{2}\left(\Gamma_\phi L(\sigma_i^z)\rho + \Gamma_- L(\sigma_i^-)\rho\right), \tag{7.28}$$

where we define the Lindblad operators $L(\hat{A})\rho = (2\hat{A}\rho\hat{A}^\dagger - \hat{A}^\dagger\hat{A}\rho - \rho\hat{A}^\dagger\hat{A})/2$. We consider a damping rate for the cavity of $\kappa = 2\pi \times 10$ kHz, and a decoherence and decay rate for a single transmon qubit of $\Gamma_\phi = \Gamma_- = 2\pi \times 20$ kHz. We compute a numerical simulation for the XYZ dynamics for two transmon qubits, following the scheme as in Fig. 7.8a, using for the XY exchange gate steps the outcome of the evolution obtained from Eq. (7.28), and perfect single-qubit gates. We plot our result in Fig. 7.10. The dynamics for the density operator ρ, encoding the evolution of the

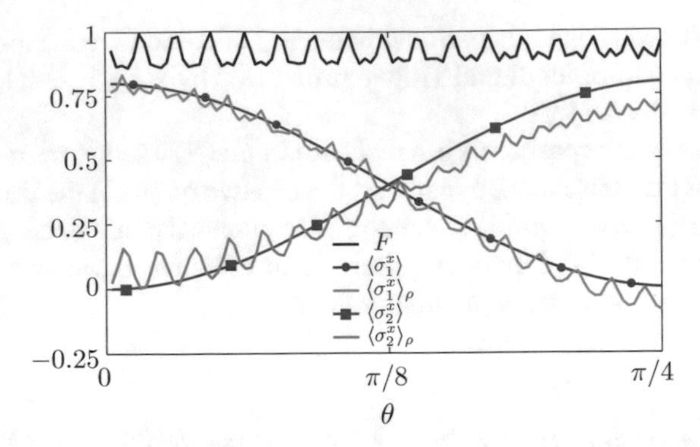

Fig. 7.10 Evolution of the emulated Heisenberg Hamiltonian for two superconducting transmon qubits, initialized in state $1/\sqrt{5}(|\uparrow\rangle + 2|\downarrow\rangle) \otimes |\downarrow\rangle$. The fidelity $F = \mathrm{Tr}(\rho|\Psi_I\rangle\langle\Psi_I|)$ represents the performance of the protocol for the simulated phase θ. The ideal spin evolution $\langle\sigma_i^x\rangle$ for both transmon qubits is depicted versus average values $\langle\sigma_i^x\rangle_\rho$ which are given through the qubit Hamiltonian H_t. Reprinted with permission from [104], Copyright (2014) American Physical Society

two transmon qubits, is contrasted to the ideal quantum dynamics $|\Psi\rangle_I$, evolving with the Hamiltonian in Eq. (7.23), where $J = g_{01}^2\omega_1/(\omega_1^2 - \omega_r^2) \approx 2\pi \times 6$ MHz. It can be appreciated that the simulation fidelities $F = Tr(\rho|\Psi_I\rangle\langle\Psi_I|)$ obtained are good for nontrivial evolutions. We point out that the application of the XYZ Hamiltonian on an initial state, corresponding to an eigenstate of the ZZ operator, would be just equivalent to the one of the XY exchange dynamics. To show characteristic behavior of the XYZ interaction, we considered an initial state which does not have this feature. One can as well appreciate the standard short-time fidelity fluctuations due to the spurious terms of the dispersive exchange Hamiltonian. Making use of a larger detuning of the qubits from the cavity, the contribution of the non-dispersive part of the interaction can be reduced, increasing the total protocol fidelity.

Summarizing, we have proposed a digital quantum simulation of spin systems with circuit QED platforms. We have analyzed a prototypical model: the Heisenberg interaction. Moreover, we have studied the feasibility of the protocol with current technology of transmon qubits coupled to microwave cavities. These protocols may be generalized to many-body spin systems, paving the way towards universal digital quantum simulation of spin models with superconducting circuits.

7.5.2 Digital Quantum Simulations of Quantum Field Theories with Superconducting Circuits

Our current knowledge of the most fundamental processes in the physical world is based on the framework of interacting quantum field theories [108]. In this context, models involving the coupling of fermions and bosons play a prominent role.

In these systems, it is possible to analyze fermion-fermion scattering mediated by bosons, fermionic self-interactions, and bosonic polarization. In this section, we will study [105] a quantum field theory model with the following assumptions: (i) $1+1$ dimensions, (ii) scalar fermions and bosons, and described by the Hamiltonian ($\hbar = c = 1$)

$$H = \int dp\, \omega_p (b_p^\dagger b_p + d_p^\dagger d_p) + \int dk\, \omega_k a_k^\dagger a_k + \int dx\, \psi^\dagger(x)\psi(x) A(x). \quad (7.29)$$

Here, $A(x) = i \int dk\, \lambda_k \sqrt{\omega_k / 4\pi}(a_k^\dagger e^{-ikx} - a_k e^{ikx})$ is a bosonic operator, with coupling constants λ_k, and $\psi(x)$ is a fermionic field, $b_p^\dagger (b_p)$ and $d_p^\dagger (d_p)$ are its corresponding fermionic and antifermionic creation(annihilation) mode operators for frequency ω_p, while $a_k^\dagger (a_k)$ is the creation(annihilation) bosonic mode operator associated with the frequency ω_k. We propose a protocol for the scalable and efficient digital-analog quantum simulation of interacting fermions and bosons, based on Eq. (7.29), making use of the state-of-the-art circuit QED platforms. In this fast-evolving quantum technology, one has the possibility of a strong coupling of artificial atoms with a one-dimensional bosonic continuum.

In order to map the proposed model to the circuit QED setup, we consider a further assumption in Eq. (7.29): (iii) one fermionic and one antifermionic field modes [109] that interact via a bosonic continuum. Accordingly, the interaction Hamiltonian is given by

$$H_{\text{int}} = i \int dx dk \lambda_k \sqrt{\frac{\omega_k}{2}} \left(|\Lambda_1(p_f, x, t)|^2 b_{\text{in}}^\dagger b_{\text{in}} + \Lambda_1^*(p_f, x, t) \Lambda_2(p_{\bar{f}}, x, t) b_{\text{in}}^\dagger d_{\text{in}}^\dagger \right. \quad (7.30)$$
$$\left. + \Lambda_2^*(p_{\bar{f}}, x, t) \Lambda_1(p_f, x, t) d_{\text{in}} b_{\text{in}} + |\Lambda_2(p_{\bar{f}}, x, t)|^2 d_{\text{in}} d_{\text{in}}^\dagger \right) \left(a_k^\dagger e^{-ikx} - a_k e^{ikx} \right).$$

The fermionic and antifermionic creation and annihilation operators obey anticommutation relations $\{b_{\text{in}}, b_{\text{in}}^\dagger\} = \{d_{\text{in}}, d_{\text{in}}^\dagger\} = 1$, and the bosonic creation and annihilation operators satisfy commutation relations $[a_k, a_{k'}^\dagger] = \delta(k - k')$. Here, we have spanned the field $\psi(x)$ in terms of two comoving anticommuting modes as a first order approximation, neglecting the remaining anticommuting modes. These are given by the expressions,

$$b_{\text{in}}^\dagger = \int dp\, \Omega_f(p_f, p) b_p^\dagger e^{-i\omega_p t} \quad (7.31)$$

$$d_{\text{in}}^\dagger = \int dp\, \Omega_{\bar{f}}(p_{\bar{f}}, p) d_p^\dagger e^{-i\omega_p t}, \quad (7.32)$$

where $\Omega_{f,\bar{f}}(p_{f,\bar{f}}, p)$ are the fermion and antifermion wavepacket envelopes with average momenta p_f and $p_{\bar{f}}$, respectively.

Thus, the fermionic field reads

$$\psi(x) \simeq \Lambda_1(p_f, x, t) b_{\text{in}} + \Lambda_2(p_{\bar{f}}, x, t) d_{\text{in}}^\dagger, \quad (7.33)$$

where the coefficients can be computed by considering the anticommutators $\{\psi(x), b_{\text{in}}^{\dagger}\}$ and $\{\psi(x), d_{\text{in}}\}$ as follows

$$\Lambda_1(p_f, x, t) = \{\psi(x), b_{\text{in}}^{\dagger}\} = \frac{1}{\sqrt{2\pi}} \int \frac{dp}{\sqrt{2\omega_p}} \Omega(p_f, p) e^{i(px-\omega_p t)}, \quad (7.34)$$

$$\Lambda_2(p_{\bar{f}}, x, t) = \{\psi(x), d_{\text{in}}\} = \frac{1}{\sqrt{2\pi}} \int \frac{dp}{\sqrt{2\omega_p}} \Omega(p_{\bar{f}}, p) e^{-i(px-\omega_p t)}, \quad (7.35)$$

where we have considered $\psi(x)$ in the Schrödinger picture.

With this proposal, we think that emulating the physics of a discrete number of fermionic field modes coupled to a continuum of bosonic field modes will significantly enhance the quantum simulations of full-fledged quantum field theories.

We now use the Jordan-Wigner transformation [110, 111] that maps fermionic mode operators onto tensor products of spin operators: $b_l^{\dagger} = \prod_{r=1}^{l-1} \sigma_l^- \sigma_r^z$, and $d_m^{\dagger} = \prod_{r=1}^{m-1} \sigma_m^- \sigma_r^z$, where $l = 1, 2, \ldots, N/2, m = N/2+1, \ldots, N$, with N the total number of fermionic and antifermionic modes. Thus, Hamiltonian (7.30) is reduced to just three different kinds of couplings: single and two-qubit gates interacting with the bosonic continuum $H_1 = i\sigma_j \int dxdk\ g_k(a_k^{\dagger} e^{-ikx} - a_k e^{ikx})$, $H_2 = i(\sigma_j \otimes \sigma_\ell) \int dxdk\ g_k(a_k^{\dagger} e^{-ikx} - a_k e^{ikx})$, with $\sigma_q = \{\sigma_x, \sigma_y, \sigma_z\}$ for $q = 1, 2, 3$, and couplings that involve only bosonic field modes, $H_3 = i \int dxdk\ g_k(a_k^{\dagger} e^{-ikx} - a_k e^{ikx})$. Therefore, the quantum simulator should produce a way of generating multiqubit entangling gates and coupling qubit operators to a bosonic continuum through a digital-analog method [109].

Circuit QED platforms consisting of the coupling between coplanar waveguides (CPW) and transmon qubits [112–114] are an appropriate setup to implement our digital-analog simulator model. We depict in Fig. 7.11a, a scheme of our setup, which is based on a microwave transmission line supporting a continuum of electromagnetic field modes (open line) that interacts with three qubits in the transmon regime.

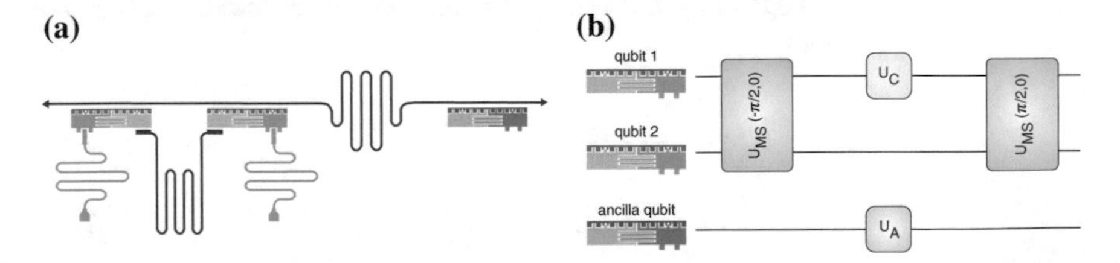

Fig. 7.11 **a** Scheme of our protocol for emulating fermion-fermion scattering in QFTs. An open line supporting a bosonic continuum is coupled to three superconducting transmon qubits. A second, one-dimensional stripline waveguide, forming a cavity, contains a single bosonic mode of the microwave field and couples with two superconducting transmon qubits. Each of the qubits can be locally addressed through external flux drivings generating fluxes Φ_{ext}^{j} and $\bar{\Phi}_{\text{ext}}^{j}$ to adjust the coupling and its associated frequencies. **b** Sequential protocol of multiple and single qubit gates, in a single digital step, acting on transmon qubits to produce two-qubit gates interacting with the continuum. Reprinted with permission from [105], Copyright (2015) American Physical Society

Moreover, we consider a microwave stripline resonator with a single bosonic mode coupled only with two of the qubits. We point out that two superconducting transmon qubits may interact at the same time with both CPWs, while the ancilla transmon qubit will interact only with the open transmission line.

In our proposal [105], we take into account tunable couplings among each transmon qubit and the CPWs, as well as tunable transmon qubit energies via applied magnetic fluxes. More specifically, our method for emulating fermion-fermion scattering will be based on the capacity to turn on/off each CPW-qubit coupling with tunable parameters. The latter may be performed by using controlable coupling superconducting qubits, [112, 113] and typical techniques of band-pass filter [115] to apply in the open line, in the sense that just a finite bandwidth of bosonic field modes plays a role in the evolution. In this respect, to decouple a superconducting qubit from the open transmission line may be achieved by shifting the qubit frequency outside of the permitted bandwidth. Moreover, our proposal may be extrapolated to many fermionic field modes by considering more superconducting transmon qubits, as shown in Fig. 7.12.

In our proposal, the transmon qubit-continuum and the transmon qubit-resonator couplings are expressed through the interaction Hamiltonian

$$\begin{aligned} H_{\text{int}} = \; & i \sum_{j=1}^{3} \sigma_j^y \int dk\, \beta(\Phi_{\text{ext}}^j, \bar{\Phi}_{\text{ext}}^j) g_k (a_k^\dagger e^{-ikx_j} - a_k e^{ikx_j}) \\ & + i \sum_{j=1}^{2} \alpha(\Phi_{\text{ext}}^j, \bar{\Phi}_{\text{ext}}^j) g_j \sigma_j^y (b^\dagger - b), \end{aligned} \tag{7.36}$$

where $a_k^\dagger(a_k)$ and ω_k are the creation(annihilation) operator and the free energy associated with the kth continuum field mode, respectively. In addition, $b^\dagger(b)$ denotes the creation(annihilation) bosonic operator in the microwave cavity, and σ^y is the corresponding Pauli operator. The couplings g_k and g_j are a function of specific properties of the CPW as for example the photon frequencies and its impedance. Moreover, x_j denotes the jth transmon qubit position, and the function $\beta(\alpha)$ can be changed over the interval $[0, \beta_{\text{max}}]([0, \alpha_{\text{max}}])$ via applied magnetic fluxes Φ_{ext}^j and $\bar{\Phi}_{\text{ext}}^j$, which are externally driven on the jth superconducting qubit. We point out that these external magnetic fluxes allow as well to modify the qubit frequency.

We show now how the interaction Hamiltonian (7.36) can emulate the evolution associated with Hamiltonian (7.30). We plot in Fig. 7.11b the quantum gates needed for emulating two-qubit operations interacting with the continuum in a single digital step [9, 109] to be performed by our digital-analog emulator. In this proposal, making use of a superconducting circuit framework, each unitary operator will be associated with the dynamics under the Hamiltonian (7.36) for corresponding external fluxes Φ_{ext}^j and $\bar{\Phi}_{\text{ext}}^j$. More concretely, the operators acting on the first two superconducting transmon qubits are, sequentially applied, a Mølmer-Sørensen [116] operator $U_{\text{MS}}(\pi/2, 0)$ performed through the cavity [117], a local rotation $U_C = \exp[-\phi\sigma_1^y \int dk\ g_k(a_k^\dagger e^{-ikx} - a_k e^{ikx})]$ that produces an interaction between the spin matrices and the bosonic field continuum, and the inverse Mølmer-Sørensen operator $U_{\rm MS}(-\pi/2, 0)$. The combination of the three gates will produce the corresponding two-qubit gate coupled with the continuum of bosonic modes, $H_2 = i(\sigma_j \otimes \sigma_\ell) \int dk\ g_k(a_k^\dagger e^{-ikx} - a_k e^{ikx})$.

The U_c operator will be employed independently on each superconducting transmon qubit to produce the corresponding single-qubit gates interacting with the bosonic continuum. Moreover, the auxiliary transmon qubit permits to produce the operators involving only the bosonic field modes by using a gate,

$$U_A = \exp[-\phi\sigma_A^z \int dk g_k(a_k^\dagger e^{-ikx} - a_k e^{ikx})], \tag{7.37}$$

where σ_A^z is the corresponding Pauli operator. The necessary operator is achieved by initializing the auxiliary qubit in an eigenstate of σ_A^z. An equivalent sequence of operators can be applied on more superconducting qubits for scaling the model in order to emulate couplings involving many fermionic field modes.

The path for scaling this proposal to many fermionic field modes is to take into account more superconducting transmon qubits interacting both with the resonator and with the open line, as we plot in Fig. 7.12. When one considers N superconducting qubits, N fermionic field modes can be emulated. Therefore, our protocol can realize a large number of fermionic field modes coupled to the bosonic field continuum. This proposal will represent a significant step forward towards an advanced quantum simulation of full-fledged quantum field theories in controllable superconducting qubit setups.

Making use of the proposed method, one can extract information of relevant features of quantum field theories, as for example pair creation and annihilation of fermions as well as self-interaction, mediated via a bosonic field continuum. This quantum simulation is based on unitary gates related to Hamiltonian (7.30). In this respect, as opposed to standard perturbative techniques in quantum field theories, the realization of our proposal will be associated with an infinite number of Feynman diagrams and a finite number of fermionic field modes. Accordingly,

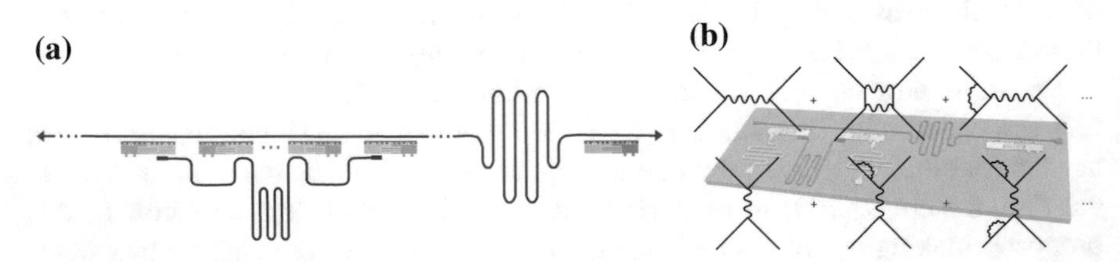

Fig. 7.12 **a** Schematic representation for the realization of N fermionic field modes interacting with a bosonic continuum. Each fermionic field mode is mapped onto a nonlocal spin operator implemented among N superconducting transmon qubits. **b** Feynman diagrams related to the quantum simulation of two fermionic field modes interacting with a bosonic continuum in a circuit QED setup, as described in the text. Reprinted with permission from [105], Copyright (2015) American Physical Society

this path towards full-fledged QFTs is at variance from standard techniques, because it needs the addition of more fermionic field modes instead of more perturbative Feynman diagrams. On the other hand, the fact that we consider a continuum of bosonic field modes in circuit QED makes our protocol nearer to the targeted theory.

To conclude, we have introduced a method for a digital-analog quantum emulation of fermion-fermion scattering and quantum field theories with circuit QED. This quantum platform benefits from strong coupling between superconducting transmon qubits with a microwave cavity and a continuum of bosonic field modes. Our method is a significant step forward towards efficient quantum simulations of quantum field theories in perturbative and nonperturbative scenarios.

7.6 Conclusion

We have presented the topic of analog and digital quantum simulations in the light of circuit QED technologies. In particular, we have discussed the basic concepts of circuit network theory and their applications to electric circuits operating at the quantum degeneracy regime imposed by the superconducting state.

We have shown how circuit QED with a transmon or a flux qubit represents a building block for circuit QED lattices of interacting microwave photons aiming at simulating Hamiltonians of condensed matter physics such us the Bose-Hubbard model, the Jaynes-Cummings-Hubbard model, models with nearest-neighbor Kerr nonlinearities that exhibit Bose-Hubbard features, the bosonic Kagome lattice, and the Rabi-Hubbard model. The latter has interesting predictions provided by the counter-rotating terms that appear in the system Hamiltonian.

Regarding the digital approach of quantum simulations, we have presented two recent developments, i.e., the simulation of spin systems, and the digital/analog simulation of quantum field theories exemplified by the fermion-fermion scattering mediated by a continuum of bosonic modes. These theoretical efforts are based on the state-of-the-art in circuit QED with transmon qubits, and may pave the way for experimental developments in the near future.

Acknowledgements We acknowledge funding from the Basque Government IT472-10; Spanish MINECO FIS2012-36673-C03-02; Ramón y Cajal Grant RYC-2012-11391; UPV/EHU UFI 11/55; UPV/EHU Project EHUA14/04; Chilean FONDECYT 1150653; PROMISCE; SCALEQIT EU; Spanish MINECO/FEDER FIS2015-69983-P and Basque Goverment IT986-16 projects.

References

1. V. Kendon, K. Nemoto, W.J. Munro, Philos. Trans. R. Proc. A **368**, 3609 (2010)
2. J.I. Cirac, P. Zoller, Nat. Phys. **8**, 264 (2010)
3. P. Hauke, F.M. Cucchietti, L. Tagliacozzo, I. Deutsch, M. Lewenstein, Rep. Prog. Phys. **75**, 082401 (2012)

4. T. Schätz, C.R. Monroe, T. Esslinger, New J. Phys. **15**, 085009 (2013)
5. I. Buluta, F. Nori, Science **336**, 108 (2009)
6. I.M. Georgescu, S. Ashhab, F. Nori, Rev. Mod. Phys. **86**, 153 (2013)
7. T.H. Johnson, S.R. Clark, D. Jaksch, EPJ Quant. Technol. **1**, 10 (2014)
8. R.P. Feynman, Int. J. Theor. Phys. **21**, 467 (1982)
9. S. Lloyd, Science **273**, 1073 (1996)
10. M.H. Devoret, J.M. Martinis, in *Superconducting Qubits*, ed. by D. Esteve, J.-M. Raimond, J. Dalibard. Quantum Entanglement and Information Processing, Les Houches Session LXXIX (Elsevier, Heidelberg, 2004), p. 443
11. F.K. Wilhelm, J. Clarke, Nature **453**, 1031 (2008)
12. M.H. Devoret, R.J. Schoelkopf, Science **339**, 1169 (2013)
13. A. Blais, R.-S. Huang, A. Wallraff, S.M. Girvin, R.J. Schoelkopf, Phys. Rev. A **69**, 062320 (2004)
14. A. Wallraff, D.I. Schuster, A. Blais, L. Frunzio, R.-S. Huang, J. Majer, S. Kumar, S.M. Girvin, R.J. Schoelkopf, Nature **431**, 162 (2004)
15. I. Chiorescu, P. Bertet, K. Semba, Y. Nakamura, C.J.P.M. Harmans, J.E. Mooij, Nature **431**, 159 (2004)
16. J.J. García-Ripoll, E. Solano, M.A. Martin-Delgado, Phys. Rev. B **77**, 024522 (2008)
17. L. Tian, Phys. Rev. Lett. **105**, 167001 (2010)
18. E.J. Pritchett, C. Benjamin, A. Galiautdinov, M.R. Geller, A.T. Sornborger, P.C. Stancil, J.M. Martinis, arXiv:1008.0701
19. Y. Zhang, L. Yu, J.-Q. Liang, G. Chen, S. Jia, F. Nori, Sci. Rep. **4**, 4083 (2014)
20. F. Mei, V.M. Stojanović, I. Siddiqi, L. Tian, Phys. Rev. B **88**, 224502 (2013)
21. O. Viehmann, J. von Delft, F. Marquardt, Phys. Rev. Lett. **110**, 030601 (2013)
22. A. Kurcz, A. Bermudez, J.J. García-Ripoll, Phys. Rev. Lett. **112**, 180405 (2014)
23. R. Barends et al., Nat. Commun. **6**, 7654 (2015)
24. Y. Salathé et al., Phys. Rev. X **5**, 021027 (2015)
25. M. Tinkham, *Introduction to Superconductivity*, 2nd edn. (Krieber, Malabar, 1985)
26. T.A. Orlando, K.A. Delin, *Foundations of Applied Superconductivity* (Addison-Wesley, 1991)
27. M.H. Devoret, in *Quantum Fluctuations in Electrical Circuits*, ed. by S. Reynaud, E. Giacobino, J. Zinn-Justin. Quantum Fluctuations, Les Houches Session LXIII (Elsevier, Heidelberg, 1997), p. 351
28. G. Burkard, R.H. Koch, D.P. DiVincenzo, Phys. Rev. B **69**, 064503 (2004)
29. S.E. Nigg, H. Paik, B. Vlastakis, G. Kirchmair, S. Shankar, L. Frunzio, M.H. Devoret, R.J. Schoelkopf, S.M. Girvin, Phys. Rev. Lett. **108**, 240502 (2012)
30. J. Koch, T.M. Yu, J. Gambetta, A.A. Houck, D.I. Schuster, J. Majer, A. Blais, M.H. Devoret, S.M. Girvin, R.J. Schoelkopf, Phys. Rev. A **76**, 042319 (2007)
31. T.P. Orlando, J.E. Mooij, L. Tian, C.H. van del Wal, L.S. Levitov, S. Lloyd, J.J. Mazo, Phys. Rev. B **60**, 15398 (1999)
32. E.T. Jaynes, F.W. Cummings, Proc. IEEE **51**, 89 (1963)
33. D. Braak, Phys. Rev. Lett. **107**, 100401 (2011)
34. J. Majer, J.M. Chow, J.M. Gambetta, J. Koch, B.R. Johnson, J.A. Schreier, L. Frunzio, D.I. Schuster, A.A. Houck, A. Wallraff, A. Blais, M.H. Devoret, S.M. Girvin, R.J. Schoelkopf, Nature (London) **449**, 443 (2007)
35. P.J. Leek, S. Filipp, P. Maurer, M. Baur, R. Bianchetti, J.M. Fink, M. Göppl, L. Steffen, A. Wallraff, Phys. Rev. B **79**, 180511 (2009)
36. A. Blais, J. Gambetta, A. Wallraff, D.I. Schuster, S.M. Girvin, M.H. Devoret, R.J. Schoelkopf, Phys. Rev. A **75**, 032329 (2007)
37. L. DiCarlo, J.M. Chow, J.M. Gambetta, L.S. Bishop, B.R. Johnson, D.I. Schuster, J. Majer, A. Blais, L. Frunzio, S.M. Girvin, R.J. Schoelkopf, Nature **460**, 240 (2009)
38. G. Haack, F. Helmer, M. Mariantoni, F. Marquardt, E. Solano, Phys. Rev. B **82**, 024514 (2010)
39. R.C. Bialczak, M. Ansmann, M. Hofheinz, E. Lucero, M. Neeley, A.D. O'Connell, D. Sank, H. Wang, J. Wenner, M. Steffen, A.N. Cleland, J.M. Martinis, Nat. Phys. **6**, 409 (2010)

40. T. Yamamoto, M. Neeley, E. Lucero, R.C. Bialczak, J. Kelly, M. Lenander, M. Mariantoni, A.D. O'Connell, D. Sank, H. Wang, M. Weides, J. Wenner, Y. Yin, A.N. Cleland, J.M. Martinis, Phys. Rev. B **82**, 184515 (2010)
41. J.M. Fink, R. Bianchetti, M. Baur, M. Göppl, L. Steffen, S. Filipp, P.J. Leek, A. Blais, A. Wallraff, Phys. Rev. Lett. **103**, 083601 (2009)
42. M. Neeley, R.C. Bialczak, M. Lenander, E. Lucero, M. Mariantoni, A.D. O'Connell, D. Sank, H. Wang, M. Weides, J. Wenner, Y. Yin, T. Yamamoto, A.N. Cleland, J.M. Martinis, Nature **467**, 570 (2010)
43. L. DiCarlo, M.D. Reed, L. Sun, B.R. Johnson, J.M. Chow, J.M. Gambetta, L. Frunzio, S.M. Girvin, M.H. Devoret, R.J. Schoelkopf, Nature **467**, 574 (2010)
44. J.P. Groen, D. Ristè, L. Tornberg, J. Cramer, P.C. de Groot, T. Picot, G. Johansson, L. DiCarlo, Phys. Rev. Lett. **111**, 090506 (2013)
45. C. Ciuti, G. Bastard, I. Carusotto, Phys. Rev. B **72**, 115303 (2005)
46. J. Bourassa, J.M. Gambetta, A.A. Abdumalikov Jr., O. Astafiev, Y. Nakamura, A. Blais, Phys. Rev. A **80**, 032109 (2009)
47. D. Hagenmüller, S. De Liberato, C. Ciuti, Phys. Rev. B **81**, 235303 (2010)
48. P. Nataf, C. Ciuti, Phys. Rev. Lett. **104**, 023601 (2010)
49. T. Niemczyk, F. Deppe, H. Huebl, E.P. Menzel, F. Hocke, M.J. Schwarz, J.J. Garcia-Ripoll, D. Zueco, T. Hümmer, E. Solano, A. Marx, R. Gross, Nat. Phys. **6**, 772 (2010)
50. P. Forn-Díaz, J. Lisenfeld, D. Marcos, J.J. García-Ripoll, E. Solano, C.J.P.M. Harmans, J.E. Mooij, Phys. Rev. Lett. **105**, 237001 (2010)
51. F. Beaudoin, J.M. Gambetta, A. Blais, Phys. Rev. A **84**, 043832 (2011)
52. H.-P. Breuer, F. Petruccione, *The Theory of Open Quantum Systems* (Clarendon, Oxford, 2006)
53. P. Nataf, C. Ciuti, Phys. Rev. Lett. **107**, 190402 (2011)
54. G. Romero, D. Ballester, Y.M. Wang, V. Scarani, E. Solano, Phys. Rev. Lett. **108**, 120501 (2012)
55. J. Casanova, G. Romero, I. Lizuain, J.J. García-Ripoll, E. Solano, Phys. Rev. Lett. **105**, 263603 (2010)
56. S. De Liberato, Phys. Rev. Lett. **112**, 016401 (2014)
57. A. Crespi, S. Longhi, R. Osellame, Phys. Rev. Lett. **108**, 163601 (2012)
58. A. Nunnenkamp, J. Koch, S.M. Girvin, New J. Phys. **13**, 095008 (2011)
59. S. Felicetti, M. Sanz, L. Lamata, G. Romero, G. Johansson, P. Delsing, E. Solano, Phys. Rev. Lett. **113**, 093602 (2014)
60. I. Carusotto, C. Ciuti, Rev. Mod. Phys. **85**, 299 (2013)
61. M.P.A. Fisher, P.B. Weichman, G. Grinstein, D.S. Fisher, Phys. Rev. B **40**, 546 (1989)
62. M.J. Hartmann, F.G.S. Brandão, M.B. Plenio, Nat. Phys. **2**, 849 (2006)
63. M.J. Hartmann, M.B. Plenio, Phys. Rev. Lett. **99**, 103601 (2007)
64. A.D. Greentree, C. Tahan, J.H. Cole, L.C.L. Hollenberg, Nat. Phys. **2**, 856 (2006)
65. D.G. Angelakis, M.F. Santos, S. Bose, Phys. Rev. A **76**, 031805(R) (2007)
66. J. Cho, D.G. Angelakis, S. Bose, Phys. Rev. Lett. **101**, 246809 (2008)
67. J. Koch, A.A. Houck, K. Le Hur, S.M. Girvin, Phys. Rev. A **82**, 043811 (2010)
68. S. Mittal, J. Fan, S. Faez, A. Migdall, J.M. Taylor, M. Hafezi, Phys. Rev. Lett. **113**, 087403 (2014)
69. F. Grusdt, F. Letscher, M. Hafezi, M. Fleischhauer, Phys. Rev. Lett. **113**, 155301 (2014)
70. T. Kitagawa, M.A. Broome, A. Fedrizzi, M.S. Rudner, E. Berg, I. Kassal, A. Aspuru-Guzik, E. Demler, A.G. White, Nat. Commun. **3**, 882 (2012)
71. J. Cho, D.G. Angelakis, S. Bose, Phys. Rev. A **78**, 062338 (2008)
72. A. Kay, D.G. Angelakis, Eur. Phys. Lett. **84**, 20001 (2008)
73. A.A. Houck, H.E. Türeci, J. Koch, Nat. Phys. **8**, 292 (2012)
74. D. Underwood, W.E. Shanks, J. Koch, A.A. Houck, Phys. Rev. A **86**, 023837 (2012)
75. A. Hosseinkhani, B.G. Dezfouli, F. Ghasemipour, A.T. Rezakhani, H. Saberi, Phys. Rev. A **89**, 062324 (2014)
76. S. Schmidt, J. Koch, Ann. Phys. (Berlin) **525**, 395 (2013)

77. J. Raftery, D. Sadri, S. Schmidt, H.E. Türeci, A.A. Houck, Phys. Rev. X **4**, 031043 (2014)
78. S. Haroche, J.-M. Raymond, *Exploring the Quantum* (Oxford University Press, New York, 2006)
79. M. Leib, M.J. Hartmann, New. J. Phys. **12**, 093031 (2010)
80. M. Leib, M.J. Hartmann, Phys. Scr. **T153**, 014042 (2013)
81. J. Tangpatinanon, D.G. Angelakis, Unpublished
82. C. Eichler, Y. Salathe, J. Mlynek, S. Schmidt, A. Wallraff, Phys. Rev. Lett. **113**, 110502 (2014)
83. D.G. Angelakis, S. Bose, S. Mancini, Eur. Phys. Lett. **85**, 20007 (2009)
84. D.G. Angelakis, L. Dai, L.C. Kwek, Eur. Phys. Lett. **91**, 10003 (2010)
85. I. Carusotto, D. Gerace, H.E. Türeci, S. De Liberato, C. Ciuti, A. Imamoglu, Phys. Rev. Lett. **103**, 033601 (2009)
86. M.J. Hartmann, Phys. Rev. Lett. **104**, 113601 (2010)
87. T. Grujic, S.R. Clark, D. Jaksch, D.G. Angelakis, New J. Phys. **14**, 103025 (2012)
88. T. Grujic, S.R. Clark, D. Jaksch, D.G. Angelakis, Phys. Rev. A **87**, 053846 (2013)
89. N. Schetakis, T. Grujic, S.R. Clark, D. Jaksch, D.G. Angelakis, J. Phys. B At. Mol. Opt. Phys. **46**, 224025 (2013)
90. D.G. Angelakis, A. Kay, New. J. Phys. **10**, 023012 (2008)
91. E.S. Kyoseva, D.G. Angelakis, L.C. Kwek, Eur. Phys. Lett. **89**, 20005 (2010)
92. J. Jin, D. Rossini, R. Fazio, M. Leib, M.J. Hartmann, Phys. Rev. Lett. **110**, 163605 (2013)
93. F. Verstraete, V. Murg, J. Cirac, Adv. Phys. **57**, 143 (2008)
94. N. Schuch, I. Cirac, D. Pérez-García, Ann. Phys. **325**, 2153 (2010)
95. H.-P. Breuer, F. Petruccione, *The Theory of Open Quantum Systems* (Oxford University Press, Clarendon, 2006)
96. H. Zheng, Y. Takada, Phys. Rev. A **84**, 043819 (2011)
97. M. Schiró, B. Öztop, H.E. Türeci, Phys. Rev. Lett. **109**, 053601 (2012)
98. G. Zhu, S. Schmidt, J. Koch, New J. Phys. **15**, 115002 (2013)
99. D. Ballester, G. Romero, J.J. García-Ripoll, F. Deppe, E. Solano, Phys. Rev. X **2**, 021007 (2012)
100. J.S. Pedernales, R. Di Candia, D. Ballester, E. Solano, New J. Phys. **15**, 055008 (2013)
101. T.O. Wehling, A.M. Black-Schaffer, A.V. Balatsky, Adv. Phys. **76**, 1 (2014)
102. B.P. Lanyon, J.D. Whitfield, G.G. Gillet, M.E. Goggin, M.P. Almeida, I. Kassal, J.D. Biamonte, M. Mohseni, B.J. Powell, M. Barbieri, A. Aspuru-Guzik, A.G. White, Nat. Chem. **2**, 106 (2009)
103. B.P. Lanyon, C. Hempel, D. Nigg, M. Müller, R. Gerritsma, F. Zähringer, P. Schindler, J.T. Barreiro, M. Rambach, G. Kirchmair, M. Hennrich, P. Zoller, R. Blatt, C.F. Roos, Science **334**, 57 (2011)
104. U. Las Heras, A. Mezzacapo, L. Lamata, S. Filipp, A. Wallraff, E. Solano, Phys. Rev. Lett. **112**, 200501 (2014)
105. L. García-Álvarez, J. Casanova, A. Mezzacapo, I.L. Egusquiza, L. Lamata, G. Romero, E. Solano, Phys. Rev. Lett. **114**, 070502 (2015)
106. S. Filipp, M. Göppl, J.M. Fink, M. Baur, R. Bianchetti, L. Steffen, A. Wallraff, Phys. Rev. A **83**, 063827 (2011)
107. C. Rigetti, J.M. Gambetta, S. Poletto, B.L.T. Plourde, J.M. Chow, A.D. Córcoles, J.A. Smolin, S.T. Merkel, J.R. Rozen, G.A. Keefe, M.B. Rothwell, M.B. Ketchen, M. Steffen, Phys. Rev. B **86**, 100506(R) (2012)
108. M.E. Peskin, D.V. Schroeder, *Quantum Field Theory* (Westview Press, 1995)
109. J. Casanova, A. Mezzacapo, L. Lamata, E. Solano, Phys. Rev. Lett. **108**, 190502 (2012)
110. P. Jordan, E. Wigner, Z. Phys. **47**, 631 (1928)
111. A. Altland, B. Simons, *Condensed Matter Field Theory* (Cambridge University Press, Cambridge, 2010)
112. J.M. Gambetta, A.A. Houck, A. Blais, Phys. Rev. Lett. **106**, 030502 (2011)
113. S.J. Srinivasan, A.J. Hoffman, J.M. Gambetta, A.A. Houck, Phys. Rev. Lett. **106**, 083601 (2011)
114. J.M. Chow et al., Nat. Commun. **5**, 4015 (2014)
115. D.M. Pozar, *Microwave Engineering* (Wiley, 2012)
116. K. Mølmer, A. Sørensen, Phys. Rev. Lett. **82**, 1835 (1999)
117. A. Mezzacapo, L. Lamata, S. Filipp, E. Solano, Phys. Rev. Lett. **113**, 050501 (2014)

Chapter 8
Dirac Dynamics in Waveguide Arrays: From *Zitterbewegung* to Photonic Topological Insulators

F. Dreisow, M.C. Rechtsman, J.M. Zeuner, Y. Plotnik, R. Keil, S. Nolte, M. Segev and A. Szameit

Abstract Simulating the evolution of a nonrelativistic quantum-mechanical particle in a periodic potential by propagating an optical wave packet in an array of evanescently coupled waveguides has received continuous and increasing attention in recent years. However, one can also simulate the evolution of a *relativistic* quantum particle in free space, as described by the spinor-type Dirac equation, by carefully designing the periodic optical potential. In this chapter, the optical simulation of various phenomena based on the Dirac equation will be discussed, such as Klein tunneling, *Zitterbewegung*, relativistic gauge fields, and photonic topological insulators, which all can be realized in the framework of paraxial optics in periodic media, without requiring specially synthesized materials with subwavelength controlled properties.

F. Dreisow · J.M. Zeuner · S. Nolte · A. Szameit (✉)
Abbe Center of Photonics, Institute of Applied Physics, Friedrich-Schiller-Universität Jena, Max-Wien-Platz 1, 07743 Jena, Germany
e-mail: alexander.szameit@uni-jena.de

F. Dreisow
e-mail: f.dreisow@uni-jena.de

J.M. Zeuner
e-mail: julia.zeuner@uni-jena.de

S. Nolte
e-mail: stefan.nolte@uni-jena.de

M.C. Rechtsman · Y. Plotnik · M. Segev
Physics Department, Solid State Institute Technion, 32000 Haifa, Israel
e-mail: mcrworld@gmail.com

Y. Plotnik
e-mail: yonatanplotnik@gmail.com

M. Segev
e-mail: msegev@techunix.technion.ac.il

R. Keil
Institut für Experimentalphysik, Universität Innsbruck, Technikerstrasse 25, 6020 Innsbruck, Austria
e-mail: robert.keil@uibk.ac.at

D.G. Angelakis (ed.), *Quantum Simulations with Photons and Polaritons*, Quantum Science and Technology, DOI 10.1007/978-3-319-52025-4_8

8.1 Introduction

Integrated waveguide arrays for simulating single particle nonrelativistic quantum-mechanics effects have received significant attention in the last two decades [1]. The underlying idea of mapping nonrelativistic quantum mechanics onto an optical model system is the conceptual similarity between the Schrödinger equation for matter waves and the scalar paraxial wave equation for optical waves. This formal correspondence allowed for the observation of various classical analogues of nonrelativistic phenomena commonly associated with the evolution of electrons in periodic potentials, such as optical Bloch oscillations [2], optical dynamic localization [3], and Anderson localization in disordered lattices [4]. It has been a common belief that the use of optical waveguides as a model system for quantum mechanics carries the intrinsic drawback of being limited to nonrelativistic phenomena and that the observation of optical analogues of relativistic phenomena requires subwavelength structured media like photonic crystals or metamaterials [5–7]. However, only recently it has been realized that, by carefully designing the underlying periodic potential, paraxial light propagation is capable of simulating the evolution of a relativistic quantum particle, as described by the spinor-type Dirac equation [8, 9]. Thus, optical analogues of such important phenomena as Klein tunneling [8, 10], *Zitterbewegung* [11], relativistic gauge fields [12], and photonic topological insulation [13] can be realized in the framework of paraxial optics in periodic media, without requiring specially synthesized media with subwavelength controlled properties. Such optical simulations offer various benefits, such as the direct measurement of the wave function and the long coherence time of light beams.

The first part of the chapter deals with the fabrication technology of the photonic lattices, where ultrashort femtosecond (fs) laser pulses are tightly focused into a bulk material, resulting in permanent waveguides within the bulk. In the second part, various relativistic emulations in one-dimensional lattices are presented, including the *Zitterbewegung* of a photonic wave packet, Klein tunneling and pair creation. In the third part of this chapter we focus on the implementation of the graphene geometry in form of photonic honeycomb lattices (known as *photonic graphene*), which is the basis for several new phenomena such as relativistic pseudo-magnetic fields, topological edge states and photonic topological insulators.

8.2 Fabrication Technology

For the detailed investigation of relativistic effects in photonic lattices, the latter have to meet a variety of conditions. The coupling between the individual lattice sites has to be highly homogeneous to prevent disorder effects. Furthermore, the properties of every single waveguide should be tuneable, which in particular allows for the insertion of artificial defects. Moreover, a diversity of topologies has to be realizable in order to investigate the light evolution in one- and two-dimensional periodic lattices as well as in non-periodic configurations such as waveguide interfaces and junctions. Additionally, the fabricated arrays should be stable and permanent. It turns out that the fs laser writing technique is an excellent approach to fulfil these requirements [14].

When ultrashort laser pulses are tightly focused into the bulk material, nonlinear absorption takes place leading to optical breakdown and the formation of a micro plasma, which induces a permanent change of the material's molecular structure. In the particular case of fused silica as the processed material (refractive index $n_{\text{bulk}} = 1.45$), the density is locally increased [15], yielding an increase of the refractive index [16]. This phenomenon is commonly explained by the fact that bulk fused silica consists of molecular ring structures composed of several $Si\,O_2$ molecules, which are partially broken by the fs laser radiation (see Fig. 8.1a). Subsequent recombination to smaller ring structures, which are composed of fewer $Si\,O_2$ molecules with a higher packing density, yields the observed densification and, hence, a local increase of the refractive index. When the spherical aberrations caused by the sample surface are small, the dimensions of these induced changes are approximately given by the size of the focal region of the writing objective and can be calculated using

$$w_0 = M^2 \frac{\lambda}{\pi \text{NA}}, \quad b = M^2 \frac{\lambda}{\pi \text{NA}^2}. \tag{8.1}$$

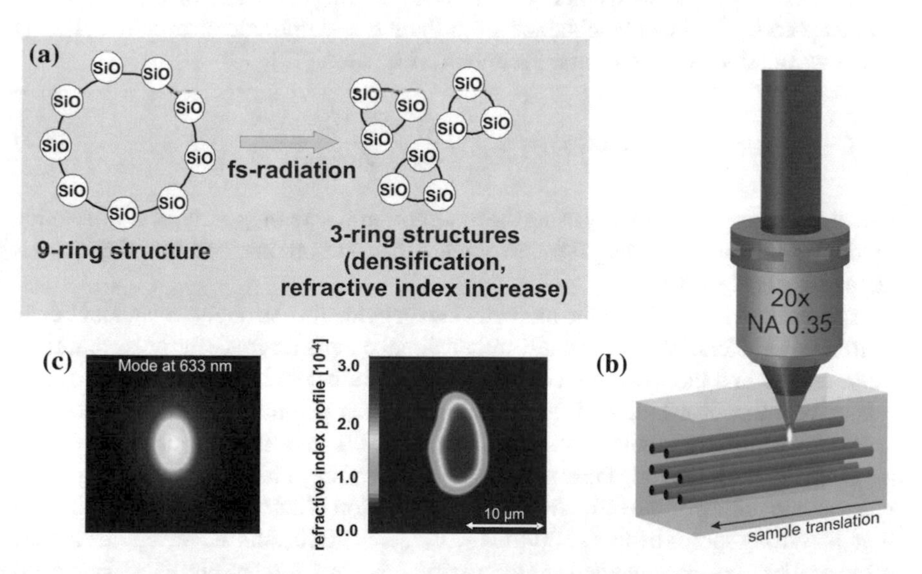

Fig. 8.1 Writing waveguides in fused silica using fs laser pulses. **a** In the focal region of the writing objective, the high light intensities partially break the silica bonds. After reconfiguration this results in densification and a locally increased refractive index. **b** Moving the sample transversely to the beam results in an elongated index increase: a waveguide is created. **c** The guided mode intensity $|A(x, y)|^2$ at $\lambda = 633$ nm (*left*) and the cross section of the refractive index distribution (*right*) of an individual waveguide

Here, λ is the optical wavelength, w_0 is the FWHM (full width at half maximum) radius of the focal spot and b is the associated Rayleigh length. The quantity M^2 characterizes the difference between a real laser beam and a diffraction-limited Gaussian beam, while NA is the numerical aperture. By moving the sample transversely with respect to the beam, a continuous modification is obtained and a waveguide is created [17]. Such a guide can be written along arbitrary paths since the only limiting factor in the placement of the focus is the focal length of the writing objective [18]. Furthermore, all structural changes are permanent and stable after fabrication.

Writing waveguides is a very active field of research. Main topics are investigations on inscriptions in a variety of bulk media, on directly influencing the refractive index distribution of individual guides by beam shaping, minimizing propagation losses, detailed analysis of the multiple fabrication parameters, and more. We refer the interested reader to recent reviews on this topic and the references therein [19–22].

The waveguides used for the experiments presented in this chapter were usually fabricated by a Ti:Sapphire laser system (RegA/Mira, Coherent Inc.) with a repetition rate of 100 kHz at a laser wavelength of 800 nm. The beam was typically focused into polished fused-silica samples by a 20× microscope objective with a numerical aperture of 0.35. This relatively weak focusing ensures that the created refractive index change is almost independent on the writing depth inside the material, so that even large waveguide arrays are still homogeneous in the vertical dimension. A sketch of the writing setup is shown in Fig. 8.1b. The resulting refractive index change (for a fixed polarization) can be obtained from the mode-amplitude distribution A(x, y) of the waveguides via the inverse Helmholtz-Equation [23],

$$n^2(x, y) = n_{\text{eff}}^2 - \frac{\lambda^2}{4\pi^2} \frac{\nabla^2 A(x, y)}{A(x, y)} \tag{8.2}$$

which can be approximated by using the assumption $n_{\text{eff}} = n_{\text{bulk}}$. A typical waveguide mode and the corresponding refractive index distribution obtained by using Eq. 8.2 are shown in Fig. 8.1c.

The guiding properties of the individual waveguides are strongly dependent on the writing parameters. Within the multi-dimensional parameter space the pulse duration, pulse energy and the writing speed are the key parameters influencing the resulting refractive index distribution. The optimal processing parameters with respect to the described setup are a pulse duration of about 150fs and a pulse energy of 0.3 μJ, respectively [24]. Keeping these parameters fixed by only changing the writing speed v, a specific tuning of the refractive index modulation is achieved. It could be shown that for writing speeds of about 500 μm/s $< v_{\text{writing}} <$ 1500 μm/s the waveguide losses, measured by a cut-back method, are approximately constant (see Fig. 8.2a). However, the refractive index modulation decreases for increasing writing speeds due to the smaller overlap between successive pulses, as shown in Fig. 8.2b. This can be used in particular to specifically introduce defects in a waveguide array with constant waveguide spacing.

As a particular feature, the light intensity distribution can be directly monitored inside the samples using a fluorescence microscopy technique [25, 26]. During the

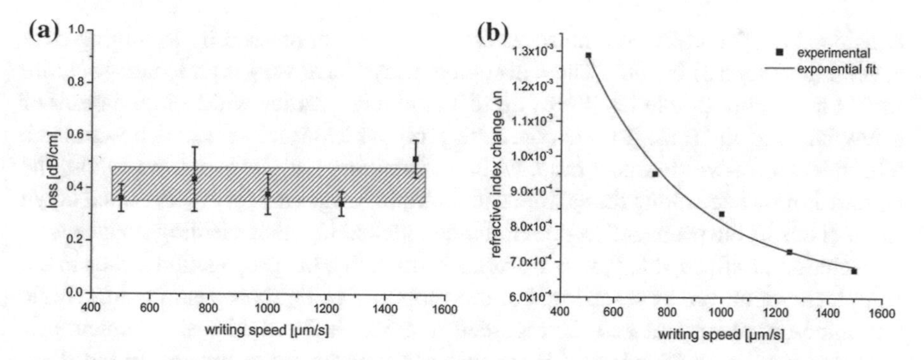

Fig. 8.2 Characteristics of fs laser-written waveguides. **a** The losses are approximately independent of the writing speed. **b** The refractive index change is a function of the writing speed. Figures from [24]

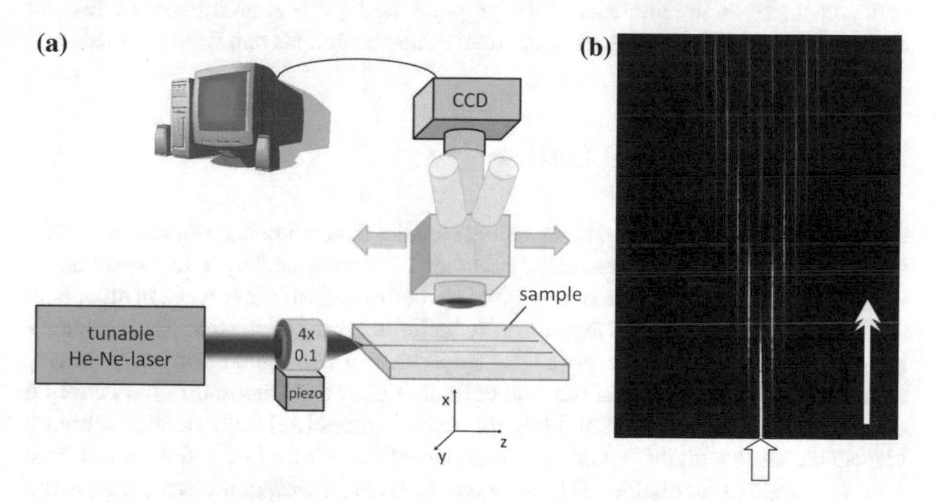

Fig. 8.3 Fluorescence microscopy. **a** Experimental setup to observe the fluorescence radiation. **b** Observed fluorescence pattern of a straight waveguide array

writing process non-bridging oxygen hole centers (NBOHCs) are generated due to breaking of molecular bonds. The NBOHCs have a broad absorption band around 2.0 eV [27] and exhibit an absorption maximum at 620 nm [28]. Hence, when launching red light from a HeNe laser at $\lambda = 633$ nm (corresponding to 1.95 eV) into the waveguides, these color centers are excited and the resulting fluorescence (at $\lambda = 650$ nm) can be directly observed. Since the color centers are formed exclusively inside the waveguides, this technique yields a high signal-to-noise ratio. In contrast to fluorescent polymers (see e.g. [29]), the bulk material causes almost no background noise. The experimental setup is sketched in Fig. 8.3a. The laser beam is launched into the waveguide array using a $4\times$ microscope objective (NA $= 0.25$). The fluorescence distribution representing the diffraction pattern is imaged onto a CCD camera using a $5\times$ objective (NA $= 0.13$). Most of the scattered HeNe-laser light is blocked by a 650 nm longpass filter which improves the image quality considerably. The electronic noise of the camera is suppressed by averaging over several (usually 10) images. The waveguide arrays have very high length-to-width aspect ratio with sample lengths of up to 100 mm and lattice widths in the order of a few hundred microns. To overcome the problem to visualize a lattice with such a high length-to-width aspect ratio, while maintaining a high spatial resolution, the camera is translated along the sample and multiple images are recorded, scaled down to 20 pixels in the propagation direction and stitched together yielding images with a resolution of 40 μm × 0.5 μm per pixel. Additionally, the propagation losses in the samples are removed by rescaling. For this purpose, the light propagation in a single waveguide is monitored and the recorded intensity is fitted with an exponentially decaying function, which is used to normalize the fluorescence images. In Fig. 8.3b, a fluorescence microscope image of light propagation in a waveguide array after a single waveguide excitation is shown. Importantly, the dependence of the fluorescence intensity on the intensity of the propagating light is approximately linear, so that not only qualitative, but also quantitative measurements can be performed.

8.3 One-Dimensional Lattices

The evolution of a Dirac particle in a single spatial dimension is governed by its two chiral components. Mathematically, this can be represented by a two-component wavefunction [30]. There are two avenues to realize such a spinor wave in an optical setting. One can realize a waveguide array including Bragg reflection gratings yielding counter-propagating waves [31] or implement a waveguide array with binary substructure. The latter again has two options. Either one can modify the coupling constants or alternating high and low refractive indices [32] both yielding a binary waveguide array with the potential to simulate spinor waves. Let us first concentrate on a pure index modulation. Therefore, we consider a waveguide array consisting of high refractive index change on even waveguides and low index change on odd waveguides (see Fig. 8.4a). The index change is assumed to be small such that we can assume an unaffected and homogeneous coupling constant κ and an index mismatch 2δ. This leads to the following discrete Schrödinger equation for the amplitudes in the mth channel a_m:

$$i\frac{da_m}{dz} + \kappa(a_{m+1} + a_{m-1}) + (-1)^m \delta a_m = 0\,. \tag{8.3}$$

This equation is most accurate in the limit where the guided modes of the waveguides are tightly bound and well separated from one another. Applying Bloch's theorem allows the computation of the eigenvalue spectrum. With the plane wave ansatz $a_m = \phi e^{i(\beta_z z - m\beta_x \Lambda)}$ one finds the eigenvalues β_z (which represent the longitudinal wave number) to be

$$\beta_z^{\pm} = \pm\sqrt{\delta^2 + 4\kappa^2 \cos^2(\beta_x \Lambda)} \tag{8.4}$$

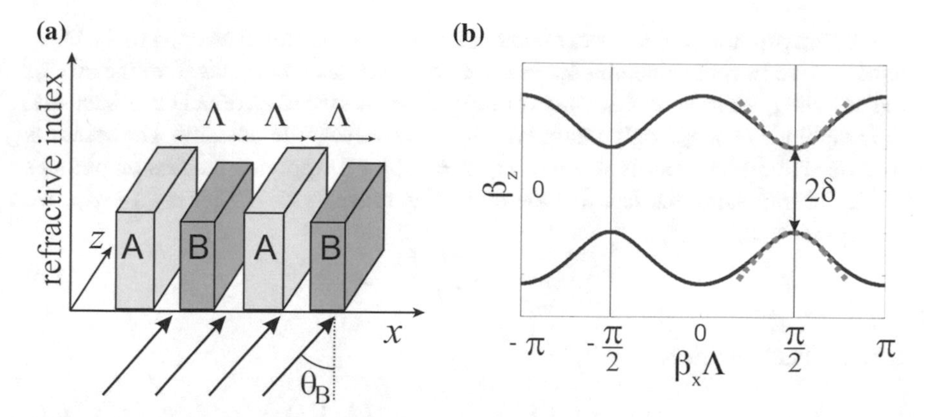

Fig. 8.4 The figure shows **a** a sketch of the binary waveguide array. It consists of an alternating sequence of high (A) and low (B) refractive indices. The arrows mark a broad excitation under the Bragg angle θ_B. **b** shows the band structure of the system. It is split in two opposite bands, which are separated by a gap of 2δ. The *dotted curves* shows the hyperbolic approximation close to the edge of the band ($\beta_x = \pi/2\Lambda$)

with the waveguide spacing Λ. The setting supports two bands denoted with the positive and negative sign. The non-vanishing index mismatch δ opens a gap between these bands, which has the smallest spacing at the edge of the Brillouin zone $\beta_x\Lambda = \pi/2$ with β_x as the transverse wave number. Note here, that the Brillouin zone has only half the width $-\pi/2 \leq \beta_x\Lambda \leq \pi/2$ compared to the homogeneous array $-\pi \leq \beta_x\Lambda \leq \pi$. The reason is that the Wigner-Seitz cell contains two waveguides and, therefore, in reciprocal space the unit cell has only half the size.

In the vicinity of the edges of the band structure $\beta_x\Lambda \approx \pm\pi/2$ one can approximate $\beta_z^\pm$ as two hyperbolas with opposite sign. For the approximation we introduce the shifted transverse and normalized wavenumber q, which is defined as $q = \beta_x\Lambda - \pi/2$. Hence we can write

$$\beta_z^\pm \approx \pm\sqrt{\delta^2 + \kappa^2 q^2} \tag{8.5}$$

as the dispersion relation of the waveguide system (Fig. 8.4b). One can immediately recognize the energy momentum dispersion relation of a free relativistic particle. In this analogy, q corresponds to the particle's momentum and β_z to its energy. The quantities δ and κ correspond to mass and speed of light, respectively.

We will show now explicitly that the system governed by Eq. 8.3 indeed follows the dynamics of the Dirac equation. We perform the analysis at $q \approx 0$. Hence, we demand the excitation width of the light field to be rather broad, covering at least several lattice sites. Therefore, the spectrum $G(\beta_x\Lambda)$ can be assumed to be narrow. The average momentum will be zero if the light beam is sent under the Bragg angle $\theta_\mathrm{B} = \lambda/4n_0\Lambda$ into the waveguides. As this results in a phase shift of $\pi/2$ between adjacent waveguides, we can set the waveguide amplitudes to $a_{2m} = (-1)^m\psi_1$ and $a_{2m-1} = -i(-1)^m\psi_2$, where we will see soon that ψ_1 and ψ_2 are the two components of the Dirac spinor ψ. For a better demonstration of the required analogy to the Dirac equation we introduce continuous transverse coordinates normalized to the size of the unit cell $\xi \leftrightarrow m = x/2\Lambda$, which is allowed here as we use a broad beam such that the amplitudes change only marginally with the waveguide position. The spinor is then depending on ξ and evolves with z: $\psi = \psi(\xi, z)$. Applying the new amplitudes to Eq. 8.3 and using the fact that the derivation reads as $\partial a/\partial\xi \rightarrow a_{m+1} - a_m$, we finally arrive at:

$$i\frac{\partial \psi}{\partial z} = -i\kappa\sigma_x\frac{\partial \psi}{\partial \xi} + \delta\sigma_z\psi \tag{8.6}$$

with the Pauli matrices

$$\sigma_x = \begin{pmatrix} 0 & 1 \\ 1 & 0 \end{pmatrix} \quad \text{and} \quad \sigma_z = \begin{pmatrix} 1 & 0 \\ 0 & -1 \end{pmatrix} . \tag{8.7}$$

Equation 8.6 is the optical version of the well known Dirac equation, where the coupling constant κ equals the velocity of light $\kappa \leftrightarrow c$ and δ is proportional to the particle's mass $\delta \leftrightarrow mc^2/\hbar$. Note that the evolution of the optical beam takes place along the spatial coordinate z instead of time t. Moreover, the field amplitudes ψ_1 and ψ_2 are describing the occupation amplitudes of light in the two sublattices of the binary array.

8.3.1 Photonic Zitterbewegung

Let us now describe the wave dynamics of a relativistic particle. It was Erwin Schrödinger who analyzed the dynamics underlying Dirac's equation [33]. He found that a free relativistic electron performs a fast oscillating motion due to the interference of positive and negative energy states in the two branches of the energy momentum relation. The motion also known as *Zitterbewegung* has a very high frequency ($\sim 10^{21}$Hz) and an amplitude of the oscillation of the order of the Compton wavelength ($\lambda_C = h/m_e c =\approx 10^{-12}$m) of an electron. Therefore so far it has not been possible to directly observe the *Zitterbewegung* for free particles as no physical method exists to achieve this temporal resolution. The first observation of the *Zitterbewegung* was reported in a simulator using trapped ions [34] only recently, 80 years after the prediction of Schrödinger, which demonstrates that the observation of the *Zitterbewegung* is a challenging task.

An oscillating motion of the light beam is also expected in our optical setting. To show this we derive the expectation value for the transverse coordinate depending on the longitudinal coordinate $\xi(z)$. Therefore we start with a field excitation as required above with $E(\xi, 0) = G(\xi)e^{i\pi\xi}$, with a broad envelope $G(\xi)$ tilted at the Bragg angle θ_B. We introduce the Fourier transform of the excitation as $\hat{G}(g) = (2\pi)^{-1}\int d\xi G(\xi)e^{-i\xi q}$. One can start the derivation by using this expression and the definition of the center of mass as the expectation value $\xi(z) = \int d\xi\xi(|\psi_1|^2 + |\psi_2|^2)/\int d\xi(|\psi_1|^2 + |\psi_2|^2)$. A detailed calculation, see e.g. [9] leads to the result:

$$\xi(z) = \xi(0) + 4\pi\kappa^3 z \int dq \left(\frac{q}{\beta_z^{\pm}}\right)^2 |\hat{G}(q)|^2 + 2\pi\kappa\delta^2 \int dq \left(\frac{1}{\beta_z^{\pm}}\right)^3 |\hat{G}(q)|^2 \sin^2(\beta_z^{\pm} z) \tag{8.8}$$

If the excitation $G(\xi)$ is sufficiently broad, then the spectrum $\hat{G}$ is narrow and allows the approximation $\beta_z^{\pm} \approx \pm\delta$. This results in

$$\xi(z) = \xi(0) + v_0 z + \frac{\kappa}{2\delta}\sin(2\delta z) \tag{8.9}$$

The first term of this expression describes the initial position of the excitation, i.e., the position of the particle. The second term shows that the particle performs a (small) linear transverse translation with velocity v_0 arising from the initial momentum of the excitation. The last term describes an oscillatory motion around the mean position, which is the *Zitterbewegung*. Its amplitude is defined by $R_{\mathrm{ZB}} = \kappa/2\delta = \hbar/2mc$ and the oscillation frequency is given by $\omega_{\mathrm{ZB}} = 2\delta = 2mc^2/\hbar$. The frequency is determined by the energy gap of 2δ in the dispersion relation Eq. 8.5, whereas the amplitude can additionally be adjusted for a fixed frequency by the coupling constant κ. Even though the expression in Eq. 8.9 shows a perfect harmonic oscillation we would like to emphasize that this term is usually damped due to the finite excitation width and thus due to the extended excitation in reciprocal space.

Let us now have a look on the complete dynamics observed in a real waveguide array. The experiments have been reported in [11], where binary waveguides were fabricated using the femtosecond laser inscription technique. The investigation was performed at the wavelength of $\lambda = 633$ nm allowing for an observation of the propagation pattern by the help of fluorescence waveguide microscopy [25] in fused silica glass. The lattice period was set to $\Lambda = 16\,\mu$m leading to a Bragg angle $\theta_{\mathrm{B}} = 0.39°$ and a coupling constant $\kappa = 0.14\,\mathrm{mm}^{-1}$. The width of the excitation beam was 105 μm which covers approximately 7 lattice sites. Hence, the excitation spectrum can be treated as sufficiently small. The binary character was implemented by tuning the writing speeds during the laser inscription process achieving slight index change differences and therefore obtaining the necessary detunings δ [24, 35].

We observed three different regimes of *Zitterbewegung*. First we analyze the far relativistic regime using a small mass by employing a small detuning $\delta = 0.5\kappa$. From Eq. 8.9 we expect a small frequency and large amplitude. Figure 8.5a depicts the experimental result together with numerical simulation of the respective discrete Schrödinger equation (Fig. 8.5b). Figure 8.5c shows the analysis of the center of mass position, which is directly related to expectation value by the relation

$$\langle m(z)\rangle = \frac{\sum_{m=1}^{N} m|a_m|^2}{\sum_{m=1}^{N}|a_m|^2} = 2\xi + \frac{1}{2}, \tag{8.10}$$

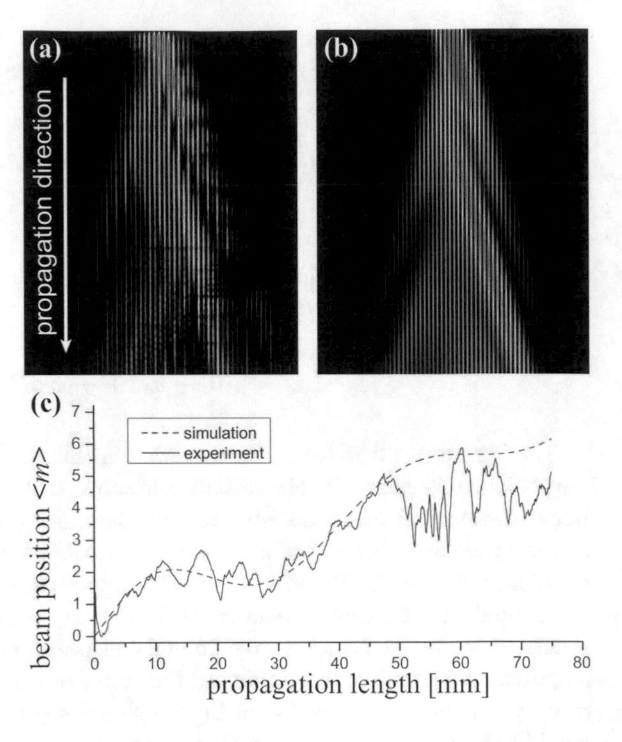

Fig. 8.5 Observation of the far relativistic regime of the *Zitterbewegung* ($\delta = 0.5\kappa$). **a** Measured propagation image, **b** corresponding calculation using Eq. (3), **c** evolution of the beam center of mass $\langle m(z) \rangle$ obtained from the fluorescence image shown in (**a**) (*solid line*) and from the numerical calculation in (**b**) (*dashed line*)

where N is the number of lattice sites. One can clearly see the oscillations of the beam accompanied by a transverse drift, as predicted by Eq. 8.9. The oscillation period $\Lambda_{ZB} = 2\pi/\omega_{ZB} = \pi/\delta = 45$ mm corresponding to approximately 2 cycles in the 80 mm long sample. The *Zitterbewegung*'s amplitude is approximately one waveguide. There is a good agreement between experimental results and numerical simulations demonstrating the asset of our Dirac modeling system. In the relativistic regime, we set $\delta = 1.1\kappa$ in the same order as κ and we expect a short oscillation period Λ_{ZB} and a moderate amplitude R_{ZB}. In Fig. 8.6 one can see an oscillatory motion with 4 cycles and a period of $\Lambda_{ZB} = 20$ mm. The *Zitterbewegung*'s amplitude is weaker with only 0.45 waveguides. The drift velocity is also reduced as one obtains a total drift of 2.5 waveguides after 80 mm propagation length compared to a total drift of 6 waveguides in Fig. 8.5. This is explained by the fact that the drift velocity v_0 decreases as the ratio δ/κ increases. The weak relativistic regime (Fig. 8.7) is obtained at $\delta = 2.1\kappa$, where one attains an even shorter oscillation period $\Lambda_{ZB} = 10.7$ mm. The amplitude is even smaller with only 0.25 waveguides (4 μm) only. The drift also decreases to less than one waveguide. This is in agreement to the fact that for sufficiently large masses (detunings) the evolution becomes classical and no *Zitterbewegung* is observed.

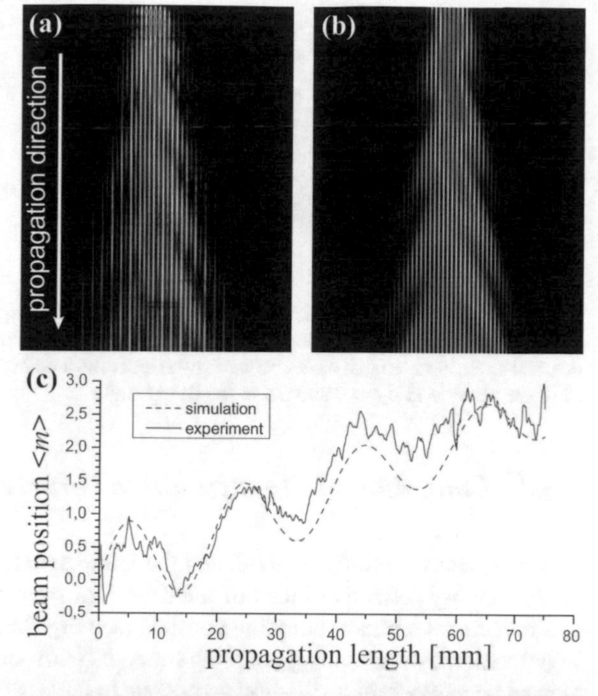

Fig. 8.6 Same as Fig. 8.5 but the relativistic regime ($\delta = 1.1\kappa$). The period is shorter and the amplitude is smaller

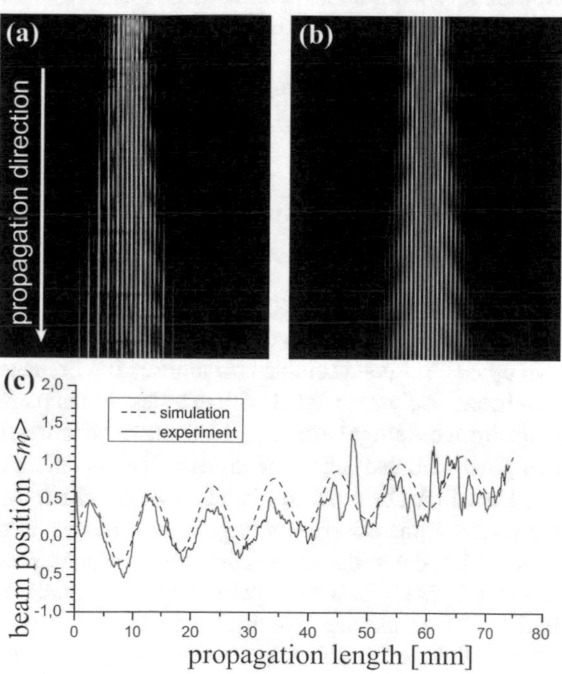

Fig. 8.7 Same as Fig. 8.5 but the weak relativistic regime ($\delta = 2.1\kappa$). The period is shortest and the amplitude almost vanishes

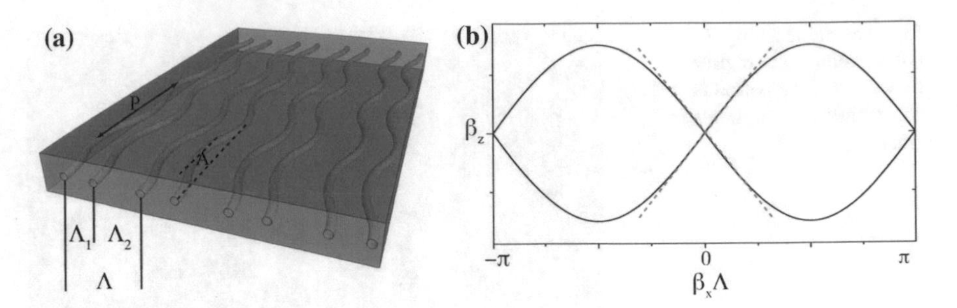

Fig. 8.8 **a** Shows a sketch of the design of the sample, the waveguides are periodically curved and have alternating spacings. **b** Shows the corresponding band structure. Compared to the binary index detuned array, here, no gap opens between two bands and a diabolic point exists. Note that the point of interest is here at $\beta_x = 0$ and not at the Bragg angle

8.3.2 Emulation of Massless Dirac Particles

In the previous section, we modelled the Dirac equation in a waveguide lattice by using a binary refractive index of the sites. However, as an alternative avenue one can implement a binary coupling constant (see Fig. 8.8a for a sketch of the setting). In particular, by introducing coupling strengths with equal magnitude but alternating signs, one can obtain a situation analogous to a massless Dirac equation [36]. This results in the discrete Schrödinger equation:

$$i\frac{da_m}{dz} + (-1)^m \kappa(a_{m+1} - a_{m-1}) = 0\ . \tag{8.11}$$

Let us first discuss the solutions of this equation. The dispersion relation reads [36]:

$$\beta_z^{\pm} = \pm 2\kappa \sin(\beta_x \Lambda) \tag{8.12}$$

and is displayed in Fig. 8.8b. Around $\beta_x = 0$ the two bands intersect. In the vicinity of this intersection, the dispersion curve can be approximated with a linear slope. The two opposite signs resemble two intersecting branches forming a so-called diabolic point commonly associated with massless Dirac particles. The diabolic point has its name from the singularity at $\beta_x = 0$ where no uniquely defined group velocity exists and gives amongst other phenomena rise to conical diffraction. Conical diffraction has been first observed in biaxial crystals where the light spreads in a cone [37]. On a screen one observes a ring, which changes in diameter with changing screen position but the width of the ring stays constant. For our one-dimensional problem this means that light would spread into two branches propagating under a fixed angle in the positive and negative direction with constant beam width.

This light evolution can also be interpreted in terms of relativistic quantum mechanics. The two beams propagating under opposite angles mimic the wave packets of a massles Dirac fermion and its anti-particle. They move away from each other, which can be directly seen from the relativistic energy momentum relation $p = \pm\sqrt{E^2 - m^2c^4}/c$, which simplifies for the massless case to the linear relation $p = \pm E/c$.

As the system describes massless particles, we expect an equation comparable to Eq. 8.6 but without the last term on the right hand side. For the derivation we follow the same formalism as for the *Zitterbewegung*. We rewrite the amplitudes to $a_{2m} = \psi_1$ and $a_{2m-1} = \psi_2$, where ψ_i are the components of the spinor ψ. Note that now the definition of the spinor is different than in the scenario of massive particles since we are interested in the central region of the dispersion relation and a flat phase is required.

Undergoing the transition from discrete to quasi-continuous coordinates as above $\xi \leftrightarrow m = x/2\Lambda$, and $\partial a/\partial\xi \to a_{m+1} - a_m$, we finally arrive at

$$i\frac{\partial\psi}{\partial z} = -i\kappa\sigma_x\frac{\partial\psi}{\partial\xi}\,. \tag{8.13}$$

This equation is precisely the Dirac equation for the massless particle. To observe the respective dynamics in a waveguide array we still need to achieve alternating positive and negative coupling constants. To this end, we utilize an effective coupling constant in sinusoidally curved waveguide arrays, which reads as [38]

$$\kappa_{m,\mathrm{eff}} = \kappa_m(\Lambda_m)J_0\left(A\frac{4\pi^2 n_0}{P\lambda}\Lambda_m\right)\,. \tag{8.14}$$

Here, $J_0(x)$ is the Bessel function of order 0, A is the modulation amplitude, P is the modulation period, Λ_m is the spacing between waveguides m and $m+1$, $\kappa_m(\Lambda_m)$ the coupling strength between a pair of straight waveguides with that separation and n_0 is the refractive index of the substrate material (see Fig. 8.8a). The coupling constant is effective as it describes the light propagation only after every full period. Depending on its argument, the Bessel function can be either positive or negative. One finds that $J_0(\mu) > 0$ for $0 < \mu < 2.405$ and $J_0(\mu) < 0$ for $2.405 < \mu < 5.52$. Hence, we can obtain the alternating signs of $\kappa_{m,\mathrm{eff}}$ just by changing the spacing of adjacent waveguides Λ_m for fixed P and A. For our binary waveguide array we need alternating coupling constants, hence, we use two different lattice constants Λ_1 and Λ_2. To achieve symmetry in the coupling constants they need to fulfill the following conditions with fixed parameters for the curvature:

$$\kappa_{\mathrm{eff}}^{+} = \kappa(\Lambda_1)J_0(\mu(\Lambda_1)) \tag{8.15}$$

$$\kappa_{\mathrm{eff}}^{-} = -\kappa_{\mathrm{eff}}^{+} = \kappa(\Lambda_2)J_0(\mu(\Lambda_2))\,. \tag{8.16}$$

For the experimental implementation [36], we fixed the amplitude and the period of the sinusoidal waveguide curving to $A = 10.1\,\mu\mathrm{m}$ and $P = 6$ mm, respectively. Together with the alternating lattice periods $\Lambda_1 = 13.9\,\mu\mathrm{m}$ and $\Lambda_2 = 18.1\,\mu\mathrm{m}$ we obtained effective coupling constants $\kappa_{\mathrm{eff}}^{\pm} \approx \pm 0.2\ \mathrm{cm}^{-1}$. The waveguide array was

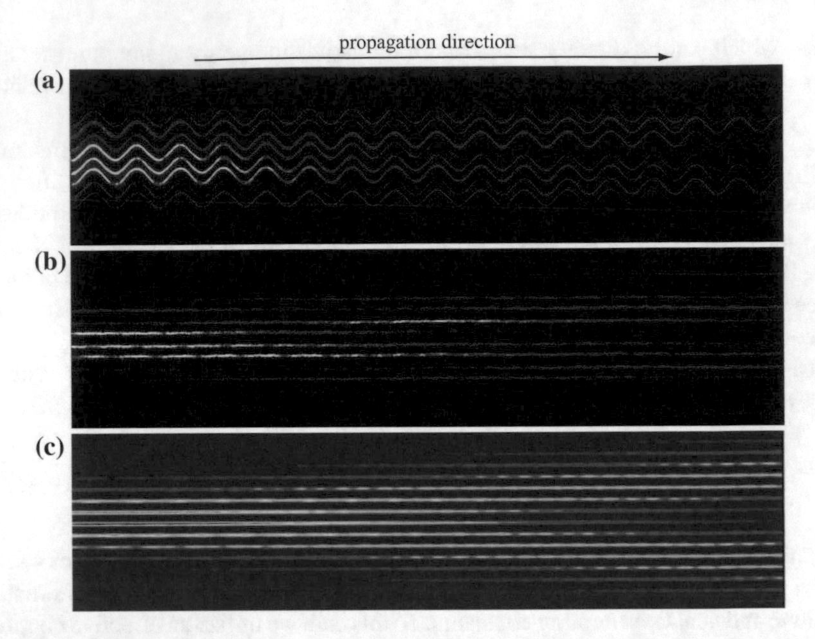

Fig. 8.9 The images show the observation of the massless particle evolution. **a** The experimental observation shows clearly the oscillation of the waveguides. **b** Digitally straightening the waveguides make the conical diffraction visible. The beam splits in two parts corresponding to particle and anti-particle. **c** Corresponding numerical simulation

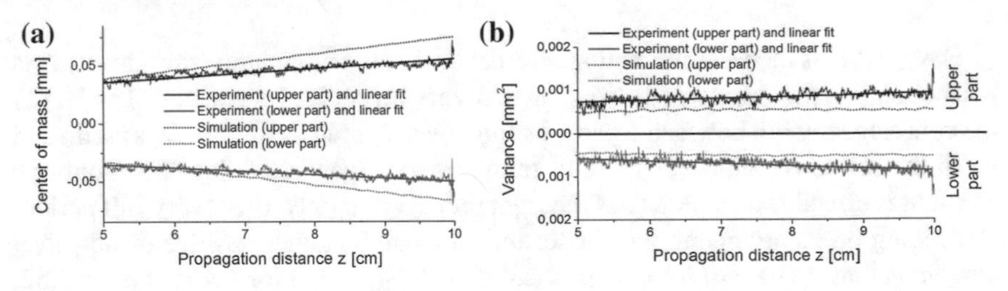

Fig. 8.10 The figure shows the analysis of the center of mass (**a**) and the variance (**b**) of the upper and lower part in Fig. 8.9b, c

excited with a broad Gaussian beam with flat phase covering 5 lattice sites. The experimental and theoretical propagation images (Fig. 8.9) show clearly the separation of the beam into two non-spreading parts moving constantly away from each other. To further substantiate the effect of conical diffraction the center of mass and the variance are plotted in Fig. 8.10. Whereas the absolute value of the barycenter increases constantly for both beams the variance stays rather the same during propagation. This effect is characteristic for the approximately linear shape of the band structure in the origin. The constant slope with opposite sign in both bands is responsible for the propagation under a fixed angle and the vanishing local curvature leads to the non-diffracting beams with constant variance.

8.3.3 Photonic Klein Tunneling

Klein tunneling describes the phenomenon where relativistic Fermions tunnel through a strong repulsive potential without exponential damping as expected by a quantum tunneling processes [39]. This work was extensively discussed [40–42], and although there has been experimental evidence in Graphene heterojunctions [43, 44] it has remained disputed [46] until the successful simulation of Klein tunneling, first in trapped ions [45] and a little later in waveguide arrays [10]. The latter work will be discussed in the following.

In order to model Klein tunneling in waveguide arrays, we again employ our setting with a binary index detuned waveguide array that mimics Dirac dynamics according to Eq. 8.6. The required potential step is implemented by an offset in the refractive index between the left and the right side of the lattice. Hence, our model system will be composed of two different regions, where in each the refractive index of the waveguides is alternating around a different average value. The wave travels from the higher index region ($x < 0$) towards the lower index region ($x > 0$) (see Fig. 8.11a, b) corresponding to a free particle impinging a potential wall.

The solution of this problem are forward and backward traveling waves at $x < 0$:

$$\psi(x < 0) = \begin{pmatrix} ck \\ E - mc^2 \end{pmatrix} e^{\frac{ikx}{\hbar}} + B \begin{pmatrix} -ck \\ E - mc^2 \end{pmatrix} e^{-\frac{ikx}{\hbar}} \tag{8.17}$$

and a forward travelling wave for $x > 0$:

$$\psi(x > 0) = D \begin{pmatrix} cp \\ E - V - mc^2 \end{pmatrix} e^{\frac{ipx}{\hbar}} . \tag{8.18}$$

Here, E is the energy of the particle related to its momentum k by the energy momentum dispersion relation $E^2 = c^2k^2 + m^2c^4$ and $p = \pm\sqrt{(E - V)^2 - m^2c^4}/c$ is the momentum within the potential wall. B and D are constants. To satisfy continuity at $x = 0$ one requires

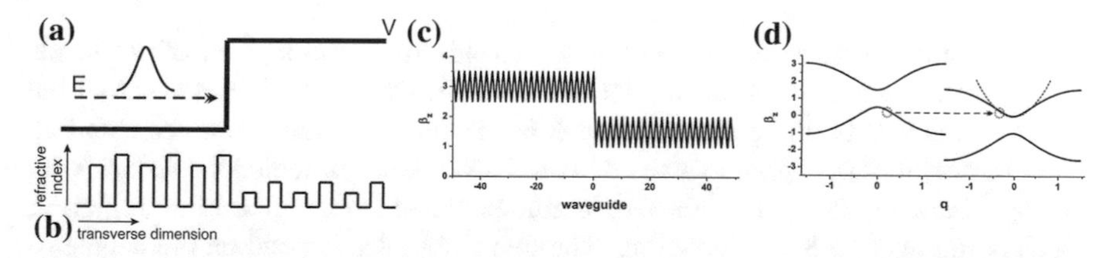

Fig. 8.11 **a** Sketch of the Klein tunneling process. A relativistic particle wave with energy E travels towards a potential barrier $V > E$. The relativistic particle experiences no exponential damping after crossing the potential barrier. **b** refractive index profile of a optical analogous waveguide array simulating Klein tunneling, the potential step is achieved by constantly shifting the refractive indices **c** corresponding propagation constants and **d** band structure showing the typical band in both regions. The bands on the *right side* are shifted to lower values as the refractive indices are weaker

$$ck(1 - B) = cpD \tag{8.19}$$

$$(E - mc^2)(1 + B) = (E - V - mc^2)D\,, \tag{8.20}$$

which yields

$$\frac{1 - B}{1 + B} = \frac{p}{k}\frac{(E - mc^2)}{(E - V - mc^2)} \equiv \frac{1}{\eta}\,. \tag{8.21}$$

The reflection and transmission coefficients R and T read then as

$$R = \left(\frac{1 - \eta}{1 + \eta}\right)^2\,, \quad T = \frac{4\eta}{(1 + \eta)^2}\,. \tag{8.22}$$

It can be easily seen that if the energy of the particle is smaller than the potential step $E < V - mc^2$, η will become negative and therefore $R > 1$ and $T < 0$. In many articles and books this fact is referred to as the Klein paradox. It is not, however, what Klein wrote down. Klein noted in his original manuscript that Pauli had pointed out to him that for $x > 0$, the particle momentum within the potential wall is p, whereas the group velocity is $v_g = dE/dp = pc^2/(E - V)$. So if the transmitted particle moved from left to right, v_g has to be positive which implies that p had to be assigned its negative value $p = -\sqrt{(E - V)^2 - m^2c^4}/c$. This result finally teaches us that the even for infinite height of the potential step ($V \to \infty$) the transmission is non-zero. This is what we learn from Kleins work: According to the Dirac equation fermions can pass through strong repulsive potentials without the expected exponential damping according to a conventional quantum tunneling process.

To investigate Klein tunneling in waveguide arrays we design our system similar to that for *Zitterbewegung* but with an offset Δ_m, which is defined as $\Delta_m = 0$ for $m < 0$ (region I) and $\Delta_m = \Delta$ for $m \geq 0$ (region II) [47]. The discrete Schrödinger equation reads

$$i\frac{da_m}{dz} + \kappa(a_{m+1} + a_{m-1}) + \left[(-1)^m\delta + \Delta_m\right]a_m = 0\,. \tag{8.23}$$

For the uniform binary array, we obtained a band structure composed of two bands that are hyperbolic near the band edge $\beta_x\Lambda = \pi/2$. Due to the introduction of the potential step we obtain again two of such bands but with their mean value shifted by Δ. Hence, for appropriate values of Δ the lower band in region I overlaps with the upper band in region II. Light is now able to tunnel from region I into region II, which is analogue to Klein tunneling. The tunneling rates depend on the degree of relativity and on the height of the potential step. The transmission is highest when the overlap of two bands is maximal.

The experimental analysis was performed in femtosecond laser written waveguide arrays [10]. The propagation constants are set according to Fig. 8.11c, resulting in a band structure shown in Fig. 8.11d. The incident beam has Gaussian shape covering approximately 9 waveguides and is tilted around the Bragg angle. For simplicity we

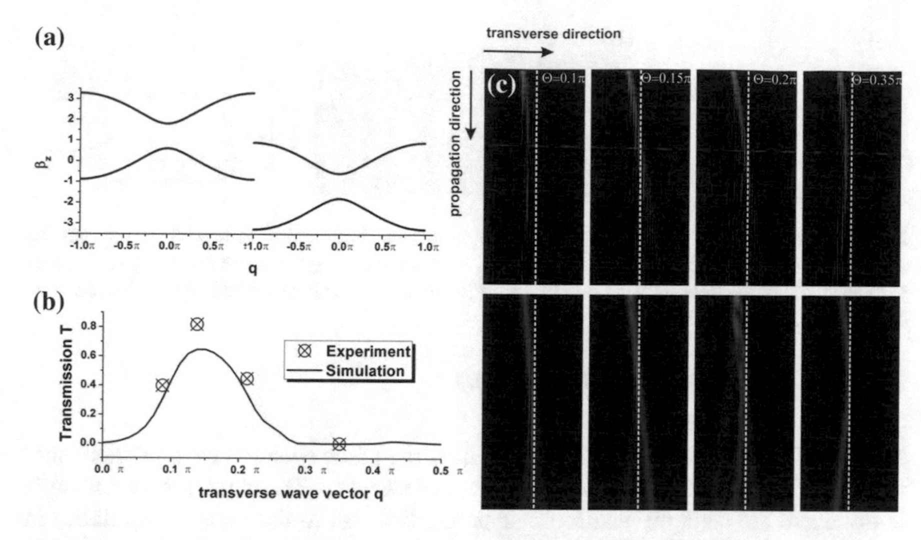

Fig. 8.12 Well overlapping bands shown in (**a**) allow for a broad region of Klein tunneling. The potential step is rather large. **b** measured and simulated transmission obtained from (**b**). The tunneling region spans from 0 to 0.3π, **b** (*upper row*) shows the measured beam propagation for different angles of incidence, meaning a probing of Klein tunneling at different positions of the band structure. (*lower row*) illustrates the corresponding simulations

use here the dimensionless incident angles q, which are shifted towards the edge of the band $q = \beta_x \Lambda - \pi/2$. The first sample was designed with a detuning $\delta = 0.15$ mm^{-1} in both regions and an offset $\Delta = 0.6$ mm^{-1}. This is a rather high potential step so that the lower band in region I overlaps completely with the upper band in region II (Fig. 8.12a). As we have a large overlapping region in the experiments we find a large range for incident angles around the Dirac point ($q = 0$) where tunneling occurs. For $q = 0.15\pi$ the transmission reaches values of up to $\approx$80%. Far away from the Dirac point ($q = 0.35\pi$) the tunneling amplitude is negligibly small and the light is completely reflected at the potential step at $m = 0$. At the Dirac point the transmission drops to zero, which can be attributed to different momenta belonging to same energies in region I and II. A complete scan of the incident angles is shown in Fig. 8.12b, c, where experimental images are compared with numerical solutions of the coupled mode equations. The transmission is obtained by integrating the light intensity in the regions I and II, respectively. The transmission is normalized to the total power in the array. For further analysis one can derive an analytical expression for the tunneling rates depending on the potential step and detuning [47]. If the overlap of the bands is much smaller ($\delta = 0.125$ mm^{-1} and $\Delta = 0.25$ mm^{-1}, see Fig. 8.13) one obtains a smaller region of Klein tunneling. Whereas in the previous experiment the tunneling range was $[0, 0.3\pi]$, the smaller overlap yields only half the tunneling range $[0, 0.15\pi]$, which is consistent to the quantum-mechanical pictures, where both, energy and momentum conservation have to be fulfilled.

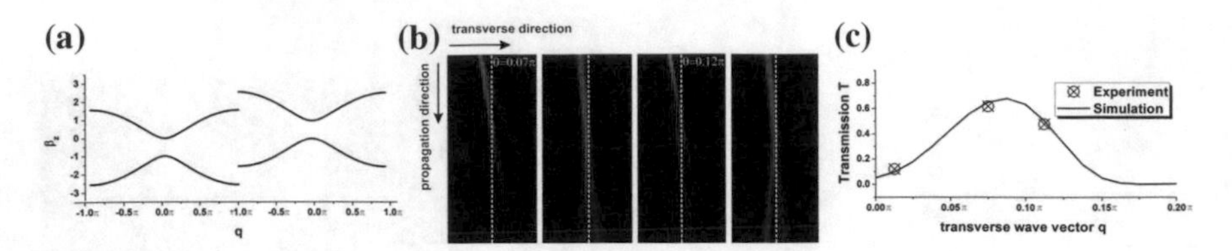

Fig. 8.13 Same as Fig. 8.12 but for a smaller potential step. The bands are not well overlapping (**a**) and therefore the tunneling is rather ineffective. Two exemplary angles where tunneling is observed are shown in (**b**) together with the corresponding simulations. Only in a small region (0–.15π) one can achieve tunneling (**c**)

8.3.4 Emulation of Pair Creation

Beyond *Zitterbewegung* and Klein tunneling, the Dirac equation even predicts phenomena which do not conserve the number of particles. The paradigmatic example is of course the electron-positron pair production due to the vacuum instability in an external electric field [48, 49]. The negative energy levels described by the Dirac equation are filled up to the vacuum state and, therefore, allow the possibility to pull electrons out of the vacuum. This leaves a hole in the reservoir of negative energy states, resulting in the creation of a positron as the anti-particle to the electron. So far there are two known mechanisms that initiate this process. By applying an electrostatic field the electron can directly tunnel through the energy gap between negative and positive energy levels. This effect is known as the Schwinger mechanism [40]. However, enormous field strengths are required to observe this effect and therefore no observation in any lab has been performed yet. On the other hand one can drive a dynamic process by an external AC field, known as dynamical pair production [50]. For this process at least an experimental proposal exists, in which two counter-propagating ultrastrong laser fields are used. But also this implementation is challenging and has not been achieved so far. The observation in a waveguide array as a model system allows this experiment to be simulated in a conventional laboratory setup (Fig. 8.14).

In optical waveguide arrays we can simulate an external field by applying a curvature with the profile $x_0(z)$ to the waveguides [3, 38, 51], such that an external field E_x is proportional to $\ddot{x_0}(z)$ and equivalent to a local shift of the waveguide's propagation constants. This can be written in terms of complex coupling constants, employing a phase term [52]:

$$i\frac{da_m}{dz} + \kappa e^{i\phi}a_{m+1} + \kappa e^{-i\phi}a_{m-1} + (-1)^m\delta a_m = 0\,. \tag{8.24}$$

The phase ϕ is depending on the longitudinal coordinate and is defined as $\phi(z) = 2\pi n_0 \Lambda/\lambda \times dx_0(z)/dz$. For straight waveguides, i.e., $\phi = 0$, the system is equivalent to the two-band lattice described in Sect. 8.3.1, in which *Zitterbewegung* can be observed. In the general scenario, however, Eq. 8.24 results in the modified Dirac equation

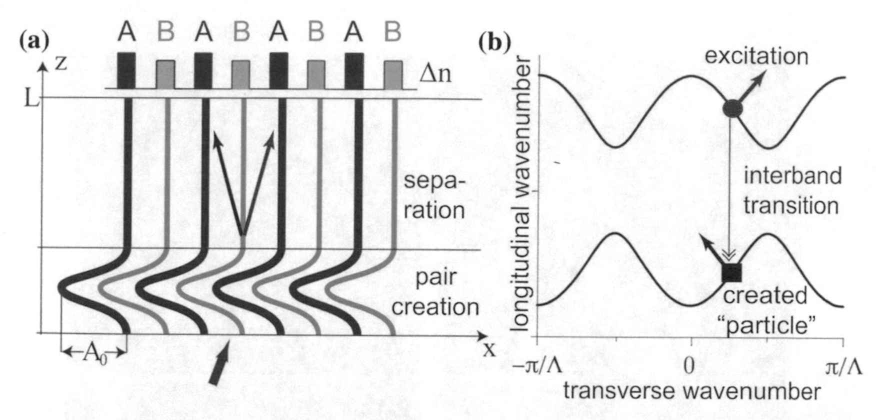

Fig. 8.14 The figure shows how dynamic pair production can be observed in the optical model system. **a** shows the sketch of the sample. The lattice consists of high and low index waveguides A and B. The index profile is shown at the top. Below the longitudinal design is depicted. The sample contains an initial section of a single cycle oscillation mimicking the external AC driving field. Here the pair is created. A subsequent section separates the beams. The beam propagating to the right is the original undistorted beam. The beam propagating to the *left* determines the pair creation rate. **b** shows the corresponding band structure. Initially the wave packet is excited in the upper band marked with the *red dot*. Due to the external field an interband transition to the lower band occurs, which corresponds to the pair creation. The created particle and the excitation have opposite propagation angles, indicated by an *arrow*. After a certain propagation distance the beams can be distinguished in real space

$$i\frac{d\psi}{dz} + i\kappa\sigma_x\frac{d\psi}{d\xi} + 2\kappa\phi(z)\sigma_x\phi - \delta\sigma_z\phi = 0\,, \tag{8.25}$$

where $\dot{\phi}(z) \sim \ddot{x_0}(z) \sim E_x$. The pair creation process can be explained in a simple optical picture. One can understand this process as a tunneling from the upper band (negative energy) to the lower band (positive states) caused by the external electric field. This effect changes the occupation probabilities in both bands. As both bands exhibit opposite slopes, the propagation direction of the light beams is opposite as well. Propagating the beam further beyond the AC field area yields a spatial separation of the respective fractions, such that they can be distinguished. The separated droplet represents the positive energy state and, therefore, the created electron. At the output the both beams are integrated and normalized to the total power yielding the band occupation in each bands. The occupation of the lower band (positive energy) determines the pair creation probability.

For the experimental realization [53], we generated the required AC pulse by implementing a sinusoidally curved region at the beginning of the waveguide arrays. The period of the cycle was 6.67 mm and the peak-to-peak amplitude A_0 was varied from 0 µm to 48 µmm in 13 realizations to study the influence of the external field strength. A straight section until the end of the sample with total length of 90 mm followed afterwards. The waveguide spacing was $\Lambda = 12\,\mu$m yielding a coupling

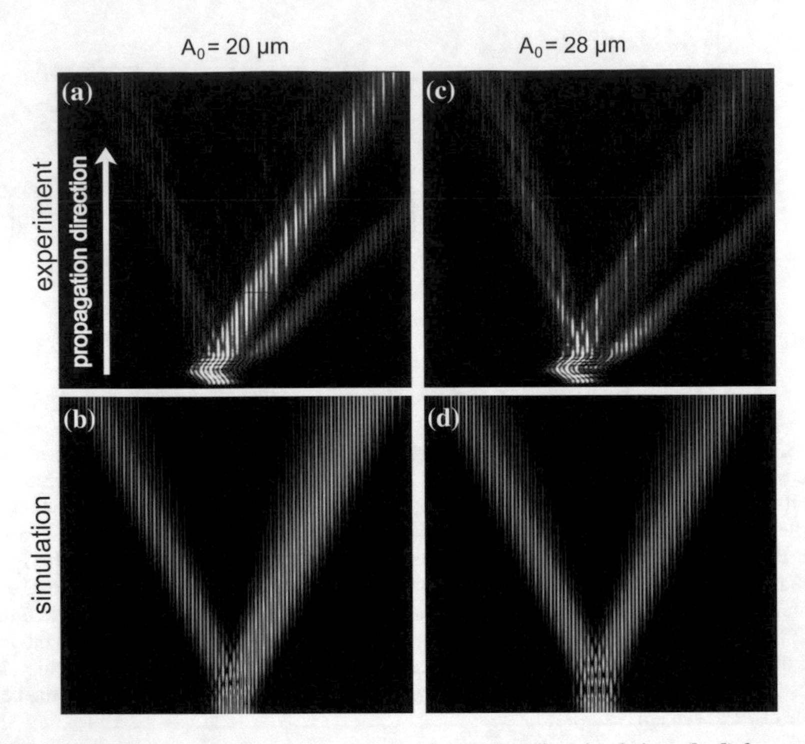

Fig. 8.15 Exemplary propagation images (**a**, **c**) and corresponding simulations (**b**, **d**) for a superlattice with a small detuning of $\delta = 0.4\kappa$. The images show results for the amplitudes $A_0 = 20\,\mu\text{m}$ (**a**, **b**) and $A_0 = 28\,\mu\text{m}$ (**c**, **d**). The tunneling is incomplete for both samples. The simulations are performed in the reference frame, where the waveguides appear straight

constant of $\kappa = 0.24$ mm^{-1}. Two different samples were fabricated to study the influence of the particle's mass. To excite an electron from the negative energy states, we slightly tuned the incident angle of the broad Gaussian beam (covering 9 waveguides) to precisely half the Bragg angle, which insures the excitation of only one band as well as fulfilling the close-to Bragg angle approximation. For vanishing external field ($A_0 = 0\,\mu\text{m}$) we see a propagation to the right side of the sample. When increasing the field strength, an inter-band transition is observed, which is manifested as a breakup of the beam to the left side of the sample. Consequently, the output is split into two beams. For the first sample (Fig. 8.15), which has an index detuning of $\delta = 0.4\kappa$, a maximum of the tunneling rate of 40% was achieved for an amplitude of $A_0 = 28\,\mu\text{m}$. This fractional tunneling coincides with the theoretical predictions and models the situation when the frequency of the driving AC field was not properly adjusted to the particle's mass. Note, that in the experiments a third beam is visible, which is a remainder of the injected laser light that did not couple into the waveguides. This beam is not relevant to the physics of the simulated pair creation process and it is not observed in the numerical simulations, which are based on a discretized (tight-binding) model of light transport.

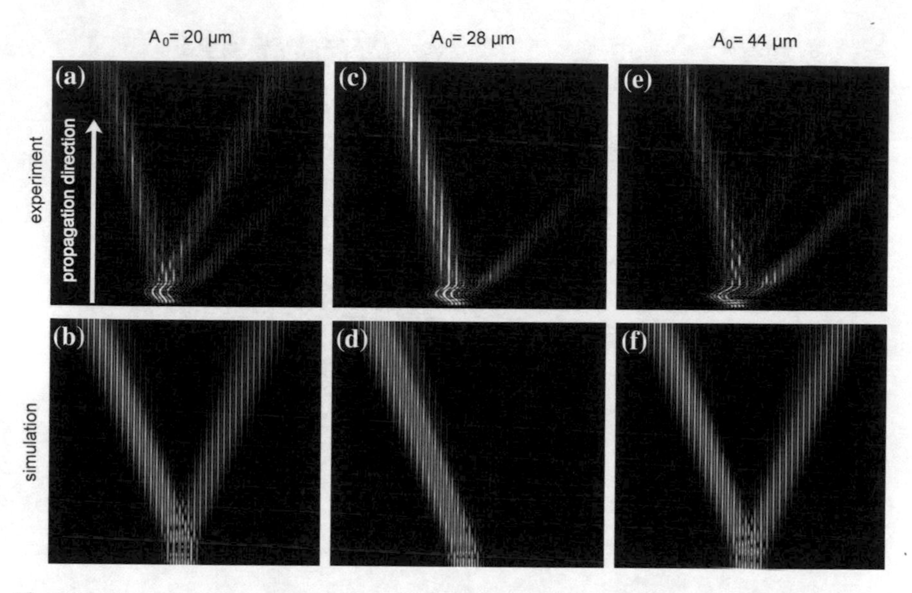

Fig. 8.16 Same as. Figure 8.15 but for a optimized photon energy (oscillation frequency) to the rest energy (detuning). Samples for $A_0 = 20\,\mu\text{m}$, $A_0 = 28\,\mu\text{m}$ and $A_0 = 44\,\mu\text{m}$ are shown (*top* experiment, *bottom* simulation). An almost complete tunneling (98%) is observed for (**c**, **d**). Increasing the driving field reduces the tunneling again

The tunneling rate is significantly increased when the particle's mass is properly adjusted. An optimization of the tunneling was achieved by tuning the particle's mass δ with a fixed frequency of the driving field. For a particle mass of $\delta = 0.8\kappa$ we achieved a transition probability of almost 100% (see Fig. 8.16). The probability decreases again at higher amplitudes, which means for the pair production process that, for a certain mass, the frequency of the driving field and the amplitude have to fulfill a particular relation to obtain maximum pair creation probability. The results are summarized in Fig. 8.17 where the tunneling rates for all 13 field strengths are plotted. One can clearly see that for small and large mass the pair creation probability is a non-monotonic of the AC field amplitude, i.e., it reaches a maximum and decrease afterwards. One can additionally see that the tunneling probability for the large mass approaches almost 100% for $A_0 = 28\,\mu\text{m}$ (black curve and data points).

To summarize this part of the chapter, we have found that waveguide arrays provide a convenient way to observe relativistic quantum mechanics described by the Dirac equation. We have shown that binary waveguide lattices can either describe massive or massless particles depending on whether the binary character is applied to the refractive index of the waveguides or the inter-site coupling. By choice of the waveguide configuration, a variety of scenarios can be investigated, ranging from free dynamics via scattering at potential barriers to such elusive phenomena as pair creation. It is even possible to simulate Dirac particles with a spatially dependent mass, which directly relates to impurity spins in antiferromagnetic materials and their mutual correlation [54].

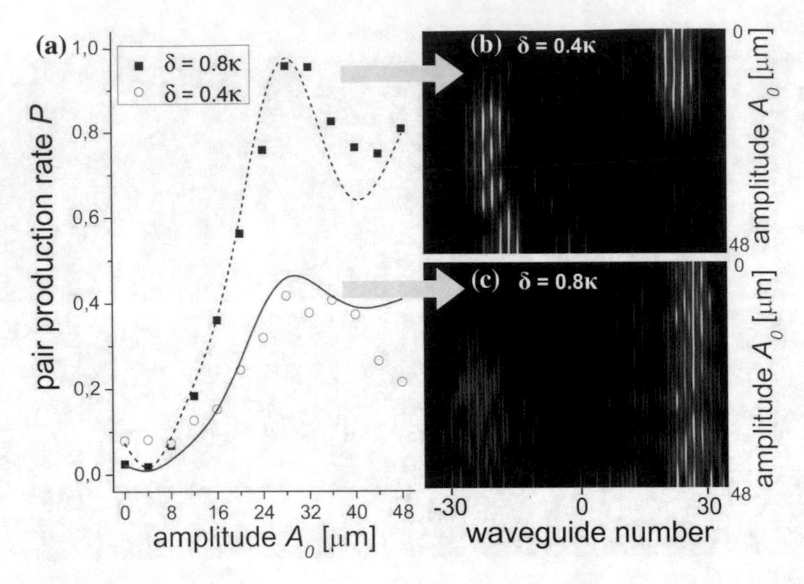

Fig. 8.17 Measurements and simulations are summarized for 13 different amplitudes. **a** shows the pair production probability versus driving field amplitude for both realizations ($\delta = 0.4\kappa$ and $\delta = 0.8\kappa$). A maximum close to 100% is obtained for the larger detuning only. **b**, **c** Output cross sectional images depending on the amplitude. For a low detuning a pair production rate of 50% cannot be exceeded, while for the high detuning one approaches almost unity for $A_0 = 28\,\mu$m

8.4 Two-Dimensional Lattices: Photonic Graphene

The first part of this chapter focused on the realization of Dirac physics in one-dimensional lattices. However, perhaps the most well-known realization of this relativistic dispersion relation in a lattice geometry is two-dimensional: i.e., it arises in graphene [55, 56]. Graphene is a two-dimensional honeycomb lattice of carbon atoms; it is composed of two interpenetrating triangular lattices, and therefore has a two-member basis. Its lattice vectors may be written as:

$$\boldsymbol{r}_1 = \Lambda(\sqrt{3}, 0)\ , \quad \boldsymbol{r}_2 = \Lambda(\sqrt{3}, 3)/2\ , \tag{8.26}$$

and the positions of the elements in the unit cell may be written as:

$$\boldsymbol{\rho}_1 = (0, 0)\ , \quad \boldsymbol{\rho}_2 = \Lambda(0, 1)\ , \tag{8.27}$$

where Λ is the nearest-neighbor spacing, and $\sqrt{3}\Lambda$ is the lattice constant. The physics of graphene has been discussed extensively [56]; the present discussion is devoted to photonic graphene, namely, a lattice of evanescently-coupled waveguides wherein the diffraction of light is governed by the same equations as the motion of electrons in graphene (see Fig. 8.18a, b). The honeycomb system has been termed photonic graphene due to this correspondence [12]. The notion of photonic graphene was

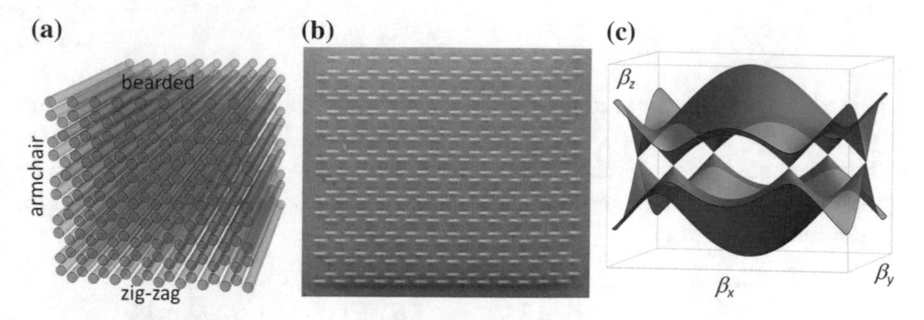

Fig. 8.18 **a** Schematic diagram of the honeycomb photonic graphene lattice with lattice terminations, namely the bearded, zig-zag and armchair edges indicated; **b** Microscope image of the input facet of a photonic lattice showing elliptical waveguides arranged in the honeycomb geometry; **c** Spatial band structure (longitudinal wavenumber, β_z, of the paraxial wavefunction envelope, plotted as a function of transverse wavenumbers β_x and β_y). The band structure describes the diffraction properties of wavepackets propagating in the photonic lattice

introduced in 2007 [5] in the context of the observation of lattice solitons in the honeycomb lattice. Other systems that obey the same mathematical properties of graphene have been discussed in the context of photonic crystal slabs [57], molecular assemblies [58], microwave resonators [59], Bose-Einstein condensates [60], and exciton-polariton condensates [61]. The focus of this discussion, however, will be on photonic lattices: arrays of coupled waveguides.

Similarly to one-dimensional lattices, when the waveguides are single-mode, the system can be accurately modelled by the following discrete Schrödinger equation:

$$i\frac{da_m}{dz} + \beta_m a_m + \sum_{<n>} \kappa_{m,n} a_m = 0\,, \tag{8.28}$$

where κ_{nm} is the coupling constant between waveguides n and m. The summation is taken over neighboring waveguides up to some cutoff distance in the transverse plane, which is determined by the evanescent coupling decay length. Unless stated otherwise, we consider only nearest-neighbor coupling ($\kappa_{n,m} = \kappa$ for neighboring waveguides, and 0 otherwise). In order to write down z-independent equations for static solutions, we take $a_m = \phi_m \exp\{i\beta_z z\}$, and therefore arrive at:

$$-\beta_z\phi_m + \beta_m\phi_m + \sum_{<n>} \kappa_{m,n}\phi_m \equiv \sum_n H_{m,n}\phi_m\,. \tag{8.29}$$

This represents an eigenvalue equation for the static solutions to these linear coupled-mode equations. Here, the matrix H is a Hamiltonian whose eigenvalues and eigenvectors represent the propagation constants and mode profiles of the static solutions of Eq. 8.28. When the system is periodic, we may use Bloch's theorem to represent the solutions of Eq. 8.29 in reciprocal space. Namely, we take:

$$\phi_m^{(A)} = Ae^{i\boldsymbol{\beta}\cdot\boldsymbol{R}_n^{(A)}} \ , \quad \phi_m^{(B)} = Be^{i\boldsymbol{\beta}\cdot\boldsymbol{R}_n^{(B)}} \ , \tag{8.30}$$

where $\phi_m^{(A)}$ represents the amplitude on the mth waveguide on the A-sublattice (similarly for $\phi_m^{(B)}$); $\boldsymbol{\beta} = (\beta_x, \beta_y)$ is the Bloch wave vector; $\boldsymbol{R}_n^{(A)}$ is the position of the mth waveguide on the A-sublattice (similarly for $\boldsymbol{R}_n^{(B)}$); and A, B are defined by Eq. 8.30. Plugging this into Eq. 8.28, we obtain the following dispersion relation for β_z as a function of $\boldsymbol{\beta}$:

$$\beta_z = \pm\sqrt{\kappa^2\left(1 + 4\cos^2\left[\frac{\beta_y \Lambda}{2}\right] + 4\cos\left[\frac{\beta_y \Lambda}{2}\right] + \cos\left[\frac{\beta_x \sqrt{3}\Lambda}{2}\right]\right)} \ . \tag{8.31}$$

This band structure diagram is plotted in Fig. 8.18c: it consists of two bands (due to the fact that the unit cell contains two waveguides), and the bands cross at singular points that reside at the Brillouin-zone vertices. These are the Dirac points—there are two unique such points per Brillouin zone. They exhibit conical dispersion and are described (locally in reciprocal space) by a two-dimensional Dirac Hamiltonian [56], namely:

$$H_{\text{Dirac}} = \frac{3\kappa}{2}\boldsymbol{p}\cdot\boldsymbol{\sigma} \ , \tag{8.32}$$

where $\boldsymbol{p}$ is the momentum measured from the Dirac point (or $\boldsymbol{\beta} - \boldsymbol{\beta_0}$, where $\boldsymbol{\beta_0}$ is a vertex point of the Brillouin zone where the Dirac point lies), and $\boldsymbol{\sigma}$ is a vector composed of Pauli matrices, (σ_x, σ_y). Note that the dispersion relation of this Hamiltonian only describes wave propagation in the vicinity of a given Dirac point, not elsewhere in the spectrum.

8.4.1 New Edge States in Photonic Graphene

The discussion up to this point has described the spectral properties of extended Bloch states propagating in a honeycomb photonic lattice. However, some of the most fascinating graphene physics arises not in the bulk of the crystal, but on the edges. Although there are an infinite number of ways to terminate the honeycomb lattice, the three simplest—and most widely referred to—are the so-called *zig-zag*, *bearded*, and *armchair* edges (labeled in Fig. 8.18a). In order to examine the effect of terminating the lattice, a unit cell with a 'strip' geometry must be considered (which is periodic in the x-direction but is terminated in the y-direction). The strip geometry incorporates Bloch periodic boundary conditions in one direction, implying an infinite system size in the x-direction, but a finite termination in the vertical direction. The finite termination (which is also called discrete Dirichlet boundary conditions) allows for the incorporation of boundary effects, such as the occurrence of states localized at the edges.

Edge calculations reveal that both the zig-zag and bearded edges exhibit states that are localized to the edges (i.e., decay exponentially from the edges in the vertical direction in Fig. 8.18a and are extended horizontally), but in the unperturbed honeycomb lattice, the armchair edge does not [62, 63]. In coupled mode theory, the bearded edge exhibits a state that resides in the central 2/3 of the Brillouin zone in the x-direction, whereas the zig-zag edge exhibits an edge state that resides in the complementary 1/3 of the zone. These states are well known from graphene physics, and arise as a result of the band crossings at the Dirac points [62]. Edge states that arise as a result of band crossings are called *Shockley states* [64], and can be derived using the bulk-edge correspondence principle via the Zak phase [65]—further discussion of these derivations can be found in Refs. [66].

The edge states of photonic graphene in the limit of nearest-neighbor hopping have completely flat dispersion (in other words, $\beta_z(\beta_x)$ is completely flat, meaning the edge states do not move or broaden as a function of z) [62]. However, this is not the case if next-nearest-neighbor coupling is included. A more general calculation which accounts for such non-nearest neighbour couplings is given by the full Schrödinger equation

$$i\frac{\partial\psi(x,y,z)}{\partial z}+\frac{\lambda}{4\pi n_0}\nabla_\perp^2\psi(x,y,z)+\frac{2\pi\Delta n}{\lambda}\psi(x,y,z)=0 \tag{8.33}$$

with $\psi(x, y, z)$ as the envelope function of the electric field and Δn as the refractive index change of the waveguides. The full Schrödinger equation is diagonalized to find the eigenvalues and eigenstates as a function of $\boldsymbol{\beta}$. The edge states for the bearded and zig-zag edges are shown for the complete diagonalization of Eq. 8.33 in Fig. 8.19a. Here, the eigenstates for a honeycomb strip terminated on one side

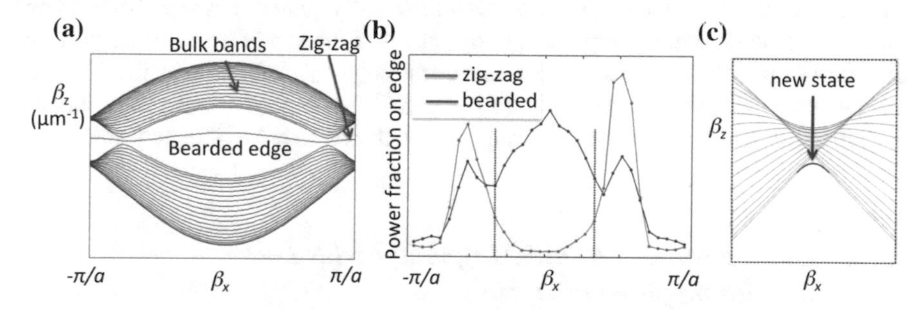

Fig. 8.19 **a** Edge band structure indicating the presence of edge states residing at the center of the edge Brillouin zone (and occupying 2/3 thereof) for the bearded edge, as well as the outer part of the Brillouin zone (occupying the remaining 1/3) for the zig-zag edge; **b** Experimental results showing the fraction of optical power remaining on the bearded edge (*blue*) and the zig-zag edge (*green*); **c** Edge band structure for the bearded edge, zoomed in around the van Hove singularity at the zone boundary. An edge state (classified as a Tamm state) that is not predicted by tight-binding theory emerges from the band as a result of a self-induced defect. The range of β_x and β_z for this *plot* is indicated by a box with a *dashed outline* in (**b**)

with the bearded edge and the other side with the zig-zag edge are shown. Note that the band structure is composed of bulk modes that permeate the entire bulk of the lattice, as well as edge states that reside only on the edges of the lattice. The bearded edge state occupies approximately the central two thirds of the zone and the zig-zag edge state occupies the outer third, in accordance to the exact values in the nearest-neighbor tight-binding limit. In order to experimentally probe the presence or absence of an edge state at a given transverse wavevector (parallel to the edge), β_x, an elliptical beam is launched at the edge in question. The parallel tilt of the beam controls β_x. Figure 8.19b shows the fraction of power remaining on the edge as a function of β_x: a large value indicates the presence of a highly confined edge state, and a low value indicates bulk excitation and no edge state. For the zig-zag edge, it is clear that the edge state resides in the outer part of the spectrum, as explained directly from Fig. 8.19a. However, for the bearded edge, there exist edge states both at the center of the Brillouin zone (the well-known state) and at the zone boundaries (an anomalous state). This indicates the presence of a novel type of edge state that was not known in the graphene literature, and was found only in the context of photonic graphene [67, 68]. The fact that it was not seen before is, in some sense, not a surprise: the bearded edge is not chemically stable in graphene itself. The state resides at a *van-Hove singularity* (where there exists a high degeneracy due to a saddle point in the bulk band structure). Figure 8.19c shows the edge dispersion zoomed in at the Brillouin zone boundary in the upper band: notice the presence of an edge state emerging from the bulk bands there (there is another edge state in the lower band).

The mechanism for the formation of this edge state may be explained as follows. The bearded edge contains two 'missing bonds' on the edge (see Fig. 8.18a). The absence of these bonds provides a small correction to the self-energy term when deriving the coupled mode equations, Eq. 8.28, from the full continuous Schrodinger equation, Eq. 8.33. This is explained in fuller detail in Ref. [67]. This term acts as an effective defect on the edge. It has been previously shown that small defects on the edge can lead to the emergence of edge states at the van Hove singularity [69]: therefore despite the lack of any real defect on the edge, this 'effective' defect induces an edge state. Therefore, it can be classified as a Tamm state [70] (also since it does not arise from any form of band crossing, and is therefore not a Shockley state).

8.4.2 Generation of Pseudo-Magnetic Fields Due to Inhomogeneous Strain

In 1998, Kane and Mele [71] showed that if graphene is strained from its honeycomb lattice geometry in a particular way, the resulting description of the electron wave dynamics around the Dirac point was equivalent to a charged, massless relativistic particle in a magnetic field (it was later termed a *pseudomagnetic field*). Building on this work, Guinea et al. [72] showed that a particular functional form of the strain would lead to a constant pseudomagnetic field (directed perpendicular to the graphene plane, that is, in z-direction). Specifically, the standard two-dimensional Dirac Hamiltonian picks up a gauge field:

$$H_{\text{Dirac}} = \frac{3\kappa}{2}\boldsymbol{p}\cdot\boldsymbol{\sigma} \rightarrow \frac{3\kappa}{2}(\boldsymbol{p}+\boldsymbol{A}(x,y))\cdot\boldsymbol{\sigma}\,, \quad \boldsymbol{A}(x,y) = \frac{B_0}{2}(-y,x) \tag{8.34}$$

where $\boldsymbol{A}(x, y)$ is a vector potential (i.e., gauge field) that corresponds to a constant magnetic field in the symmetric gauge. The strength of the magnetic field, B_0, is a function of the strain tensor: a full derivation of how the pseudomagnetic field comes about as a result of strain is discussed in Refs. [56, 72]. The pseudomagnetic field was first shown experimentally by straining graphene in a bubble-geometry [73]. A very important distinction with real magnetic fields must be noted here: an imposed external magnetic field breaks time-reversal symmetry, where a pseudomagnetic field does not. The reason is that the pseudomagnetic field acts in opposite directions for the two unique Dirac cones in the graphene spectrum, thus preserving time-reversal symmetry for the system as a whole.

When a magnetic field is imposed on a two-dimensional system, the result is the appearance of *Landau levels* in the eigenvalue spectrum—this is no different for a pseudomagnetic field. Landau levels are regions of very high degeneracy (and thus high density-of-states), which can be thought of as the result of quantization of the cyclotron orbit of charged particles rotating in a magnetic field. Since photons propagating through a honeycomb photonic lattice obey the same governing equations as electrons in graphene, Landau levels should appear in the photonic spectrum of propagation constants, β_z for a honeycomb photonic lattice with the same strain as discussed in Ref. [72]. This effect was explored in Ref. [12], observing Landau levels in a photonic spectrum for the first time. Figure 8.20a, b show a single Dirac cone in the spectrum with and without an applied constant magnetic field, respectively. Note that all of the states in the spectrum cluster in highly degenerate spectral regions. The positions of the Landau levels in the spectrum are given by $\beta_z = \pm\omega_0\sqrt{|N|}$, where N is the integer Landau level index, and $\omega_0 = 3\sqrt{B}\kappa\Lambda/2$.

The calculation showing the mode profiles of the Landau levels is given in Ref. [56]; note, however, that the derivation assumes an infinite system size. Systems in nature must of course be finite, and whenever this is the case, there exist localized states lying on the edges of the structure. As is shown in Ref. [12], an excitation of the armchair edge overlaps with modes that lie band gaps between Landau levels. Recall that without strain, the armchair edge exhibits no localized modes in the spectrum and therefore light cannot be confined on this edge. As shown in Fig. 8.20c–f together with Fig. 8.20g–j, increasing the degree of strain of the lattice causes light input on the armchair edge to become localized there. This is a result of band gap guidance, but not in an ordinary band gap—instead, the light is guided in gaps lying between Landau levels. Further results in Ref. [12] show that indeed the guided states spectrally reside in the Landau level region.

To conclude this section, it is important to discuss the technological motivation behind studying aperiodic photonic structures, and in particular, those with Landau

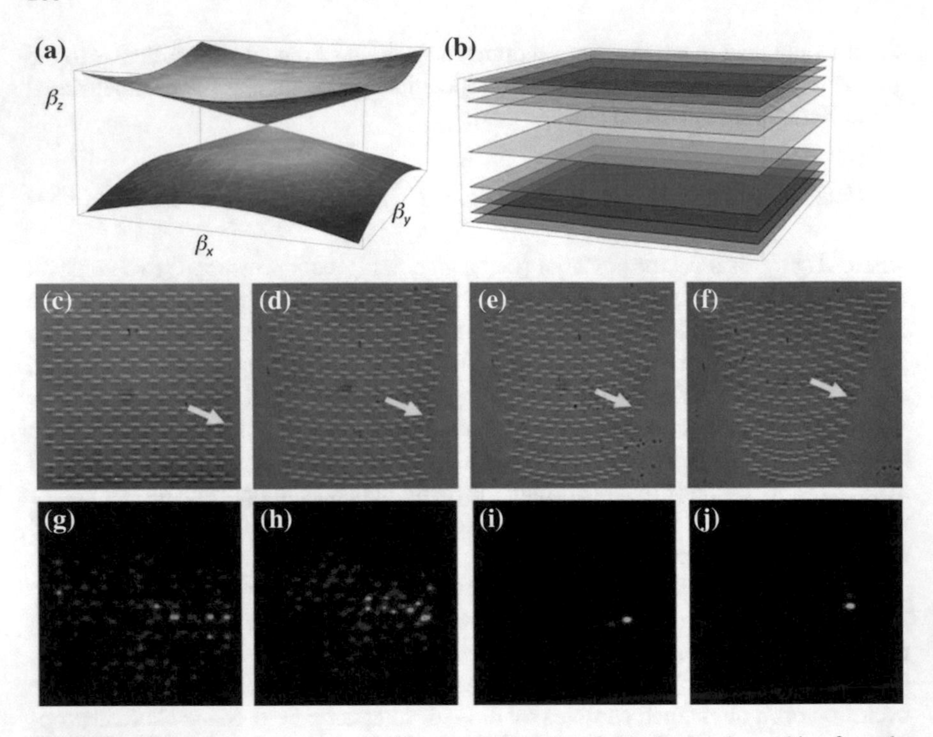

Fig. 8.20 **a** Dirac cone in an unstrained honeycomb system; **b** Landau levels resulting from the inhomogeneous strain; **c–f** Microscope images of the input facets of the arrays with increasing inhomogeneous strain (the waveguide into which light is launched is indicated by a yellow arrow); **g–j** Experimental images of output facets of the arrays after light is injected into a single waveguide on the armchair edge. Localization of the wavepacket implies that the states excited lie within band gaps between Landau levels

levels in their spectra. A Landau level is an energy level with extremely high degeneracy and thus high density-of-states. Since the efficiencies of many photonic processes are directly proportional to the ambient density of photonic states (this is called the Purcell effect [74]), the ability to manipulate this over a large-scale photonic device could be extremely important across applications where the Purcell effect plays an important role. Examples of these include slow light-based devices, where the high density-of-states at the band edge of a photonic crystal is associated with enhanced nonlinear all-optical control [75]; wavelength-scale cavity lasing [76]; and large-area lasing for biosensing [77], among many others. A high density-of-states is usually associated with a physical mechanism for why light remains confined in a given region for a long period of time compared to its inverse frequency: (1) in the case of a cavity, the mechanism is that light can be thought of as bouncing back and forth within the cavity geometry, keeping it mostly in one place; (2) in the case of slow light, the mechanism is self-evident from its name: light goes more slowly and therefore stays in one region for a longer time; (3) in the case of Landau levels, light circulates in cyclotron orbits of the pseudomagnetic field, allowing it to be confined on long time scales. This suggests that Purcell enhancement applications are most certainly achievable using pseudomagnetic fields and photonic Landau levels, especially if the device in question is significantly larger than the scale of a wavelength. Falling into this category would be directional lasing, biosensing, and even photovoltaic cells.

8.4.3 Photonic Floquet Topological Insulators

Starting in 2005, a new topological insulator revolution started with papers demonstrating the *quantum spin Hall effect* [78, 79]. A topological insulator is a material that is an insulator in the bulk, but metallic on its edges. In other words, it has a bulk band gap, but it has surface states that cross that band gap, and since the Fermi energy lies within the band gap, current flows along the surfaces. In two dimensions, the surface states are effectively one-dimensional, meaning that they are edge states. In this scenario, the edge states are completely robust to any kind of scattering as a result of defects or disorder on the edge: they pass through with perfect transmission and no reflection (and no scattering into the bulk either). The reason that they have this property is that either they are forbidden from backscattering into the backwards-propagating state (in the case of the spin-Hall effect), or there is no backwards propagating state available for them to scatter into, as in the case of the quantum anomalous Hall effect, as described by the Haldane model [80]. The reason that these states are called topologically protected has to do with the fact that their existence is associated with topological invariants of the band structure [81] (via the bulk-edge correspondence principle), but this is a topic that goes beyond the scope of the present chapter. For a thorough discussion, see Ref. [82].

In photonic graphene as discussed previously, there are edge states associated with the bearded and zig-zag edges of the structure. However, these are not the non-scattering topologically protected edge states discussed above, since they may easily backscatter into counterpropagating states. In order to eliminate the counterpropagating states, time-reversal symmetry has to be broken. Here, since the time coordinate of the usual Schrödinger equation gets replaced by the longitudinal coordinate, z (as discussed above), one must break z-reversal symmetry to eliminate the counterpropagating state. The way this is done is by making the waveguides helical in shape, rather than straight. This gives a chirality to the structure, thus establishing a preferred rotation direction (in the present work, it is clockwise). This structure is shown schematically in Fig. 8.21a. The period of the helix in the z-direction is 1 cm, and the structure is 10 cm long (meaning 10 periods of helical rotation). It was shown there that the helical rotation gives rise to a fictitious force: a circularly polarized electric field, in the reference frame co-moving with the helices. This in turn implies that the Hamiltonian describing the system is time-dependent, and therefore must be solved using Floquet boundary conditions (which are simply Bloch-periodic boundary conditions in time). This puts our structure in a class called *Floquet topological insulators* [83–85]. Further details are discussed in Ref. [13], which constituted the first realization of topological protection in optics. Topological protection was

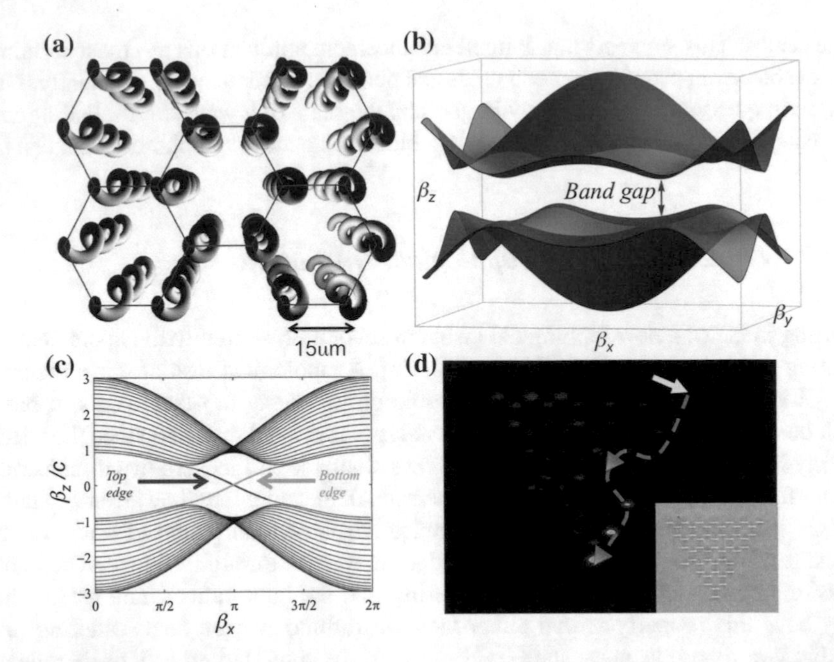

Fig. 8.21 **a** Schematic diagram of helical waveguides rotating in a counter-clockwise sense. The helicity of the waveguides breaks z-reversal symmetry, which enables transverse topological protection, as discussed in the text. **b** As a result of the helicity, the honeycomb band structure opens a band gap; **c** Edge band structure of the zig-zag edge showing the presence of topologically-protected states. The left-propagating state resides on the *top*, and the *right*-propagating state resides on the *bottom*, and thus they form a single *clockwise* edge channel propagating around the entire array. No counterclockwise channel exists in the band gap. **d** Experimental results showing that when light is launched into the *top right* waveguide (indicated by *yellow arrow*) of a honeycomb array (shaped like an equilateral *triangle*), it propagates around a lattice defect (which takes the form of a missing waveguide) without backscattering. A microscope image of structure is shown in the inset

proposed and achieved in the microwave regime earlier [86–88]. Subsequent to the photonic realization of Floquet topological insulators, a condensed matter realization was found using ultrafast optics techniques [89].

As a result of the helical rotation of the photonic lattice, the degeneracy at the Dirac cones is broken, and a band gap is opened (as shown in Fig. 8.21b). Moreover, the structure possesses topologically protected edge states (and non-zero Chern number), as can be seen in the edge dispersion plot shown in Fig. 8.21c. Note that there are two edge states crossing one another in the dispersion diagram; the right-mover (negative slope) resides on the top edge, and the left-mover (positive slope) resides on the bottom edge of the structure. In fact, these edge states taken together with those that reside on the left and right (armchair) edges constitute one chiral edge state (per momentum β) that circulates around the structure in a helical fashion. There is no counterclockwise-propagating channel; therefore there can be no scattering. This is demonstrated in Fig. 8.21d, in the following way. The honeycomb photonic lattice

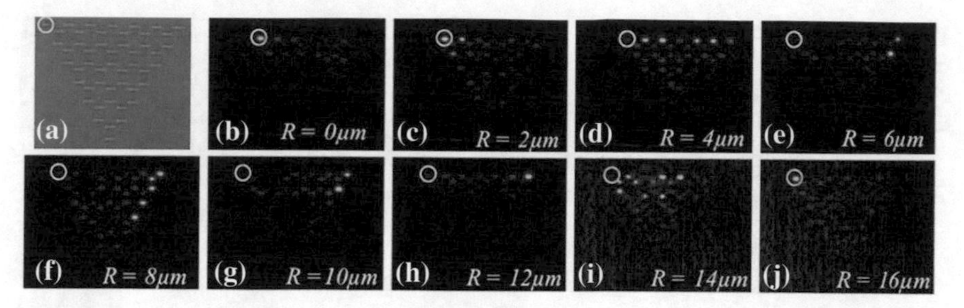

Fig. 8.22 **a** Microscope image for the input facet of an equilateral-triangle-shaped photonic topological insulator (helical honeycomb photonic lattice). **b–j** Output facets of different honeycomb photonic lattices with increasing helix radius. The displacement of the optical power along the edge from the excited waveguide (indicated by the *yellow circle*) initially increases with helix radius, but then reaches a maximum and eventually goes to zero

is arranged in an overall equilateral triangular shape. A single waveguide is removed from the right edge, to act as a defect. Light is injected in the upper-right waveguide, which has significant overlap on the edge states of the system. The light propagates through the structure and past the defect without backscattering. This can be taken as evidence of topological protection.

More evidence of topological protection can be obtained by examining the dependence of the group velocity of the edge state on the radius of the helix. In order to do so, multiple arrays were fabricated, each with a different helix radius, R. When light is injected into the top-left waveguide of the *photonic topological insulator*, in general it gets displaced in a clockwise direction along the edge. The degree to which it is displaced yields the group velocity of the state. This can be seen via the output facet of the array, as shown in Fig. 8.22a–i, each with increasing helix radius. Note that around $R = 8\,\mu\text{m}$, the group velocity reaches its maximum, and then decreases for larger helix radius. This is confirmed by results from Floquet tight-binding simulations, where the group velocity dependence is found to be in quantitative agreement with experiment. At the point where the group velocity reaches zero again the valence and conduction bands touch and the system undergoes a topological transition. This can be directly seen in Fig. 8.23. In the language of topological insulators, the Chern number of the valence band changes from -1 to 2.

In summary, photonic graphene provides a rich platform for the study of Dirac dispersion phenomena. The realization of anomalous edge states, optical pseudomagnetism and Landau levels, as well as photonic topological protection dramatically demonstrates this fact. At this stage, the effects of mean field interactions (in other words, optical nonlinearities), have just begun to be studied. For example, solitons have been predicted to exist in the topological band gap in the presence of the Kerr nonlinearity [90]. There is certain to be more very rich nonlinear Dirac physics that will arise in the optical context. Deep questions remain, such as: are nonlinear topological edge states stable against interactions? Can transport be quantized in a photonic system, much as it is in quantum Hall topological systems with non-zero Chern number? What is the nature of topological edge states in non-Hermitian

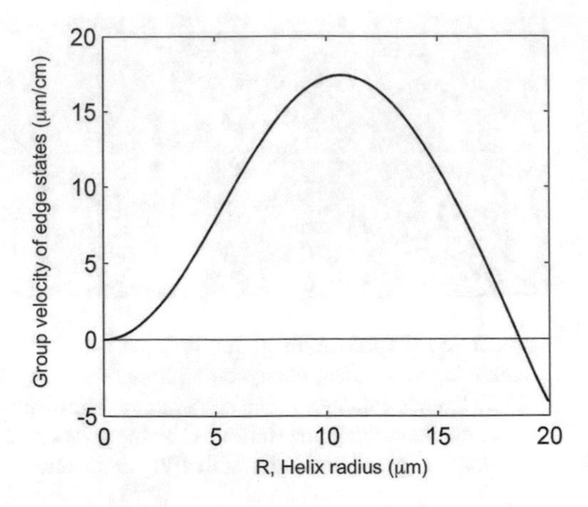

Fig. 8.23 Group velocity dependence on the helix radius, *R*, calculated from a tight-binding model

systems? These questions and others will likely yield deep topological physics in the optical domain for years to come—these are questions that optics provides a unique platform to address.

References

1. S. Longhi, Laser Photon. Rev. **3**, 243 (2009)
2. T. Pertsch et al., Phys. Rev. Lett. 83, 4752 (1999) (R. Morandotti et al. Phys. Rev. Lett. **83**, 4756 (1999))
3. S. Longhi et al., Phys. Rev. Lett. **96**, 243901 (2006)
4. T. Schwartz et al., Nat. Lond. **446**, 52 (2007) (Y. Lahini et al. Phys. Rev. Lett. **100**, 013906 (2008))
5. F.D.M. Haldane, S. Raghu, Phys. Rev. Lett. **100**, 013904 (2008) (R.A. Sepkhanov, Ya.B. Bazaliy, C.W.J. Beenakker, Phys. Rev. A **75**, 063813 (2007); O. Peleg, G. Bartal, B. Freedman, O. Manela, M. Segev, D.N. Christodoulides, Phys. Rev. Lett. **98**, 103901 (2007); O. Bahat-Treidel, O. Peleg, M. Segev. Opt. Lett. **33**, 2251 (2008))
6. X. Zhang, Phys. Rev. Lett. **100**, 113903 (2008)
7. L.G. Wang, Z.G. Wang, J.X. Zhang, S.Y. Zhu, Opt. Lett. **34**, 1510 (2009)
8. O. Bahat-Treidel et al., Phys. Rev. Lett. **104**, 063901 (2010)
9. S. Longhi, Opt. Lett. **35**, 235 (2010)
10. F. Dreisow et al., Euro. Phys. Lett. **97**, 10008 (2012)
11. F. Dreisow et al., Phys. Rev. Lett. **105**, 143902 (2010)
12. M.C. Rechtsman et al., Nat. Photon. **7**, 153 (2013)
13. M.C. Rechtsman et al., Nature **496**, 196 (2013)
14. A. Szameit, S. Nolte, J. Phys. B **43**, 163001 (2010)
15. J. Chan, T. Huser, S. Risbud, D. Krol, Opt. Lett. **26**, 1726 (2001)
16. A. Streltsov, N. Borelli, J. Opt. Soc. Am. B **19**, 2496 (2002)
17. K. Miura, J. Qui, H. Inoue, T. Mitsuyu, K. Hirao, Appl. Phys. Lett. **71**, 3329 (1997)
18. S. Nolte, M. Will, J. Burghoff, A. Tünnermann, Appl. Phys. A. **77**, 109 (2003)
19. K. Itoh, W. Watanabe, S. Nolte, C.B. Schaffer, MRS Bull. **31**, 620 (2006)
20. R.R. Gattass, E. Mazur, Nat. Phot. **2**, 219 (2008)
21. G. Della Valle, R. Osellame, P. Laporta, J. Opt. A **11**, 013001 (2009)

22. M. Ams, G.D. Marshall, P. Dekker, J.A. Piper, M.J. Withford, Laser Photo Rev. **3**, 535 (2009)
23. I. Mansour, F. Caccavale, J. Lightwave Technol. **14**, 423 (1996)
24. D. Blömer et al., Opt. Exp. **14**, 2151 (2006)
25. A. Szameit, F. Dreisow, H. Hartung, S. Nolte, A. Tünnermann, Appl. Phys. Lett. **90**, 241113 (2007)
26. F. Dreisow et al., Opt. Exp. **14**, 3474 (2008)
27. L. Skuja, T. Suzuki, K. Tanimura, Phys. Rev. B **52**, 15208 (1995)
28. M. Stevens-Kalceff, A. Stesmans, J. Wong, Appl. Phys. Lett. **80**, 758 (2002)
29. H. Trompeter, T. Pertsch, F. Lederer, D. Michaelis, U. Streppel, A. Brauer, U. Peschel, Phys. Rev. Lett. **96**, 023901 (2006)
30. H. de Vries, *Physics Quest: Understanding Relativistic Quantum Field Theory*, Physics-Quest.org (2013)
31. J. Feng, Opt. Lett. **18**, 1302–1304 (1993)
32. A.A. Sukhorukov, Y.S. Kivshar, Opt. Lett. **28**, 2345–2347 (2003)
33. E. Schrödinger, Sitzungsber. Preuss. Akad. Wiss. Phys. Math. Kl. **24**, 418 (1930)
34. R. Gerritsma et al., Nature **463**, 68 (2010)
35. F. Dreisow et al., Phys. Rev. Lett. **101**, 143602 (2008)
36. J.M. Zeuner et al., Phys. Rev. Lett. **109**, 023602 (2012)
37. H. Lloyd, Trans. R. Irish Acad. **17**, 145 (1837)
38. S. Longhi, Opt. Lett. **30**, 2137 (2005)
39. O. Klein, Z. Phys. **53**, 157 (1929)
40. F. Sauter, Z. Phys. **69**, 742 (1931)
41. F. Hund, Z. Phys. **117**, 1 (1941)
42. J. Schwinger, Phys. Rev. **82**, 664 (1951)
43. N. Stander, B. Huard, D. Goldhaber-Gordon, Phys. Rev. Lett. **102**, 026807 (2009)
44. A.F. Young, P. Kim, Nat. Phys. **5**, 222 (2009)
45. R. Gerritsma et al., Phys. Rev. Lett. **106**, 060503 (2011)
46. D. Dragoman, Phys. Scr. **79**, 015003 (2009)
47. S. Longhi, Phys. Rev. B **81**, 075102 (2010)
48. E.S. Fradkin, D.M. Gitman, ShM Shvartsman, *Quantum Electrodynamics with Unstable Vacuum* (Springer, Berlin, 1991)
49. H.K. Avetissian, *Relativistic Nonlinear Electrodynamics* (Springer, New York, 2006)
50. E. Brezin, C. Itzykson, Phys. Rev. D **2**, 1191 (1970)
51. G. Lenz, I. Talanina, M. de Sterke, Phys. Rev. Lett. **83**, 963 (1999)
52. S. Longhi, Phys. Rev. A **81**, 022118 (2010)
53. F. Dreisow, S. Longhi, S. Nolte, A. Tünnermann, A. Szameit, Phys. Rev. Lett. **109**, 110401 (2012)
54. R. Keil et al., Nat. Commun. **4**, 1368 (2013)
55. K.S. Novoselov et al., Nature **438**, 197 (2005)
56. A.H. Castro et al., Rev. Mod. Phys. **81**, 109 (2009)
57. R.A. Sepkhanov, Y.B. Bazaliy, C.W.J. Beenakker, Phys. Rev. A **75**, 063813 (2007)
58. K.K. Gomes, W. Mar, W. Ko, F. Guinea, H.C. Manoharan, Nature **483**, 306 (2012)
59. U. Kuhl et al., Phys. Rev. B **82**, 094308 (2010)
60. L. Tarruell, D. Greif, T. Uehlinger, G. Jotzu, T. Esslinger, Nature **483**, 302 (2012)
61. T. Jacqmin et al., Phys. Rev. Lett. **112**, 116402 (2014)
62. M. Kohmoto, Y. Hasegawa, Phys. Rev. B **76**, 205402 (2007)
63. Y. Kobayashi, K. Fukui, T. Enoki, K. Kusakabe, Y. Kaburagi, Phys. Rev. B **71**, 193406 (2005)
64. W. Shockley, Phys. Rev. **56**, 317 (1939)
65. J. Zak, Phys. Rev. Lett. **62**, 2747–2750 (1989)
66. P. Delplace, D. Ullmo, G. Montambaux, Phys. Rev. B **84**, 195452 (2011)
67. Y. Plotnik et al., Nat. Mater. **13**, 57–62 (2014)
68. M.C. Rechtsman et al., Phys. Rev. Lett. **111**, 103901 (2013)
69. D.C. Mattis, Ann. Phys. **113**, 184 (1978)
70. I.E. Tamm, Phys. Z. Sowjetunion **1**, 733 (1932)

71. C.L. Kane, E.J. Mele, Phys. Rev. Lett. **78**, 1932 (1997)
72. F. Guinea, M.I. Katsnelson, A.K. Geim, Nat. Phys. **6**, 30 (2010)
73. N. Levy et al., Science **329**, 544 (2010)
74. E. Purcell, Phys. Rev. **69**, 681 (1946)
75. M. Soljacic, J.D. Joannopoulos, Nat. Mater. **3**, 211 (2004)
76. O. Painter et al., Science **284**, 1819 (1999)
77. B. Zhen et al., Proc. Natl. Acad. Sci. **110**, 13711–13716 (2013)
78. C.L. Kane, E.J. Mele, Phys. Rev. Lett. **95**, 226801 (2005)
79. B.A. Bernevig, T.L. Hughes, S.-C. Zhang, Science **314**, 1757 (2006)
80. F.D.M. Haldane, Phys. Rev. Lett. **61**, 2015 (1988)
81. M.Z. Hasan, C.L. Kane, Colloquium: topological insulators. Rev. Mod. Phys. **82**, 3045–3067 (2010)
82. B.A. Bernevig, *Topological Insulators and Topological Superconductors* (Princeton University Press, 2013)
83. T. Oka, H. Aoki, Phys. Rev. B **79**, 081406 (2009)
84. N.H. Lindner, G. Refael, V. Galitski, Nat. Phys. **7**, 490 (2011)
85. T. Kitagawa, E. Berg, M. Rudner, E. Demler, Phys. Rev. B **82**, 235114 (2010)
86. F.D.M. Haldane, S. Raghu, Phys. Rev. Lett. **100**, 013904 (2008)
87. Z. Wang, Y. Chong, J.D. Joannopoulos, M. Soljacic, Phys. Rev. Lett. **100**, 013905 (2008)
88. Z. Wang, Y. Chong, J.D. Joannopoulos, M. Soljacic, Nature **461**, 772 (2009)
89. Y.H. Wang, H. Steinberg, P. Jarillo-Herrero, N. Gedik, Science **342**, 453 (2013)
90. K. Lumer, Y. Plotnik, M.C. Rechtsman, M. Segev, Phys. Rev. Lett. **111**, 243905 (2013)

Printed in the United States
By Bookmasters